A+

MW01537075

2/6 -224

48 LECTURES

IN

MODERN BIOLOGY

JOHN L. SOUTHIN
McGILL UNIVERSITY, MONTREAL CANADA

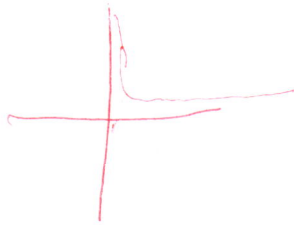

KENDALL/HUNT PUBLISHING COMPANY
2460 Kerper Boulevard P.O. Box 539 Dubuque, Iowa 52004-0539

This edition has been printed directly from camera-ready copy.

Copyright © 1993 by Kendall/Hunt Publishing Company

ISBN 0-8403-8316-9

All rights reserved. No part of this publication may be reproduced,
stored in a retrieval system, or transmitted, in any form or by any
means, electronic, mechanical, photocopying, recording, or otherwise,
without the prior written permission of the copyright owner.

Printed in the United States of America
10 9 8 7 6 5 4 3 2 1

CONTENTS

PREFACE

> Now, what I want is, Facts. Teach these boys and girls nothing but Facts. Facts alone are wanted in life. Plant nothing else, and root out everything else. You can only form the minds of reasoning animals upon Facts: nothing else will ever be of any service to them. This is the principle on which I bring up my own children, and this is the principle on which I bring up these children. Stick to the Facts, sir!
>
> C. Dickens
> "Hard Times"

Biology, even just "Modern Biology", covers such an immensity of topics that it would be impossible in the course of a single term even to *describe* them all, much less teach anything useful about them. Hence, neither "Introductory Cell and Molecular Biology", nor "Essential Biology", the two McGill University first-year courses for which these Lectures provide the accompaniment, aims at anything so ambitious as "covering" either field, but more modestly at "*un*covering" just a few of the more interesting aspects of both. Because most college courses consist of between 30 and 40 lectures, it is obvious that this book represents more a menu from which a balanced selection can to be made than a prescribed diet to be consumed *holus bolus*. It is also anticipated that even when these lectures do form part of a course each topic would benefit from some additional fleshing out by the instructor or from a regular text book.

These Lectures treat modern biology as an experimental science. In consequence the focus will not be on a memorizable collection of "facts" (many of which will be disproved later anyway) but on the experimental processes by which we arrived at our current understanding. Cicero noted some time ago that "the causes of events are more interesting than the events themselves", and the same is true in science: why we believe what we do is often more interesting than the belief. We shall also have many occasions to see that it is often more important that a scientific proposition be *interesting* even than it be true. The sad truth is that, facts, at least in Biology, don't keep much better than fish. In summary, what I hope the student or reader will come away with is an appreciation of the modern biological experiment: its design, its logic, and its relationship to other experiments. Modern biology is essentially an enormous reticulum of experiments; to look at it as just so many facts to be memorized leaves one with the ashes but none of the fire that makes this subject so exciting.

None of this is to suggest, however, that the "facts" can be wholly ignored. There is no escaping the necessity of knowing—memorizing, if you will—the meaning of all the words in, for example, V. Nabokov's *Lolita* if one is to appreciate the novel for the masterpiece it is. Yet if involvement with the book is arrested at merely knowing the meaning of its words, one trivializes a literary accomplishment of uncommon eloquence, subtlety and wit. So it is with the science of biology. Facts are always important for the moment, but the excitement of biological science rests with their discovery rather than with their subsequent labelling, cataloguing and memorizing. "If a poet is anybody", noted e.e.cummings, "he is someone to whom things made matter very little—someone who is obsessed by Making." The same is true of scientists and the facts that are the perishable by-products of their activities.

One great advantge students in an introductory biology course have over their coevals in many other scientific disciplines is that the former can often be led, in an informed way, through a series of experiments that goes right to the perimeter of our current knowledge. Sometimes, of course, it is not possible to go quite that far, since the experimental trail may lead through thickets of mathematics, chemistry or physics beyond the coping of most first year students. However, for each of the major topics treated in these Lectures I try to point out the problems that are the subject of the most intense investigation at the present time.

Because *Scientific American* is very widely distributed in public and school libraries, I have largely confined myself to this readable lay publication in giving references for additional reading. However, when a particular experiment is discussed in detail, reference is usually made to the original report in the scientific "literature". Similarly, when information in the text is too recent to have found its way into textbooks, I have occasionally given readers the opportunity to check it out for themselves by giving the original reference.

John L. Southin,
McGill University,
Montreal, Canada.

Fall, 1992.

LECTURE 1

SOME ELEMENTARY CHEMICAL PRINCIPLES: 1

The science of Biology rests on a foundation of chemistry and physics. Although it is possible to study intensively many fields within biology without frequent reference to chemical principles, the number of such sub-disciplines is decreasing all the time. In fact, many scientists see the resolution of the whole of biology into chemistry and physics to be the ultimate goal of their research. Whether that goal is attainable or even desirable need not concern us here; the fact is that much of modern biology is an applied sort of chemistry and that the study of biology requires, in consequence, some understanding of basic chemical principles. Some of this background information is presented in this introductory section, and more is given where it becomes relevant in later sections.

THE NATURE OF ATOMS

All matter is made up of **atoms**, a fact guessed at by the ancients (for example, by the Greek atomists, such as Democrates) and confirmed by modern physicists. There are only 92 different kinds of stable atoms, constituting the 92 **elements** that occur with greater or lesser frequency on earth. It is likely that these 92 are the only ones that appear anywhere in the universe, although their relative frequencies in other locations may be very much different from what we find on earth. In this century physicists have succeeded in synthesizing several dozen additional elements, but all of them are unstable and disintegrate into combinations of the familiar 92.

The 92 elements are given by international convention distinctive symbols to represent them. There capitalized one- or two-letter symbols are often related to their English or Latin name (eg. "U" stands for Uranium, "H" for Hydrogen, "Na" for Natrium (better known to English speakers as Sodium), and so on. Rest assured that the identities of the elements for which these symbols stand will become second nature within a short period of time.

Atoms have two important areas: a nucleus and an electron cloud surrounding it. The size of the nucleus is minuscule in comparison to that of the atom as a whole. The size of the electron cloud around the nucleus determines the volume of space the atom occupies; if this volume could be thought of as a football stadium, then the nucleus is about the size of the football. Despite this disparity in volume, however, the mass (weight) of the atom is determined by the nucleus, for it contains all the particles that contribute any significant weight. We now know that the atom can be further subdivided into an array of particles, all of which (except for the electrons) exist in the nucleus. For our purposes, only two of these nuclear particles, **protons** and **neutrons**, are relevant.

Protons are particles with a positive charge and a mass arbitrarily set at one. The nucleus of the simplest atom, hydrogen, contains only a single proton and nothing else. As one adds protons, one gets heavier and heavier atoms (helium has two protons, lithium has three, etc.). The number of these protons an atom possesses determines its **atomic number**, and the range will be from hydrogen (atomic number 1) through to uranium (atomic number 92). As we shall see, the chemical properties of an atom are indirectly a consequence of the number of protons within its nucleus.

Besides having protons, all atoms except hydrogen also have neutrons within their nuclei. Neutrons (as their name implies) are without electrical charge, but like protons have a mass of one. Often the number of neutrons is about equal to the number of protons, and hence the **atomic weight** of an atom will be approximately double the atomic number. These two quantities, the atomic number and the atomic weight, are usually written preceding the symbol for the element, the former as a subscript and the latter as a superscript. It is frequently the case that additional neutrons can be added to the nucleus of an atom to form **isotopes**. Varying the number of neutrons does not, of course, change the atomic number or chemical properties of the atom—both of which are dependent solely on the number of protons—but it does change the atomic weight. (Because the volume of the atom is not significantly affected by adding neutrons to the nucleus, the change of weight without a concomitant change of volume does alter the **density** of the atom—a fact that is frequently made use of in experimental procedures we shall be examining later in these lectures.) Some isotopes of particular atoms are unstable and disintegrate over time—these are the radioactive isotopes—whereas others are stable and coexist in nature in various proportions with the standard form of the element.

The fact that protons have a single positive charge means that atomic nuclei must themselves be charged, the magnitude of this positive charge being directly proportional to the number of protons. Precisely balancing this positive nuclear charge—and therefore maintaining the atom as a whole in a neutral state—are the electrons surrounding the nucleus. An electron has a negative charge of one, but no significant mass. Hydrogen, therefore, will have one electron surrounding the nucleus, helium two, lithium three, etc. It is these electrons that exclusively determine the chemical properties of atom.

It used to be thought that electrons circled the nucleus in an analogous fashion to the orbits in which planets circle the sun, but this view has had to give way as the properties of atoms become understood in greater detail, although some of the terminology has been retained. Electrons organize themselves in specific patterns, called **shells**, around the nucleus, depending upon the number of electrons associated with that particular atom. The electrons associated with hydrogen and helium travel in a roughly spherical shell close to the nucleus. This first shell of electrons will only hold a maximum of two electrons, however, so the third electron associated with the next heaviest element, lithium, must occupy the next shell more distantly removed from the nucleus. This shell, and most remaining shells, holds up to eight electrons each[*], which progressively become filled as the atomic number of the atoms increases. In this second shell, the first two electrons move in roughly spherical patterns called **orbitals**, but the remaining six electrons of the filled shell travel in three dumbbell-shaped orbitals, a maximum of two electrons in each. All orbitals in all shells can contain a maximum of two electrons. For most purposes, the exact shape of the orbits within a particular shell is of little importance in biology, but it is of great importance to know how many electrons altogether are in the final or outermost shell of the atom. It is sometimes important to know in addition whether the outermost shell has two electrons in all its orbitals, since atoms with an outer orbital containing a single electron are more reactive than those that have two electrons in all their orbitals in the outer shell.

One should not take too literally the description if the shapes of the electron orbitals within the shells, since they represent not actual tracks or paths that the electrons follow, but are actually just volumes within which the occupying electrons most probably exist. One cannot ever say with certainty that a particular is here or there, only that it has a certain probability of being found within this particular space.

Electrons that occupy orbitals that are more distant from the nucleus are considered to be at a higher energy level than those nearer the nucleus. Occasionally an electron will absorb a certain amount of energy (from, for example, ultra-violet light) that enables it to move to an orbital further from the nucleus. Such an electron is said to be **excited**; it quickly returns to its **ground state**, giving off (usually as light) exactly the same amount of energy that excited it in the first place. The amount of energy necessary to excite an electron is known as a **quantum** and is the minimum quantity of energy that can exist in nature.[**]

---

[*]In fact, many of the subsequent shells can hold more than 8 electrons; however, their most stable configuration is with just 8. Hence, for most chemical purposes it is assumed that this is the maximum number of electrons that will occur in the outer shell.

[**]Having described the atom in physical and visual terms, I should balance the picture by at least mentioning the heresy involved here. Perhaps it is best summarized by W. Heisenberg, one of the towering figures of modern physics.

> The atom of modern physics can be symbolized only through a partial differential equation in an abstract space of many dimensions. All its qualities are inferential; no material properties can be directly attributed to it. That is to say, any picture of the atom that our imagination is able to invent is for that very reason defective. An understanding of the atomic world in that primary sensuous fashion...is impossible.

LECTURE 2

SOME ELEMENTARY CHEMICAL PRINCIPLES: ll

COMBINATIONS OF ATOMS

Although a few elements are likely to be found in nature as pure substances (many of the metals, in particular), it is much more common to encounter them as constituents of **compounds**, a mixture of atoms bonded together in precise proportions and in specific ways. Sometimes even substances that we might commonly think of as being "raw" elements, such as the oxygen or the nitrogen that compose much of the atmosphere, are actually compounds composed, in each case, of two atoms of Oxygen or of Nitrogen. (Notice that elements in chemical parlance are capitalized, whereas compounds are not.) The one compound whose constituents are known to everyone is water, H_2O. This chemical **formula** indicates that a **molecule** of the compound, water, is composed precisely of two Hydrogen atoms and one Oxygen atom. It does not, however, indicate that nature of the bonding process that holds these atoms together in such precise quantities, and that is what we shall take up next.

IONIC BONDS

The two principal means by which atoms become joined to others to make molecules of compounds are by **ionic bonds** and **covalent bonds**. In both cases the bonding propensities are a consequence of the number of electrons in the outer shells of the atoms involved. Ionic bonds are somewhat easier to envision, and we shall examine them first by using as an example a compound familiar to everyone, table salt—sodium chloride or $NaCl$. As the chemical formula indicates, this compound is composed of one atom of Sodium and one of Chlorine. Since Sodium is a metal and Chlorine a gas, it is already obvious that compounds have physical properties that can differ markedly from those of their constituent elements. Sodium, atomic number 11, has 11 protons in its nucleus and therefore eleven electrons in its shells; the outer shell, in consequence, will have a single electron inhabiting it. Chlorine, atomic number 17, will have seven electrons in its outer shell.

The underlying principle that governs the ways in which atoms interact is their attempt to have their outer shell contain the most stable number of electrons—usually eight, except in the case of hydrogen. (Helium, already having two electrons in its only shell, will not react chemically with anything.) Sodium, with one electron in its outer shell, will attempt to donate it to some other atom, eliminating its unfilled outer shall altogether, whereas Chlorine, will attempt to acquire one electron from somewhere in order to complete its outer shell. It is obvious these two atoms were "meant for each other". However, in losing an electron to Chlorine, the Sodium acquires a slight positive charge, since the number of remaining electrons in its shells (10) no longer exactly balances the number of positively charged protons in its nucleus (11). By similar reasoning, the Chlorine, having gained an electron to complete its outer shell now has one more electron than it has protons in the nucleus; hence, it will now carry a slight negative charge. Such charged particles can no longer be called atoms, which are always electrically balanced and neutral in charge, but instead are called **ions**.[*] Since the sodium and chlorine ions are now equally but oppositely charged, they attract each other and are said to be held together by an **ionic bond**. Some atoms (e.g. Calcium, Ca) have two electrons in their outer shell to lose; in this case the calcium ion would form a compound with two chlorine anions in order to balance the double positive charge the calcium cation bears. The chemical formula for calcium chloride would then be $CaCl_2$. The number of electrons an atom donates or acquires in order to establish a completed outer shell—and, in consequence, the number of oppositely charged atoms it must combine with in order to regain electrical neutrality—is known as its **valence**. Sodium would have a valence of +1; Calcium, +2; Chlorine, -1, and so on.

Sometimes atoms join together in a tight group known as a **radical**, where they behave in chemical reactions as if they were a single ion with its own valence. One of the most common of these in biological chemistry is the phosphate radical (or group, as it is sometimes called) composed of a Phosphorus atom and four Oxygen atoms (PO_4). It is a anion with a valence of -3. Others we will encounter are the sulphate anion (SO_4) with a valence of -2 and the carbonate anion (CO_3) with a valence of -2.

COVALENT BONDS

The type of bond that is much more common in biological chemistry is the **covalent bond**. In this case, individual atoms neither yield nor acquire free electrons. Instead, two or more atoms share their electrons so that all the atoms in the compound have completed outer shells. A good example would be water. The Oxygen atom (atomic number 8) has six electrons in its outer shell and needs, therefore, two more for completion. Hydrogen needs one electron for completion of its single shell. Two atoms of Hydrogen, each with their single electron, join with Oxygen, thus completing the outer shell of Oxygen. The Hydrogen atoms each acquire the one they need for completion from among the six the Oxygen already had. Hence, four electrons in the water molecule are shared. This sharing of electrons, some contributed by one of the interacting atoms and some by the other, is the essence of a covalent bond.

It is sometimes the case that two interacting atoms will share more that one pair of electrons. When two pairs are shared, the bond is said to be a **double bond**; when three pairs of electrons are shared, the bond is called a **triple bond**. The interacting atoms are held more closely to-gether by triple or double bonds than by single bonds. Single bonds also permit the two interact-ing atoms to rotate around the bond freely (unless other nearby atoms prevent such movement), whereas double and triple bonds hold the interacting atoms rigidly. When a chemical formula is given in a manner that emphasizes the bonding characteristics of the molecule, single bonds are depicted as single lines (-), double bonds as double lines (=), and triple bonds as triple lines.

---

[*]Negatively charged ions are called **anions** and positively charged ones are called **cations**.

The chemistry of the element, Carbon, constitutes the branch of chemistry known as **organic chemistry**. This name derives from the fact that it was once thought (erroneously, as we shall have many opportunities to see) that only living cells could manufacture the molecules found in living cells--and all of these molecules have Carbon as a prominent constituent. Carbon has only four electrons in its outer shell. If it were to bond with Hydrogen, for example, it would do so by forming covalent bonds with four Hydrogen atoms, and the gas, methane, would be the resulting compound. Carbon, as chemists say, has a valence of four.

It is often the case when atoms are joined by covalent bonds that the two (or more) atomic nuclei have unequal attractions for the electrons they share. Generally speaking, atoms that have a greater number of protons attract the shared electrons more strongly than do those atoms with fewer protons. The atoms exerting the greater attractive force on the shared electrons are said to possess greater **electronegativity** than do those atoms exerting less attractive force. Water is a good case in point, since Oxygen is considerably more electronegative than Hydrogen. The practical consequence of this unequal sharing of electrons is that the Oxygen in a water bears a slight negative charge and the Hydrogens—which are not opposite each other (i.e., at a 180° angle) but instead are asymmetrically oriented with respect to each other at an angle of 104.5°—bear a slight positive charge. Put another way, water molecules, like miniature magnets, are **polar**. This is a common phenomenon. Large molecules often have parts of them that are polar and other parts that are not. Polarity, in other words, does not have to involve the entire molecule, as it does in the case of water.

QUANTITATIVE MEASUREMENTS

The **molecular weight** if a compound is the sum of the atomic weights of its constituent atoms. The molecular weight unit is the dalton, abbreviated "D". Hence, water would have a molecular weight of 18 D, approximately. The **gram molecular weight** of a substance is its molecular weight given in grams. A quantitative measurement commonly used in chemistry is the **mole**, abbreviated "M". A mole of any substance is simple its gram molecular weight, and a mole of any substance will contain the same number of molecules as a mole of any other substance. That is to say, a mole of water (i.e., 18 grams of water) contains the same number of molecules as a mole of ethyl alcohol (i.e., 46 grams of alcohol, C_2H_5OH). The actual number of molecules in a mole is 6.023×10^{23}, a quantity known as **Avagadro's Number**.

ACIDS AND BASES

The **acidity** of a solution (or its opposite, **alkalinity**) is frequently an important consideration in biological chemistry. Acidity is a measure of the concentration of hydrogen ions (H+) in solution in moles per litre. Water is capable of ionizing to a limited extent, producing H+ and OH- ions. Pure water will ionize so that there will be 10^{-7} moles of H^+ per litre. Such a solution is said to have a pH of 7. The pH of a solution is, therefore, the *negative* logarithm of the hydrogen ion concentration. If a solution has a high concentration of hydrogen ions—for example, is the H^+

6

is 10^{-2} or 10^{-3} moles per litre—then the solution will have a pH of 2 or 3.[**] Water, with a pH of 7, is considered neutral; a pH lower than 7 is considered acidic and higher than 7 is considered basic or alkaline. **Acids** are substances that release H^+ ions into solution, **bases** are substances that release OH- ions into solution or else remove H^+ from solution.[***]

WATER

> What in water did Bloom, water lover, drawer of water, water carrier returning to the range, admire?
> Its universality: its democratic quality.

J. Joyce
Ulysses

Perhaps...but in this account it is the chemical properties of water, the major component of living matter, which concern us more. Water molecules are polar, by which is meant they have a negative and a positive end, like a bar magnet. In particular, the Oxygen bears a slight negative charge and the hydrogens a slight positive one.[****] Because the electronegative Oxygen exerts a pull on the Hydrogen of another water molecule, that Hydrogen becomes in a sense shared between the two molecules. It is still covalently bonded to the Oxygen of its "own" molecule, but is also much more loosely bonded to the Oxygen of the other molecule. The two water molecules that share a Hydrogen in this way are said to be **hydrogen bonded**. Hydrogen bonds have only about 5% of the strength of a standard covalent bond, but when circumstances permit a large number of them to occur, the over-all cohesion can be very strong. Water, therefore, has many of the properties of a giant molecule—a syncytium composed of all the individual molecules hydrogen bonded together. For this reason, water has a high boiling point (greater energy must be expended to get molecules to escape from the liquid), a strong surface tension that resists penetration, and a high tensile strength.

The polarity of water has other important consequences. When other charged molecules or ions mix with water, the water molecules surround them, forcing them apart. For example, the dissolving of crystalline table salt (NaCl) in water occurs when water molecules surround and

---

[**]Note that *high* concentrations of H^+ characterize acidic solutions with a *low pH*; conversely, a *high* pH denotes a basic or alkaline solution with a *low* H^+ concentration. Note, too, that the pH scale is logarithmic. This means that a solution with a pH of 3 is ten times more acidic than a solution with a pH of 4.

[***]Adding OH⁻ ions to a water solution is functionally the same as removing H^+ ions, since the latter will combine with the former to form water and hence are removed from the solution.

[****]The two Hydrogens are not directly opposite each other on either side of the Oxygen. Think of the Oxygen as the head of Mickey Mouse; the two Hydrogens are positioned somewhat like his ears. The water molecule is not, therefore, symmetrical.

separate the sodium and chloride ions. An ion surrounded in this manner by a coterie of water molecules is said to be **hydrated**. By contrast, molecules that do not bear a charge (**non-polar molecules**) such as many of those composing fats and oils do not attract water and get "squeezed out" by the attraction the water molecules have for each other. Hence, these non-polar molecules do not dissolve in water. Some molecules have parts that are charged and parts that are not. The charged portions, because of their attraction for water, are said to be **hydrophilic** whereas the uncharged (non-polar) parts are said to be **hydrophobic**. (Calling them "water-loving" and "water-fearing"—hydrophilic and hydrophobic—is perhaps over-dramatizing the situation somewhat, but those are the terms that have become established.)

LECTURE 3

NUCLEIC ACIDS AND HEREDITY

INTRODUCTION

If asked to speculate on the chemical nature of the hereditary material, most biologists of the 1930s and 40s would probably have agreed that genes were some kind of fancy protein. Webster's *Third New International Dictionary* (the latest edition, 1964) apparently still clings to this belief, since it defines "gene" as "one of the elements of the germ plasm serving as specific transmitters of hereditary characteristics and usually regarded as complex self-perpetuating protein molecules in some respects comparable to viruses..."

The identification of genes with proteins was one of those beliefs that was widely accepted simply because there seemed no reasonable alternative. Although at the time little was known of protein structure, it was clear that proteins were exceedingly complex and diverse in size, composition and shape. Genes, too, must have seemed remarkable structures, since they obviously possessed the ability to self-replicate in an exact fashion, to encode genetic information, and to change function (mutate) while still retaining their former ability to replicate. Sufficient ignorance enveloped both genes and proteins to make the equation of one with the other seem eminently reasonable. Reasonable, but wrong.

To recount the development of our current ideas about the hereditary material, it is first necessary to discuss three seemingly unrelated phenomena: plant viruses, bacteriophage and bacterial transformation.

PLANT VIRUSES

In 1885 the Dutch botanist, A. Meyer, began studying a curious plant disorder now known as **tobacco mosaic disease**. A tobacco plant afflicted with this condition manifests a mottle or "mosaic" of small necrotic yellow patches over the surface of its leaves. The disease was clearly not genetic, for Meyer convincingly demonstrated that the sap expressed from a sick plant and applied to a healthy one soon initiated a systemic mosaic infection of the second plant. Since sap from this newly-diseased plant could now transmit in turn the mosaic disease to still other healthy plants and so on *ad infinitum*, the plant disease showed many of the characteristics of being bacterial in origin. At each transfer the infectious agent, the putative bacterium, would multiply and consequently would not be diluted out by any number of serial transfers. The *Zeitgeist* of the late 19th century also strongly favored this hypothesis, since Pasteur's discoveries still played a predominant role in the scientific imagination at the time.

It soon became more difficult, however, to hold to this belief. In the 1890s the Russian botanist,

D. Iwanowsky, and then independently the Dutch microbiologist, M. Beijerinck, demonstrated that the infectious cell sap could be forced through an unglazed porcelain filter with pores small enough to retard all known bacteria, yet still the sap retained full infectivity. Moreover, nothing from this filtered, infectious sap could be cultured on any media known to support bacterial growth. To some, Iwanowsky apparently among them, this meant only that the bacterial causative agent was very small and refractory to known culture techniques; to others like Beijerinck, less distracted by the dazzle of Pasteurism, it opened up an exciting new field of investigation: plant viruses.

Tobacco Mosaic Virus (TMV) we now know to be a minute particle composed of only two chemical species, protein and nucleic acid. It can be visualized in the electron microscope as a rod-like structure 3000Å (300nm) long and 180Å (18nm) in diameter. (For an explanation of these units of measurement, see the table above.) Closer examination under an electron microscope reveals that the rod is really a tightly coiled helix assembled from 2130 identical protein **capsids**. Coiled around the central core of particle, and hence out of sight of the electron microscopist, is the nucleic acid. The TMV **virion**, therefore, besides being considerably smaller than any cell, lacks almost all the characteristics we usually associate with living organisms: it does not metabolize, for it has no metabolic machinery; it does not move on its own, because it has no locomotory organs; it has no nucleus, no cytoplasm, and no delimiting membranes. In 1935, W. Stanley succeeded even in crystallizing TMV without destroying its subsequent infectivity. It does, however reproduce, but only when it succeeds in in-fecting a living cell—in this case, a tobacco cell. Although it is many hundreds of times smaller than its host cell, the virus in essence "hijacks" the metabolic machinery of this cell,

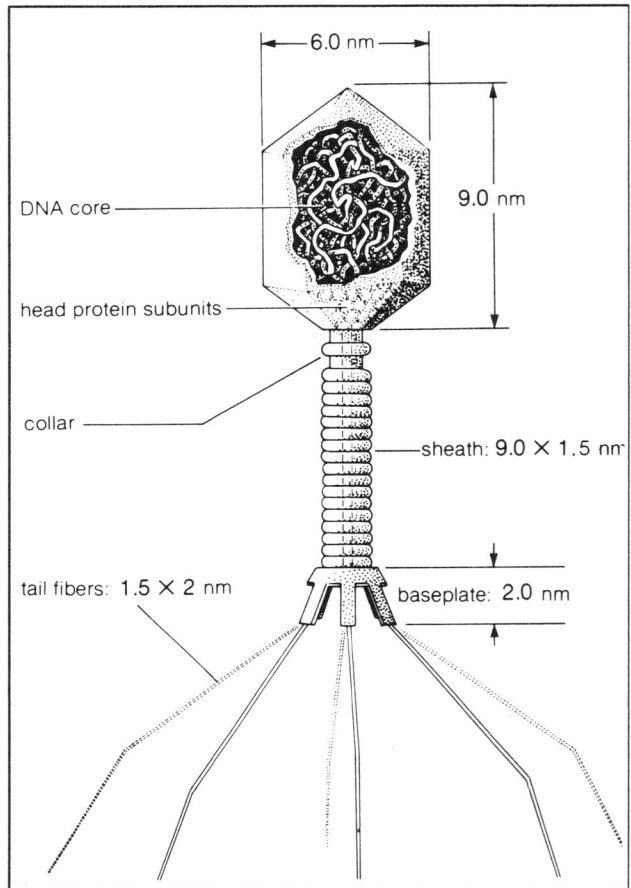

Figure 3-1. A "T-even" bacteriophage. from BIOLOGY OF THE CELL, 2ND EDITION by Stephen L. Wolfe. Copyright © 1981 by Wadsworth, Inc. Reprinted by permission of the publisher.

turning the cell into a factory for the production of progeny virus particles. In the case of TMV (and many of the viruses that will be discussed later in these notes) the host cell is killed by this effort, but different viruses can affect the host cell in a variety of ways.

Is TMV "alive"? The question recedes in importance when one realizes that the argument hinges only on what we shall *call* it, and not on any of its fundamental properties, almost all of which

10

we already know. Call it living or non-living as you please; none of its characteristics will change no matter what your choice.

BACTERIAL VIRUSES (BACTERIOPHAGE)

An English bacteriologist, E. Twort, is usually credited with having discovered a type of virus that infects bacteria and which is now known as **bacteriophage**. Twort noticed that colonies of certain bacteria (*micrococci*) growing on agar plates would occasionally and spontaneously become radically transformed in appearance. Colonies undergoing this "glassy transformation" (Twort's terminology) no longer contained any intact cells, and hence the patch of agar where they were once growing became completely sterile. If a clean needle were touched first to the transformed culture and then to a healthy one, this second culture would within several hours itself become sterile like the first. Moreover, Twort showed that this serial transfer could be continued indefinitely; whatever was causing the death of the healthy cells in the cultures must, therefore, be replenished at each transfer. Like the causative factor in tobacco mosaic disease, this factor too could pass through filters whose pores would retard the passage of any known cell.

Table 3-1. Size relationships of common measurements in biology.

Measurements Used in Microscopy

1 centimeter (cm) = 1/100 meter = 0.4 inch
1 millimeter (mm) = 1/1,000 meter = 1/10 cm
1 micrometer (μm)* = 1/1,000,000 meter = 1/10,000 cm
1 nanometer (nm) = 1/1,000,000,000 meter = 1/10,000,000 cm
1 angstrom (Å) = 1/10,000,000,000 meter = 1/100,000,000 cm
or
1 meter = 10^2 cm = 10^3 mm = 10^9 μm = 10^9 nm = 10^{10} Å

As is usual in science, a number of different hypotheses were advanced to explain these results. The transforming agent could, for example, be a bacterial poison or a destructive enzyme that was secreted by dying cells and transferred by Twort's needle from colony to colony. At each transfer it would be replenished as the cells were killed. The hypothesis favored by Twort and the one that has come to be universally accepted interprets the transforming agent as a virus which attacks bacteria and in multiplying kills the host. As "eaters of bacteria", these viruses were in time given the special name, "bacteriophage" or, by apheresis, **phage**.

Unfortunately, although Twort seems to have made all the right guesses about the cause of his "glassy transformations", few people paid his explanations or even his discovery itself any attention. A somewhat shy and retiring scientist, he lacked the personal pizazz that is often needed in order to make other busy people take one's discoveries seriously. He published his findings, of course—just a single paper in 1915—and waited for the scientific world to become excited about his findings. No one did; Twort's discovery had to be re-made by someone else

11

before it was accepted. Sometimes it is true "that the race is not to the swift nor the battle to the strong..." (*Ecclesiastes*), but still most of the time the swift do outdistance the slow and the strong clobber the weak.

Two years after Twort's paper, F. d'Herelle, a Montrealer working in Paris, independently published a series of observations similar to Twort's. A bacterial dysentery was then raging among certain military units billeted at Maisons-Lafitte and d'Herelle noted that highly filtered extracts from the feces of certain patients would occasionally cause the normally turbid cultures of the Shiga bacillus to clear overnight and become completely sterile. Again it was noted that the causative agent could be serially transferred to other cultures and was small enough to pass through the finest filters. d'Herelle at once seized on the notion that the active agent was a bacterial virus that was "eating" (i.e. destroying, or in technical parlance, **lysing**) the bacteria, and it was he who gave these viruses the name, "bacteriophage".

d'Herelle was ignorant of the earlier work of Twort's when he published his own observations in 1917. Unlike Twort, d'Herelle was not the type of scientist content to blossom under a stone, but set about to publicize his discovery and to talk up its scientific and medical potential[*]. He succeeded admirably, and for many years bacteriophage were commonly called "d'Herelle bodies"—which, it must be admitted, has considerable euphoric merit over the historically more valid term, "Twort bodies".

Bacteriophage can be as simply constructed as the TMV virion, or much more complex. A commonly used complex phage is called T4[**] , which attacks the laboratory bacterium *Escherichia coli*, a normal innocuous[***] inhabitant of the human gut. In gross morphology, the T4 resembles a frog tadpole about 2100Å in length. The expanded part, or head, of the phage is about 650Å wide and 950Å long.

---

[*]The medical potential of bacteriophage has been a tantalizing *ignis fatuus* almost from the moment of their discovery. This hope is best illustrated in the novel *Arrowsmith* by Sinclair Lewis. In reality, the body quickly builds up antibodies against the bacteriophage, rendering them useless.

[**]Bacteriophage nomenclature is a mare's nest of confusion. If T4 sounds unusual, how about MS-2, f2, Φx174, and so on. There is a thread of logic that runs through some of phage nomenclature, but the tedium of the search and the exiguous advantage attached to finding it makes understanding this logic not worthwhile.

[***]Innocuous, that is, after one grows accustomed to the strains one harbors. An encounter with a different strain of *E. coli* from the ones normally carried is thought to be the cause of "tourista" or travellers' dysentery.

Attached to the head by means of a small circular collar is the tail, which is about 350Å in diameter and 950Å long. The tail is actually composed of two parts, an inner core and an outer contractile sheath. At the end of the tail is the tail plate to which are attached six spikes 200Å long and six long tail fibers 1500Å in length. All of the parts mentioned are composed of protein, a different protein for each piece. Inside the head is found nucleic acid and a small quantity of some basic, non-structural proteins.[1]

Since phages exceed the limit of resolution of the best light microscopes, all of the structural details given above have been worked out by other means. The most useful instrument, of course, has been the electron microscope, which can resolve objects two or three orders of magnitude smaller than can be resolved with any light microscope.[*] But even before one can examine viruses under the electron microscope, one must have a means of isolating viruses free of contaminating cell debris. Often this is done in an ultra-centrifuge. In order to throw particles having the mass of viruses out of suspension, extraordinary rotational speeds have to be achieved and maintained for long periods of time.[2] Before the development of the modern Spinco (or Beckman) ultracentrifuge, an instrument capable of 40,000 rpm and developing centrifugal forces of 125,000 x g, high speed centrifugation was an extremely hazardous undertaking. Since the instruments frequently exploded, those who ventured to operate them were doomed to a mole-like existence in the basement, sheltered—sometimes inadequately—behind several feet of reinforced concrete. Nowadays almost every decent molecular biology lab has a Beckman or two sitting handy nearby, and few people have occasion to be reminded of the lethal antecedents of this now benign device.

REFERENCES

1. For a good historical introduction to bacteriophage, see chapter 1 of *Molecular Biology of Bacterial Viruses* by G. Stent (Freeman, San Francisco, 1963).

2. For a discussion of ultracentrifugation, see Beams, J., 1961. Ultrahigh-speed rotation. Sci. Amer. (Apr.): 134.

---

[*] Microscopes are limited by the wave length of the medium they use. Microscopes that employ light, which has a wave length from about 0.4 to 0.7 micrometers, have a theoretical limit of resolution of 0.2 micrometers. (For reference, bacteria are about 0.5 micrometers in diameter.) Electron microscopes employ electrons, whose wave length varies according to the speed of travel: the greater the voltage applied, the faster the electron, the smaller the wave length. The practical limit to the resolution of an electron microscope is about 0.1 nanometers (1Å).

LECTURE 4

BACTERIOPHAGE
BACTERIAL TRANSFORMATION

INTRODUCTION

Bacteriophage have become one of the most extensively used "organisms" in molecular biology research. Before any organism can be effectively used in the laboratory, however, it is necessary to know certain fundamental facts about its existence: how does it reproduce? how long does this process take? and how many progeny are produced at each round of replication? The following account attempts to answer these questions about bacteriophage.

LIFE CYCLE OF BACTERIOPHAGE

Bacteriophage are grown in the laboratory in one of two ways, depending upon the purpose of the experiments. The manner in which d'Herelle serendipidously grew his phage would now be called **liquid culture**; that is, bacteria are grown in a broth medium in a test tube, a suspension of phage is added, and the growth of the phage results in the death of the bacteria and consequently the clearing of the culture. This clearing of a normally turbid bacterial culture can, if desired, be monitored by passing light through the test tube and measuring the amount of light transmitted.

A different technique is called **solid culture**. The host bacteria are plated over an agar nutrient medium in a petri dish and allowed to grow until the expanding individual bacterial colonies grow together into a confluent "lawn" of bacteria. A dilute suspension of phage is then poured over the bacterial lawn and evenly distributed. If the petri dish is then incubated for several hours, small clear areas, called **plaques**, begin to appear in the murky bacterial lawn. Each plaque has resulted from the earlier settling of a single phage at that site. This phage particle subsequently reproduced and the progeny phage attacked neighboring cells. The repetition of these events finally gives rise to a visible plaque which abounds with phages but is devoid of intact bacterial cells. The number of plaques, therefore, becomes a measure of the number of phages present in the original suspension: on incubation each parent phage gives rise to one plaque.[*]

For most work with phage it is necessary to know more about the intracellular events that take

---

[*]"[B]acterial viruses make themselves known by the bacteria they destroy, as a small boy announces his presence when a piece of cake disappears." (from Delbruck, M., and M. Delbruck, 1946. Sci. Amer. (Nov.). Don't small girls steal cake?

place following the infection of a bacterium by a phage particle. At various times throughout these lectures we shall return to this question and add additional layers of complexity to the process. For the time being, we shall content ourselves with understanding the bare bones of the life cycle, as it was elaborated by E. Ellis and M. Delbruck, in 1939. This work is often cited as the beginning of modern phage research.

THE ELLIS-DELBRUCK EXPERIMENT[1]

In investigating the phage life cycle one would ideally like to take a single phage particle, drop it on a single bacterium, and watch what happens. However, since viruses cannot be seen or manipulated individually, some less direct means of watching what happens had to be devised. The approach used by Ellis and Delbruck was to infect simultaneously a large number of bacteria by adding a concentrated phage preparation to a small volume of concentrated *E. coli* growing in liquid culture. After allowing a few minutes for all bacteria to become infected, they added a large volume of sterile nutrient broth in sufficient quantity to dilute the culture of infected bacteria by a factor of 10^5 or 10^6. The purpose of this dilution—which is the key to the whole experiment—is to prevent any bacteria which had escaped infection during the first few minutes from becoming infected during the subsequent course of the experiment[*]. What they in effect achieved, therefore, was a dilute population of synchronously infected bacteria. Since the course of infection proceeds in each cell in step with that in every other infected cell, events happening to the population of infected cells can be extrapolated to the level of a single constituent cell.

To determine what was happening, Ellis and Delbruck took a small aliquot from the infected culture every minute after dilution and spread it on a lawn of *E. coli* growing on solid medium. After incubating these petri dishes overnight, they recorded the number of plaques on each. The results, plotted in the form of a graph, are given in Figure 4.1..

The number of plaques can be seen in the accompanying graph to remain constant up to about 35 minutes following infection. These plaques originate in one of two ways: from unabsorbed parental phage floating in the diluted bacterial cultures (in practice, there should be few if any such unabsorbed phage) and from *infected bacteria* transferred to the solid cultures. Every pre-infected bacterium that settles on the lawn of *E. coli* will, when its load of progeny phages is finally released, initiate a single plaque just as if a single phage had settled in that spot. In the solid cultures derived from samples taken from the liquid cultures during the first 35 minutes

---

[*]A better way to achieve synchrony, and a common variant of the Ellis-Delbruck procedure, is to add potassium cyanide (KCN) to the host bacterial culture at the same time as the phage are added. The effect of the cyanide is to throw the bacteria into a state of metabolic suspension. Since the replication of phage must be a metabolic event, no phage replication can occur in the presence of the cyanide and all phage no matter when they infected a bacterium are restrained in their development at exactly the same point. When the bacterial culture is diluted by fresh nutrient broth, the inhibitory effect of the KCN is instantly lifted and all phage begin replication in synchrony.

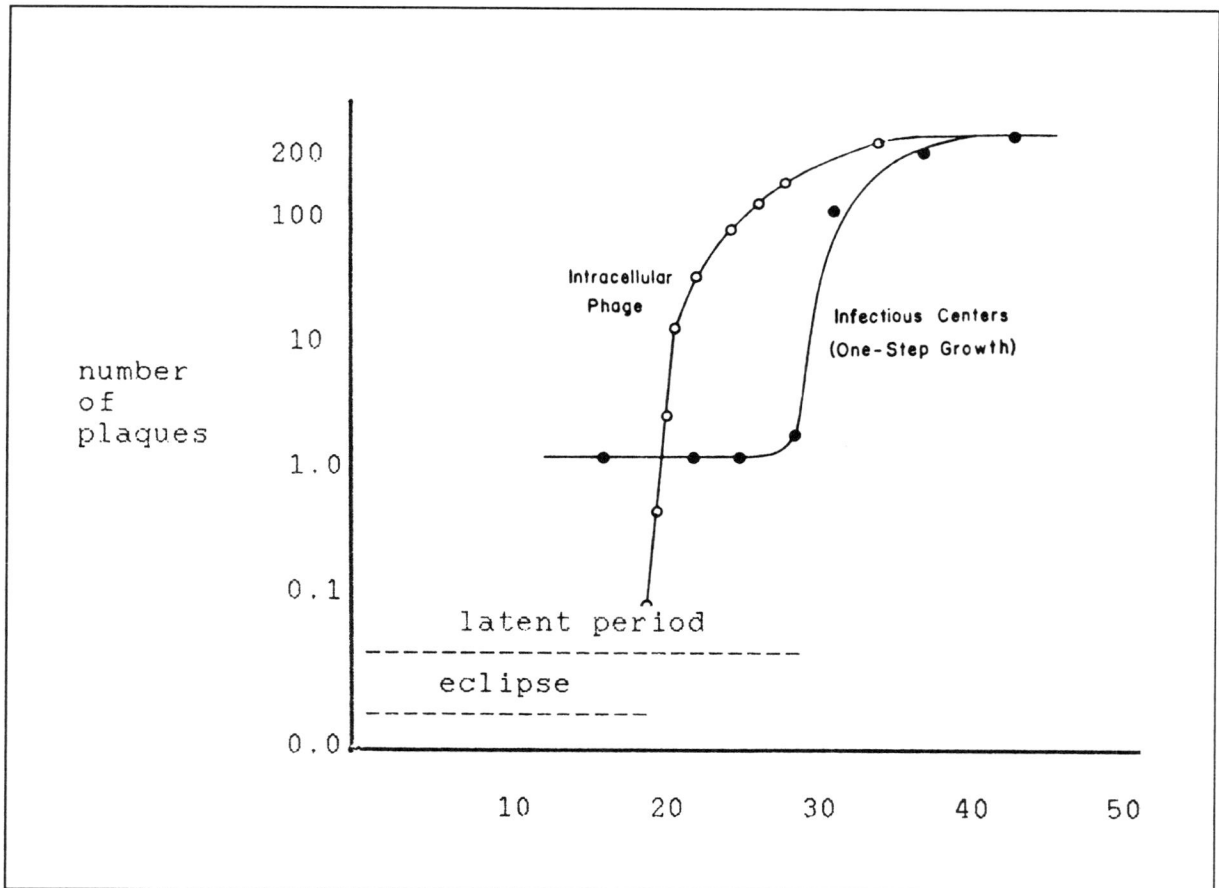

Figure 4-1. The data from the Ellis-Delbruck experiment. See text for explanation.

or so, such infected bacteria are the source of almost all of the plaques.

A great increase in the number of plaques is noted in the lawns seeded with transfers from the diluted liquid culture that were made beginning at about 35 minutes after infection. These plaques originate neither from unabsorbed parental phage nor from infected bacteria, but from progeny phage liberated from the infected bacteria. It must take, therefore, about 35 minutes (at 37°C) for a single phage to infect a cell and to complete one growth cycle, the end of which is signalled by the destruction of the host cell and the release of progeny phage.

The actual number of progeny phages released from each cell can also be deduced from the data. It turns out to be somewhere around 100.

The period of intracellular phage development between infection and lysis is called the **latent period**. The **rise period** is the time interval between when the first cell lyses and when the last cell does—think of popping corn and the time between the first "pop" and the last. Finally the **burst size** refers to the average number of progeny phage released per cell. The fact that T2 phage can reproduce themselves a hundred-fold within about half an hour makes them very

16

attractive organisms for genetic research. Fruit flies, which were once the favorite organism of geneticists, produce about 200 offspring in nine or ten days; in comparison with phage, they seem practically sterile.

DOERMAN'S CONTRIBUTION[2]

One might well wonder what is happening inside the infected bacterium during that 35-minute latent period. Does phage replication begin, for example, immediately upon infection and the latent period represent, therefore, the length of time needed for sufficient numbers of phage to accumulate and rupture the cell? Alternatively, does the latent period represent for the most part the time needed for the accumulation of phage precursors (e.g. head, tail and tail fibre protein, etc.) and only at the very end of the latent period do they suddenly assemble into mature virions? To answer these questions, A. Doerman repeated the Ellis-Delbruck experiment, but added a wrinkle of his own. The infected bacteria which are transferred to the solid culture were first artificially broken open by various harsh chemical or physical treatments and hence only the *contents* of the infected bacteria were spread over the bacterial lawn. From samples taken during the first 20 minutes or so, no plaques resulted, but thereafter a steady increase in the number of plaques was noted until, at about 35 minutes or so, the number was equivalent to what would have been achieved by the natural lysis of the infected cells. These data are also given in Figure 4.1.

This first 20 minute interval is called the **eclipse period** and represents the minimum amount of time needed for the formation of the first progeny particle. During the eclipse period, therefore, there are no infective phages detectable inside the infected bacterium, but thereafter the number of infective particles steadily increases until lysis. It is interesting to note that after infection the infecting *parental* phage must enter a non-infectious state, since it cannot be detected as a plaque-forming particle even if one attempts to rescue it by artificially breaking open the cell. The reason for this will become clear when you learn about the Hershey-Chase experiment in a subsequent lecture.

It can be noted here, parenthetically, that not all viruses lyse the host cell as a consequence of the viral multiplication. Many animal and plant viruses are secreted from the infected host cell over a long period of time. These infected cells are often clearly debilitated, but are nonetheless intact and alive all the while. Eventually, however, the host cell may succumb to the infection, but the end is a less dramatic one than occurs when an *E. coli* cell lyses. In later lectures we shall have occasion to examine in some detail how several other viruses affect cells, especially cells of higher organisms such as mammals.

BACTERIAL TRANSFORMATION[3]

The third strand we shall pick up concerns a phenomenon in bacteria called **transformation**.[*] To appreciate transformation, it is first necessary to know something about a particular bacterium, *Diplococcus pneumoniae*, which, as its name implies, causes pneumonia. These organisms usually exist as pairs of cells around which is secreted a single external cell wall. The virulent variety of *Diplococcus*, when grown in a petri dish, gives rise to colonies which have an over-all slimy appearance, and for that reason are called Type S for "Smooth". The "smooth" characteristic is founded on the ability of Type S cells to manufacture a carbohydrate substance (a mucopolysaccharide), which is the secreted outer wall. The smooth characteristic is, therefore, a hereditary property of this strain of *Diplococcus pneumoniae*.

There is also another strain of this species called Type R, for "Rough". Type R *Diplococcus pneumoniae* cannot manufacture the carbohydrate substance found in the Type S, and related somehow to this deficiency is their inability to cause pneumonia if injected into various mammals.[**] If grown in a petri dish, Type R colonies have quite a different appearance from Type S, and although one wouldn't exactly describe the colonies as "rough", they are nonetheless called Rough just to distinguish them from Smooth.

In 1928 F. Griffith discovered that heat-killed Type S if injected into a mouse simultaneously with live Type R caused the death of the mouse within just a few days. More startling still was the observation that the lungs of the dead mice teemed with live Type S cells. Live R or dead S injected alone did not produce death nor any symptoms whatever of pneumonia, nor did they ever by themselves give rise to virulent live S cells. A few years later it was shown that the mouse was merely incidental in the original observations: live R cells cultured *in vitro*[***] in the

---

[*]Despite the identical terminology, this "transformation" is entirely unrelated to the "glassy transformation" phenomenon described by Twort. Twort's terminology never caught on, and the word "transformation" has become available to describe other events. For the benefit of the more widely read student, it should be noted that the term "transformation" also crops up in studies of animal carcinogenesis. In this instance a transformed cell becomes metabolically altered and loses its "contact inhibition" when growing in tissue culture. Normal cells stop proliferating *in vitro* after they form a single confluent monolayer over the anchoring surface; transformed cells, in what is thought to be an *in vitro* manifestation of a neoplastic growth *in vivo*, keep growing and consequently form layers several cells in thickness.

[**] For one thing, they elicit a greater immune response than do Type S.

[***] In vivo means, roughly, "in life" or more aptly "in the living cell, tissue, animal, or whatever"; in vitro means "in glass" or "in a non-living environment" (such as a test-tube); in situ means "in place" or "in the same location as in the living organism". The terms are ubiquitous

presence of dead S cells would spontaneously give rise to some colonies of live S. The process whereby live R is converted into live S by the presence of dead S is called **transformation**. The transformed cells (i.e. the newly-arising Type S colonies) are in no way distinguishable from regular S colonies or cells and do not revert to Type R in the course of their further growth. It is as if the R cells that became transformed have experienced a permanent alteration of their fundamental characteristics—as if, indeed, they had suddenly acquired all the hereditary information to enable them now to make the carbohydrate cell wall substance for themselves. In fact, as we shall see, this is precisely what had happened.

REFERENCES

1. Ellis, E. and M. Delbruck, 1939. J. Gen. Physiol. 22: 365

2. Doerman, A., 1952. J. Gen. Physiol. 35: 645.

3. Griffith, F., 1928. J. Hygiene, Cambridge 27: 113.

---

in the biology literature.

LECTURE 5

THE CHEMICAL NATURE OF THE HEREDITARY MATERIAL

INTRODUCTION

The follow-up, many years later, to Griffith's discovery of bacterial transformation, provided a strong clue to the chemical identity of the hereditary material. We shall also see in this lecture the roles played by tobacco mosaic virus and by bacteriophage in unravelling this puzzle.

CHEMICAL NATURE OF THE TRANSFORMING PRINCIPLE

Transforming principle is the name given to the chemical agent or agents that caused the R *Diplococcus* cells to become Type S. If one accepts the notion that the R cells on being transformed had actually acquired the S genes needed to direct the synthesis of S-specific carbohydrate, then the transforming principle would be these genes themselves. Identifying chemically this transforming principle would be tantamount to identifying the chemical responsible for heredity.

This challenge was successfully taken up by a team of three American investigators, O. Avery, C. McLeod, and M. McCarty, in 1944.[1] Instead of mixing whole heat-killed Type S with live R, the investigators first extracted one or another class of chemicals from the dead S and then added only this extract to the growing Type R cells. Contrary to expectations, both proteins and carbohydrates from S cells were found to be wholly incapable of effecting the transformation, but the greater surprise came when S-derived nucleic acids were assayed. Nucleic acid, specifically the type of nucleic acid known as **deoxyribonucleic acid (DNA)**, a chemical which up to then had been thought to be of little significance to the cell, was found to be the only chemical in S cells capable of transforming Type R into S. Furthermore, it was only from S cells that active transforming principle could be obtained; DNA taken from any other bacterial, plant or animal source can never bring about this particular transformation.

Could it be that genes were not proteins after all, but were really composed of nucleic acid? Most scientists at the time thought not. For one thing, critics quickly pointed out, the transforming DNA Avery, McLeod and McCarty obtained from S cells was not 100% pure. A residual amount of protein still contaminated their best DNA preparations. Agreed, replied the three investigators, who then countered with the observation that DNAse (an enzyme that degrades DNA specifically) would completely inactivate transforming principle, but various proteases (enzymes that specifically degrade proteins) had no effect. Moreover, as more and more highly purified DNA preparations were obtained from S cells, the transforming ability of the DNA increased. It would be indeed strange for a chemical to become more active the less of it remained in the preparations. Interesting, thought their critics, who nonetheless remained largely unconvinced. We shall see many additional examples of the immutability of a scientist's mind once it is made up. Science advances not because dissenting investigators change their opinions under the weight of new evidence but because they die or retire and a new group with its own strongly-held but different prejudices takes over for awhile.

Table 5-1. Radioactive isotopes commonly used in biology.

ISOTOPE	EMISSION	HALF-LIFE
C^{14}	beta	5570 years
H^3	beta	12 years
P^{32}	beta	14 days
S^{35}	beta	87 days

INFECTIOUS NUCLEIC ACID FROM PLANT VIRUSES

In an earlier lecture, tobacco mosaic virus (TMV) was described as being composed only of protein and nucleic acid. In the mid-1940s, G. Schramm succeeded in chemically separating each of these components free of the other, and later H. Fraenkel-Conrat and R. Williams demonstrated that active virus particles could be reconstituted simply by mixing this purified TMV protein and nucleic acid.[2] This suggested, of course, that, given a large number of free TMV capsids and full-length pieces of TMV nucleic acid, the morphogenesis of active TMV virions was essentially a spontaneous event.[*]

Are both the protein and the nucleic acid necessary for the replication of TMV? To answer this question, Fraenkel-Conrat took healthy tobacco plants and rubbed the leaves with either the pure TMV protein or the pure TMV nucleic acid. Their first results were disappointing, for neither preparation seemed capable by itself of initiating a tobacco mosaic infection. Further work revealed, however, that if the surface of the leaf was gently abraded before the preparation was applied, TMV nucleic acid, but not TMV protein, was fully capable of producing a systemic

---

[*]At the time, the media, as is their wont, described the reconstitution of active virus from the separate components as "THE CREATION OF LIFE IN A TEST TUBE". It is no such thing, of course, since both the nucleic acid and the protein had previously been synthesized in living cells--those of the tobacco plant host. A nearer approach to this newspaper headline is the recent success in producing active polio virus by supplying polio nucleic acid (an RNA) to an *in vitro* preparation of material extracted from human cells. The polio proteins were synthesized in their entirety and correctly assembled into normal virus particles within this cell-free extract. Since it is technically feasible to manufacture the whole of the polio nucleic acid from scratch, this does represent the complete *in vitro* synthesis of an active virus. But is a virus alive?

TMV infection. From plants infected in this way large quantities of perfectly normal whole TMV particles could then be isolated.[*]

The full implications of this experiment must not be overlooked. By rubbing only the nucleic acid portion of a TMV particle on to healthy tobacco leaves, whole TMV particles—composed of both nucleic acid *and* protein—were obtained. The nucleic acid portion of TMV must, therefore, be able by itself to encode all the information needed in order to have both the TMV nucleic acid and the TMV protein synthesized during the course of the infection. In other words, the complete instructions for TMV synthesis reside in the nucleic acid, which must, in consequence, constitute the hereditary material. The original observation that the naked nucleic acid cannot unaided initiate an infection gives a clue to the role of the protein capsids: they facilitate entry of the nucleic acid into the plant cell. They also serve to protect the viral nucleic acid from damage when the particles are in the free state. If an alternative means of gaining entry to a plant cell is provided, as happens, for instance, when the leaves are rubbed gently with an abrasive to weaken the plant cell walls, naked TMV nucleic acid can then enter these cells and once inside can both replicate itself and direct the synthesis of TMV capsid protein. Whole TMV particles are then released from these cells and these can in turn infect other cells of the plant in the usual manner.

To demonstrate conclusively that it was the nucleic acid and not some contaminating intact viruses that was responsible for the infection, Fraenkel-Conrat and his co-workers performed on further test. They discovered a strain of TMV that caused a tobacco mosaic disease with recognizably different symptoms than those caused by the standard virus. This difference is the result of a minor chemical alteration in the coat protein of the variant virus[**], which is called strain HR. The investigators then combined *in vitro* the nucleic acid from HR with the protein from normal TMV to make a **hybrid virus**. When this hybrid virus infected tobacco plants, the disease that resulted was characteristic of HR—the source of the nucleic acid—and analysis of the coat protein from the progeny viruses that were a consequence of this infection showed conclusively that they carried the HR alteration. The conclusion seemed inescapable that the hereditary properties of the virus were determined only by its nucleic acid.

---

[*]The observation that TMV RNA by itself could infect was actually the consequence of a piece of work by B. Singer (Mrs. H. F.-C.) that had an entirely different objective: the purification of TMV RNA entirely free of contaminating intact TMV particles. Since it was thought that only the intact particles could be infectious, RNA preparations which retained measurable infectivity were initially thought to be contaminated with intact particles. However, it seemed impossible, despite many attempts, to eliminate a very low level of infectivity from the supposedly pure RNA preparations, which then led to the hypothesis that perhaps the pure RNA was itself all that was needed for an infection to take place. The work was then redirected to finding ways to *increase* rather than decrease this residual infectivity in pure RNA preparations.

[**]The HR variant contains two amino acids, histidine and methionine, not found in the common TMV strain.

The nucleic acid found in TMV is not DNA but a chemically related type called **RNA**, for **ribonucleic acid**. The use of RNA as hereditary material is limited to certain viruses, many of which, like flu, polio, rabies, HIV, etc., are common pathogens of humans. The demonstration that the hereditary properties of TMV reside with its RNA instead of its protein, like Avery, McLeod, and McCarty's demonstration that transforming principle was DNA, made little impact on scientific thought at the time. Viruses were strange creatures anyway, and what applied to them might very well not apply to more conventional organisms like ourselves. Viruses may not even rightfully be called organisms at all. Small wonder, therefore, that Fraenkel-Conrat's work, while undeniably interesting, could easily be dismissed as irrelevant to questions concerning the nature of the hereditary material.

DNA AS THE HEREDITARY MATERIAL OF PHAGE[3]

The third demonstration that the hereditary material was more likely nucleic acid than protein was carried out by A. Hershey and M. Chase in 1953. Although no one would care to deny the cogency of these experiments, one must nonetheless admit that their immediate acceptance as proof of the role of nucleic acid had much to do with factors extrinsic to the experiments themselves. Hershey and Chase were at the time within the inner circle of molecular biologists and were also using an "organism" (T_2 phage) that had considerable faddish appeal. Moreover, they carried out their work just at the eleventh hour before the natural birth of an idea whose time had finally come anyway.

To follow the work of Hershey and Chase it is first necessary to learn something about the use of radioactive tracers. There is probably no other single technical advance that has done more to further our understanding of cell metabolism than has the introduction of radioactive tracers into biological experiments, but it was not until the 1940s that they became available in sufficient and reliable supply to revolutionize biochemistry and usher in what is now termed molecular biology. Radiotracers are biological compounds which have incorporated unstable emitting atoms. Commonly the emission is a beta particle, although alpha, neutron and gamma emitters sometimes find special uses. The emission is detected with the appropriate instrument (e.g. geiger counter, scintillation counter, photographic film, etc.) and hence the presence of the molecules that had incorporated the label is made manifest. Since all molecules of a particular kind will usually behave more or less identically in a given separation procedure, only a few need be labelled in order to tell where the lot of them is. Trace amounts of a substance that would elude the most sophisticated chemical analysis can readily be detected by the presence of this radioactive label.

One can also use two or more atomic tracers simultaneously, even when both of them emit the same type of particle. This is possible because the energies with which the particle is emitted will likely differ in the two types of atoms. The detecting instrument can first be adjusted to detect all the particles emitted and then only the stronger of the two emissions. The presence of the weaker can then be deduced by subtraction.

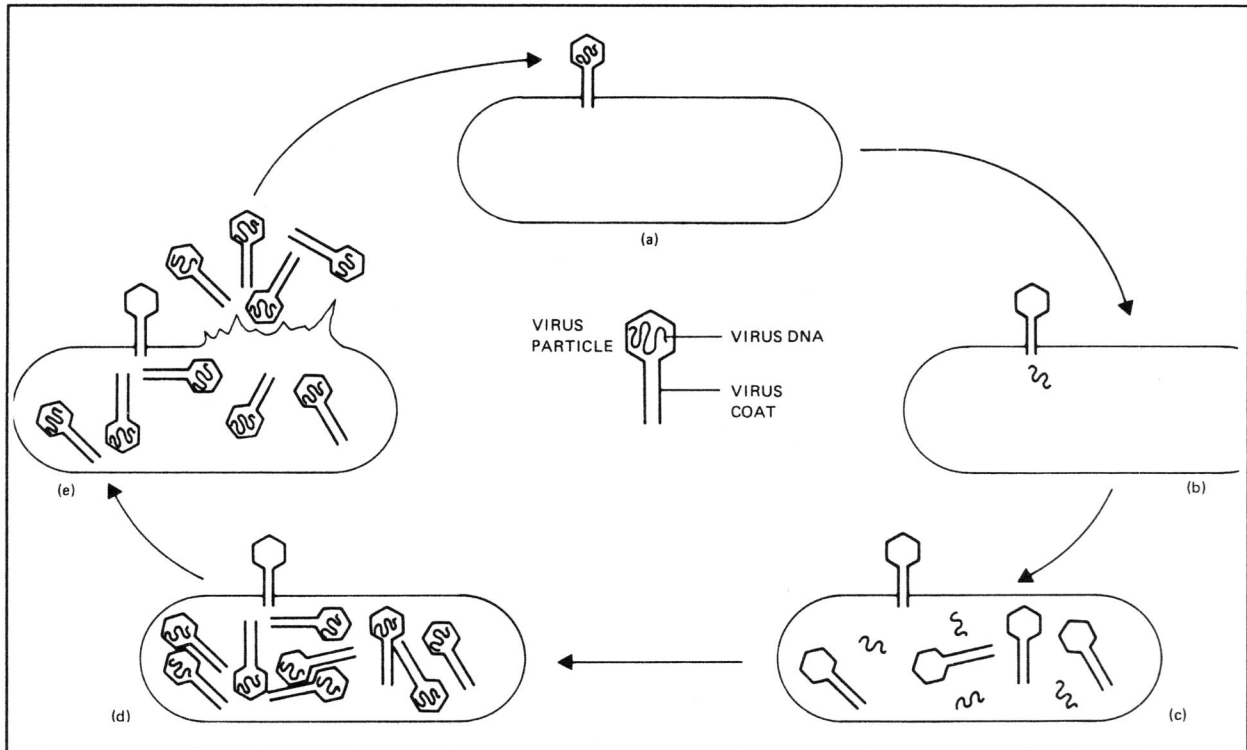

Figure 5.1 The life cycle of the bacteriophage. From CELL BIOLOGY, THIRD EDITION by John W. Kimball. Copyright © 1984 by John W. Kimball. Reprinted by permission of the author.

The experiment by Hershey and Chase has several parts. The first was the simple demonstration that the phage nucleic acid could be separated "from the phage protein by subjecting the phage to a sudden osmotic shock, say by diluting them with a large volume of distilled water. Under these conditions the phage head ruptures and the nucleic acid sprays out, leaving behind the collapsed protein structure now called a **ghost**. The ready separation of the protein and nucleic acid components of the phage gave the clue that something akin to this may happen during the process of infection.

To test this idea some means had to be devised to label differentially the nucleic acid and protein components. Fortunately it happens that nucleic acid is rich in phosphorus, an atom lacking in phage proteins, and phage protein is rich in sulphur, something not found in nucleic acid. To incorporate radioactive phosphorus (P^{32}) into phage nucleic acid or radiosulphur (S^{35})[*] into phage protein may seem at first glance a formidable task in itself. All one has to do, however, is to grow ordinary phage on bacteria themselves growing on significant quantities of P^{32} or S^{35}, and all the progeny phages will of necessity have incorporated that label into their nucleic acid or

---

[*]To be rigorously correct, the numerical superscript should precede the chemical symbol for the element. However, since scientists refer to the isotope as, for instance, "P-thirty-two", many authors write the mass number superscript following the symbol, as I do.

their protein, as the case may be. (As "eaters of bacteria", the phage derive their components from recycled bacterial constituents, just as we do from our own foodstuff.)

Hershey and Chase then took phages whose nucleic acid was labelled with P^{32} and infected them on ordinary *E. coli* growing in liquid culture. After allowing a few minutes for the bacteria to become infected, they dumped the infected bacteria into an ordinary kitchen blender and switched it on. The commotion that this set up in the culture was such that anything loosely attached to the surface of the bacterium would be stripped off, whereas anything firmly attached or incorporated inside the cell would remain unaffected. They then pelleted the bacteria by centrifugation and tested them for the presence of radiophosphorus. Most of the phosphorus remained with the cells, which in due course lysed to release the usual number of progeny phage.

When this experiment was repeated using phage labelled with S^{35}, quite a different result was obtained. Almost all the S^{35}, that is, almost all the phage protein, was successfully separated from the infected bacteria by the blender treatment. This protein would be found in the supernatant of the centrifuged culture and not with the pelleted bacterial cells. Nevertheless, these cells did lyse at the expected time and released about the same number of progeny phage as would infected cells not so treated.

Hershey and Chase's work reveals, therefore, that only the phage nucleic acid enters the bacterial cell in quantity and that most or all of the phage protein remains on the cell surface. This protein can then be stripped away without in any way affecting the outcome of the infection. This protein must be just a vehicle to ensure the injection of the phage nucleic acid and must not be part of the hereditary apparatus of the particle. Once inside the cell the phage nucleic acid alone (which is DNA, in this case) can direct the manufacture not only of more of itself but also of all the unique proteins found in the mature T2 virion.

In summary, by 1953 the scientific world was faced with three quite different lines of evidence—from plant viruses, from bacteria and from bacterial viruses—that converged on the same unambiguous conclusion: that nucleic acid and not protein was responsible for carrying hereditary information. At long last scientists had become convinced, and soon only cranks and fools believed otherwise.

REFERENCES

1. Avery, O., McCleod, M., and McCarty, M., 1944. J. Exper. Medicine 79: 137.

2. Fraenkel-Conrat, H. and R. Williams, 1955. Proc. Nat. Acad. Science 41: 690
   Fraenkel-Conrat, H., 1956. J. Amer. Chem. Soc. 78: 882

3. Hershey, A., and M. Chase, 1952. J. Gen. Physiol. 36: 39. (For an interesting discussion on how this experiment has been variously reported over the years, see Wyatt, H., 1974. Nature 249: 803)

LECTURE 6

CHEMISTRY OF NUCLEIC ACIDS

Considering the fanfare nucleic acids have been given in the previous lectures, you now should not be surprised to hear that much of molecular biology hinges on their activities. It is time to learn something about their chemistry and structure.

EARLY STUDIES ON DNA[1]

The discovery of nucleic acids is usually attributed to F. Miescher in 1868.[2] Miescher was interested in the chemistry of the cell nucleus and devised a method (extraction with dilute, cold hydrochloric acid) to isolate cell nuclei free from significant amounts of contaminating cytoplasm. The cells he chose to work with were pus cells (leucocytes) which he was able to collect fresh and in large quantities from the unlucky patients of a nearby surgical clinic. After drying the pure preparation of pus cell nuclei, he subjected the residue to various chemical analyses and discovered that a significant portion of the nucleus was composed of a heretofore unknown phosphorus-rich compound he called **nuclein**. We would now call it **nucleic acid**.

Around the turn of the century a German school of biochemists began to study nucleic acid in earnest. First A. Kossel discovered that nucleic acid was composed of four **nitrogenous bases**: guanine and adenine (both representatives of a class of compounds called **purines**), and cytosine and uracil (both of which are **pyrimidines**). Then in the 1920s P. Levine and W. Jones showed that there are really two fundamentally different kinds of nucleic acid, depending upon which type of five-carbon sugar (a **pentose**) is present in the molecule: RNA has a **ribose** sugar and DNA has **deoxyribose**. As a second difference between them, DNA has the pyrimidine **thymine** instead of **uracil**, the latter being found only in RNA.

For awhile it was thought that RNA was found only in plants and DNA only in animals; RNA was in consequence often called "plant nucleic acid" and DNA "animal nucleic acid". Later, however, it was shown that both types of nucleic acid are found in all plants and animals. The bulk of the DNA in the cell is found associated with the chromosomes in the nucleus, a fact first appreciated by R. Feulgen in 1924. RNA, by contrast, is found mainly in the cytoplasm. Cells with a big nucleus relative to the volume of their cytoplasm (e.g., pus cells, salmon sperm, thymus gland cells) will appear to contain mostly DNA. Cells where the reverse is true (e.g., many plant cells) will yield mainly RNA.

To make the identification of particular chemical bonds easier, the atoms in the ring structures of both the nitrogenous bases and the sugars are numbered as shown in the accompanying representations of their structures. To distinguish between a particular atom in a nitrogenous base and one bearing the same number in the pentose sugar, the latter numbers are given a superscript stroke (called a "prime"). Identifying a particular linkage as being, for instance, "three prime" (3') immediately indicates that it involves the number three carbon of the sugar moiety.

Individual nitrogenous bases bond chemically with the ribose or deoxyribose sugars to form a structure called a **nucleoside**. Since there are two types of sugars and five types of nitrogenous bases (A, T, G, C and U, in standard abbreviation), there are ten theoretically possible nucleosides. However, as mentioned earlier, uracil is not found in DNA nor is thymine found in RNA. Consequently, in practice one finds in nature only eight different nucleosides.*

Since neither the nitrogenous bases nor the two types of sugars contain the element phosphorus and since phosphorus formed the basis of Miescher's original identification of nucleic acid, it must be found somewhere else than as part of a nucleoside. Indeed, phosphorus, in the form of the phosphate ion, can be found attached either at the 5' or 3' carbon of the ribose or deoxyribose sugar. A nucleoside with an attached phosphate group is called either a **nucleoside phosphate** or a **nucleotide**.

Up to this point we have been treating DNA and RNA as pretty much alike. In the chemical sense, this is a justifiable approach, as both nucleic acids are compounded of the same basic building blocks, the four nucleotides, and are, moreover, very similar in three of these four. Structurally, however, DNA and RNA are very dissimilar, and as the discussion goes on to describe in detail these structures and how they were discovered, it is best to focus on DNA and RNA separately. Since only certain viruses use RNA as the hereditary material, the importance of DNA in this respect vastly overshadows that of RNA. Hence DNA structure will be discussed first and at greater length.

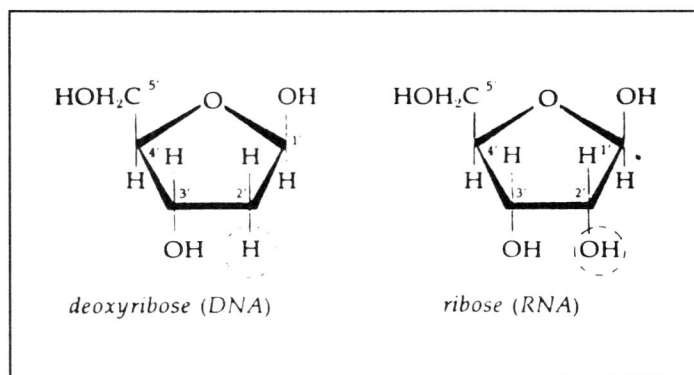

Figure 6-2. The two pentose sugars found in nucleic acids.

EARLY STUDIES ON THE STRUCTURE OF DNA

Early measures of the adenine-, guanine-, cytosine-, and thymine-containing deoxyribonucleotide frequencies in DNA suggested they were present in equivalent amounts. Without much more to recommend it than this single datum, a theory on the structure of DNA became so widely

---

*T_2 and T_4 phages do not have ordinary cytosine in their DNA; instead they have a methylated derivative, 5- hydroxymethyl cytosine. Other chemical variations are also known. Transfer-RNA (discussed later in these lectures) is particularly rich in unusual bases.

accepted that it took on in time the force of fact.[*] It held that DNA could be interpreted structurally as the four nucleotides chemically bonded through their phosphate groups into a large ring or else was a repeating polymer of this four-nucleotide unit. The **tetranucleotide**, as it was called, was only challenged when new analytical techniques became available in the 1940s to demonstrate that no such equivalence of the nucleotides exists in most natural sources of DNA.

The technique which was most useful in these new analyses was **paper chromatography**. In its simplest form, paper chromatography involves, first, the spotting of the chemical mixture to be separated on to a piece of absorbent filter paper and then the passage of a non-polar solvent-water mixture along the paper by simple capillary attraction, starting from the end that bears the spot. The components of the chemical mixture will often be carried different distances along the

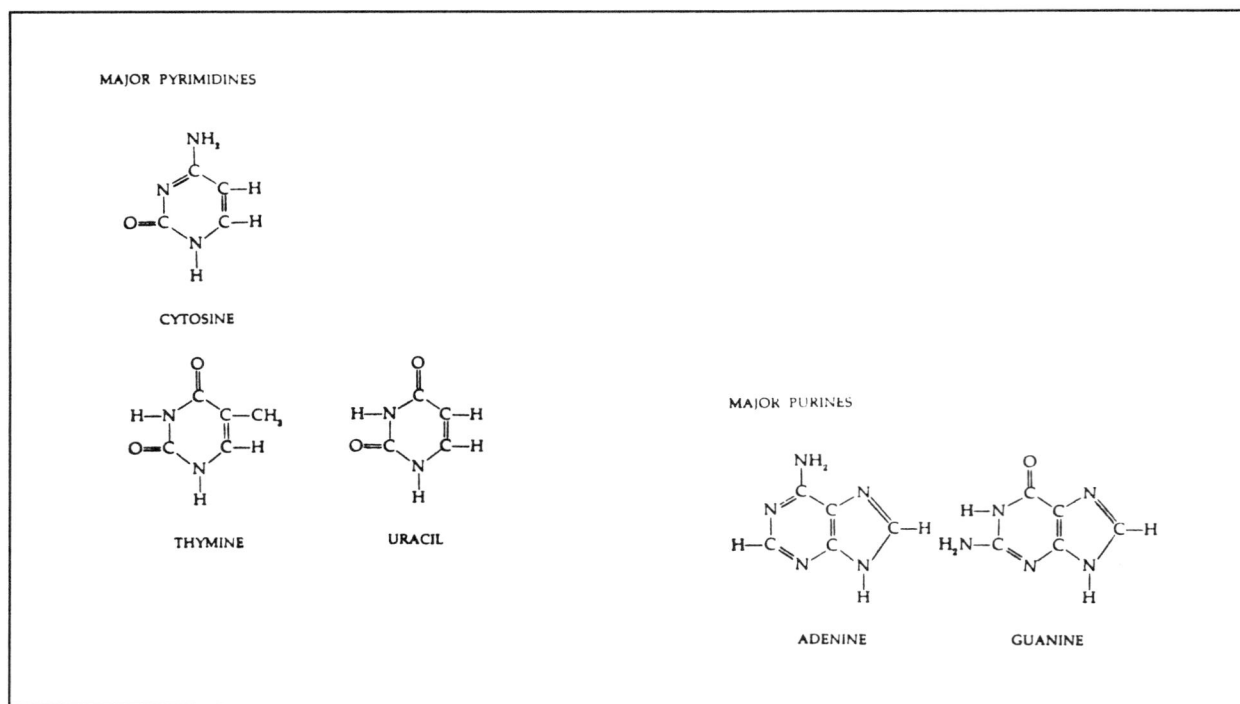

Figure 6-1. The nitrogenous bases found in nucleic acids.

paper by the solvent because each chemical compound will have a unique relative affinity for the solvent and the water that becomes bound by the cellulose fibers in the paper. Compounds that have a greater affinity for the solvent than for the stationary water will be carried further along than those that have the reverse affinities. The result will be a series of spots on the paper which

---

[*]One is reminded here of Mark Twain's sly comment: There is something fascinating about science. One gets such wholesale returns of conjecture out of such trifling investments of fact.

can be cut out, redissolved and further analyzed. Alternatively, the paper can be turned 90° and re-run with a different solvent. This is called **two-dimension chromatography**. The second solvent will often resolve two or more compounds that stayed together in the first run.

Adopting paper chromatography to the study of DNA, E. Chargaff[3] was able to show convincingly that, although the equivalences expected under the tetranucleotide theory were not found, there were other real equivalences that proved equally interesting. For one thing, Chargaff showed that for a variety of DNAs from sources widely separated on the phylogenetic scale, the total amount of purine bases was always equal to the total amount of pyrimidine bases (i.e., A + G = C + T), and also that the amount of adenine always equalled the amount of thymine, and the amount of guanine equalled the amount of cytosine (i.e., A = T and G = C). These ratios, unexpected and unexplained, became known as **Chargaff's Laws**. They spelled the end of the tetranucleotide theory of DNA structure but offered no obvious alternative in its place.

The second technical advance that played a large part in elucidating the structure of DNA was **X-ray crystallography** or **X-ray diffraction**, a physical rather than a chemical technique. It involves the passage of a beam of X-rays through a crystal and the capturing on photographic paper of the pattern formed by the deflected (or "diffracted") rays. These rays have, of course, been diffracted by collision with the assemblages of atoms making up the crystal; since crystals have fairly regular structures (or else they would not be

Table 6-1. The base compositions of DNA from various organisms.

	Adenine %	Thymine %	Guanine %	Cytosine %
Man (sperm)	31.0	31.5	19.1	18.4
Salmon	29.7	29.1	20.8	20.4
Sea urchin	32.8	32.1	17.7	17.7
Yeast	31.7	32.6	18.8	17.4
Mycobacterium tuberculosis	15.1	14.6	34.9	35.4
Escherichia coli	26.1	23.9	24.9	25.1
Vaccinia virus	29.5	29.9	20.6	20.3
E. coli bacteriophage T2	32.6	32.6	18.2	16.6*

*5-Hydroxymethyl cytosine.

crystals), it is possible to glean a lot of information about the regular arrangement of the atomic subunits by analyzing how these diffract X-rays. Still, the process is very painstaking and the amount of information gained often disappointingly small; one writer has compared it with trying to piece together the arrangement of the glass pendants in a complex chandelier by observing the pattern of light the chandelier throws on a nearby wall.[4]

It was R. Franklin in the early 1950s who did most of the important X-ray diffraction studies on DNA. She showed, first of all, that DNA could be crystallized, which in itself meant that DNA did have a regular, repeating structure. She then showed that something gave the molecule a stacked quality, similar to the effect of a vertical roll of pennies. Her pictures also showed that DNA was helical in shape, although she herself seemed for awhile reluctant to accept this conclusion.

REFERENCES

1. Mirsky, A., 1968. The discovery of DNA. Sci. Amer. (June).

2. Mirsky, A., 1968. The discovery of DNA. Sci. Amer. (June).

3. Chargaff, E., 1950. Experimentia <u>6</u>: 201.

4. A good account of the basic principles of X-ray diffraction is given in Perutz, M., 1964. The haemoglobin molecule. Sci. Amer. (Nov.).

LECTURE 7

THE WATSON & CRICK STRUCTURE OF DNA

"Before I knew of Zen," runs the proverb, "mountains were
mountains and waters were waters; when I was studying Zen,
mountains were no longer mountains and waters no longer waters;
then I achieved enlightenment—and mountains were mountains and
waters were waters." So, upon the elucidation of the structure of
deoxyribonucleic acid, the remaining problems of biology were
transformed—and were still there.

H.F. Judson[1]

MOLECULAR MODEL BUILDING

Probably R. Franklin and her collaborator, M. Wilkins, would have eventually pieced together
the structure of DNA had they continued these lines of investigation for another year or two.
They were not, however, given this opportunity, for in 1953 two other investigators, J. Watson
and F. Crick, published an essentially correct structure for DNA.[2] Watson and Crick relied
heavily on both the chemical data of Chargaff and the physical data of Franklin and Wilkins as
well as on a technique which, while not original with them, paid off most spectacularly in their
hands: molecular model building. They made models of the nitrogenous bases and then of the
whole nucleotides and tried fitting them together in ways that best accorded with the data of
Chargaff and Franklin. The result is their famous double-stranded helical model of DNA[*]. The
two strands are held together by hydrogen bonds that form between A and T and G and C, and
the adjacent nucleotides on the same strand are connected by covalent bonds that link the 5'
carbon of one nucleotide to the 3' carbon of the next through the phosphate ion (chemists call
this linkage a **3'-5' phosphodiester bond**). The molecule resembles a twisted ladder, with the

---

[*] Because of the importance DNA has assumed in modern biology, it might be justified in
this instance to go into what the *New Yorker* would style "wretched excess" in documenting this
discovery. The actual structure of DNA was discovered and the paper cut-out model assembled
on February 28, 1953. The first space-filling model was assembled on March 7, in the south-
west corner of Room 103, Austen Wing, Cavendish Labs, Cambridge University. The 128-line
paper outlining this work was published in the April 25, 1953, issue of Nature, vol. 171, pages
737-8. However, one shouldn't get the impression that figuring out the structure of DNA was
an especially "difficult" scientific achievement. Had Watson and Crick not done it when they
did, others would surely have soon hit upon the same solution to the problem. In that sense,
determining the structure of protein, for example, was a vastly more "difficult" problem to tackle.

31

inwardly projecting, hydrogen-bonded bases the rungs and the sugar-phosphate linkages the two rails.

Hydrogen bonds were discussed in Lecture 2, where it was noted that some highly electronegative atoms such as Oxygen (and also Nitrogen) can strongly attract Hydrogen atoms that formally "belong" to another atom. This partially shared Hydrogen then serves to weakly bond its covalent partner with the Oxygen (or Nitrogen). Although these hydrogen bonds individually have only about 5% the strength of a typical covalent bond, any reader of *Gulliver's Travels* knows the fabled strength of a multiplicity of individually weak bonds.

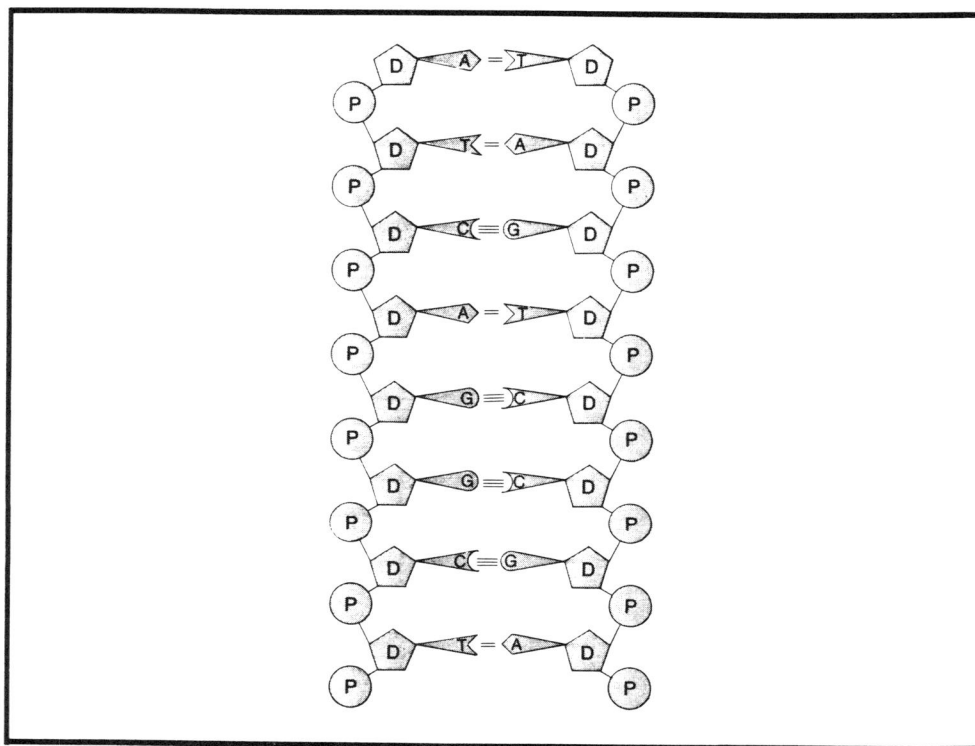

Figure 7.1 Diagram of the complimentary strands of DNA. A = adenine;T = thymine;C = cytosine;G = guanine;D = deoxyribose;P = phosphate group. From BIOLOGY: THE ESSENTIAL PRINCIPLES, SECOND EDITION by Thomas M. Graham. Copyright © 1987 by Kendall/Hunt Publishing Company. Reprinted by permission.

In DNA, two hydrogen bonds spontaneously form between adenine and thymine juxtaposed in the right orientation, and three between guanine and cytosine. Their positions are shown in the accompanying diagrams.

One feature of the structure proposed by Watson and Crick which may not be obvious on first glance is the fact that the two complementary strands of DNA "point" in opposite chemical directions. This can best be discerned by examining the pentose moieties of the two strands depicted in Figure 7.1. In one case, the oxygen in the ring structure points upward; in the complementary strand it points downward. This inversion of one strand with respect to the other is called **antiparallelism**, and in loose parlance, one says that, chemically speaking, the strands "run in opposite directions". Another manifestation of the same phenomenon can be seen at the two ends of the double helix. At the 'top' end, one strand ends in a 3' carbon and the complementary strand ends in a 5' carbon; at the opposite end, the reverse is true. This may seem like a trivial observation at present, but later we shall see that many aspects of nucleic acid function hinge on

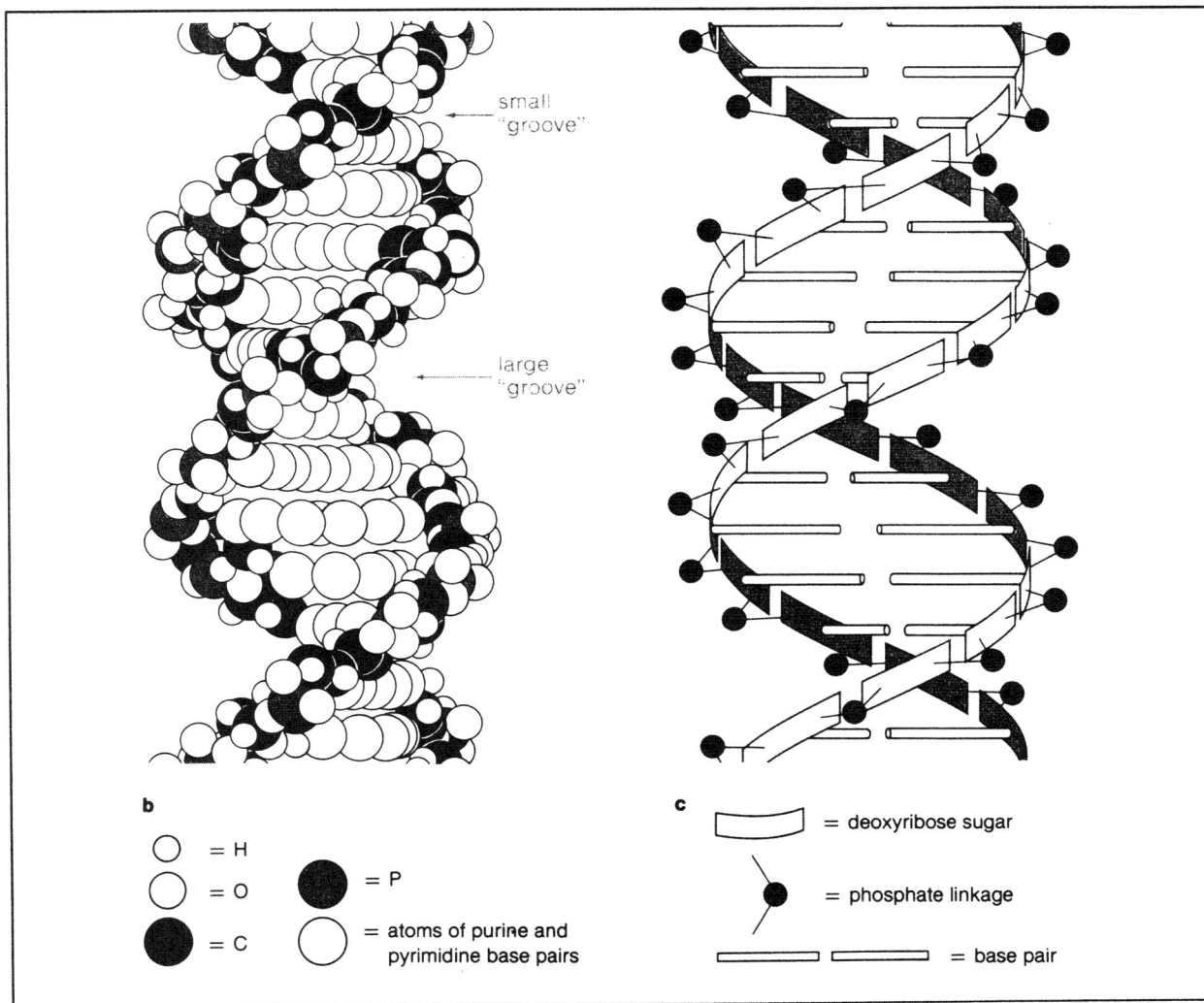

Figure 7.2 The double helical structure of DNA. From BIOLOGY OF THE CELL, 2ND EDITION by Stephen L. Wolfe. Copyright © 1981 by Wadsworth, Inc. Reprinted by permission of the publisher.

the particular directionality of the molecules in question.

Finally, it should be noted that the base-pairs found along the length of the DNA molecule can occur in any sequence as long as the complementarity of the bases comprising the pairs is maintained. DNA is, therefore, a regular polymer but one of indefinite length and sequence. These base-pairs appeared in Franklin's X-ray diffraction photographs as the "stacked" part of the molecule. Each base-pair is off-set from the one below by 36°, meaning that there will be 10 base-pairs per complete turn of the helix. Since the base-pairs are separated by a distance of 3.4Å along the molecule, one complete turn of the helix will occur every 34Å. The radius of the molecule is 10Å.[3]

A space-filling model of DNA (see Figure 7.2) shows the existence of two prominent grooves spiralling (or more correctly "helicating", if that is a word) along the molecule. The wider one is known as the **major groove** and the narrower one, the **minor groove**. Through these grooves the four possible base-pair arrangements (A-T, T-A, G-C, C-G) can be recognized from the outside of the molecule. This consideration looms important in discussions on how regulatory molecules can interact with particular base-pair sequences of an intact double-stranded molecule.

The number of DNA base-pairs that constitute the genetic contribution of a human egg or sperm is staggering to contemplate. There are about three billion base-pairs in a sperm. Even if one shortens the name of each base to a single letter (A,T,C, and G), it would take about the number of pages in 13 *sets* of the *Encyclopedia Britannica* just to write them all down.

The classical DNA structure outlined above is known as the "B structure" and is the one in which almost all natural DNA is found. There are, however, other structures DNA can assume. One of these, the **Z structure**, is found occasionally in sections of otherwise "normal" DNA and is thought to be important in regulating DNA function, but its exact significance (if any) is not yet understood. Another structure is triple-stranded. Here, the third strand has only two bases, thymine and cytosine. The former pairs with AT pairs of the duplex, and the latter with GC pairs. As far as is known, triple-stranded DNA does not occur in nature and therefore it has no biological significance. But it can be constructed in the laboratory, where it is useful in certain aspects of genetic engineering (see later Lectures).

REPLICATION OF DNA

In their historic first publication on DNA structure, Watson and Crick pointed out that they were not unaware of the potential their model held for explaining how the genetic material could store genetic information, mutate and replicate. Mutation and the storage of genetic information will be dealt with at a later time; here we shall focus on how the DNA helix replicates.

In essence Watson and Crick proposed that replication involves, first, the "unzipping" of the double helix by a breaking of the hydrogen bonds holding the two strands together, and then, closely following this first step, the formation of two "daughter" molecules as the unpaired portions of the "parent" molecule attract complementary nucleotides from the cytoplasmic milieu. (An abundance of the four deoxyribonucleotides would presumably be synthesized by the cell before DNA synthesis began.) These new nucleotides would then be joined laterally by 3'-5' phosphodiester linkages to form a continuous strand complementary to the parent strand. The picture we have, then, is of a Y-shaped replication fork which begins at one end of the parental double helix. As the fork proceeds along the length of the molecule, the single strands created

34

in its wake serve as **templates** for the assembling of the complementary free nucleotides, which then link up laterally by covalent phosphodiester bonds to yield two independent daughter double helices. When the replicating fork has proceeded the entire length of the parental molecule, one is left with two daughter double helices, each identical to the other, and both identical to the (now non-existent) parental molecule. One molecule has become two.

This mode of replication is called **semi-conservative**, since each strand of the parental molecule ends up as one of the two strands of each daughter molecule. This is in contrast to, for example, a conservative mode of replication whereby the parent molecule would be maintained intact and one daughter molecule of all-new nucleotides assembled. Another theoretical possibility would be a dispersive mode of replication, whereby the parent molecule would be degraded to the level of its individual nucleotides and these nucleotides along with others used to make two daughter molecules. These alternatives, one must be reminded, are just theoretical possibilities, for the correct overall mode of DNA replication has proved to be exactly as Watson and Crick predicted, i.e. semi-conservative. Nonetheless, we shall see anon that there have been introduced a number of important modifications to the simple "zipper" model outlined above.

Much of the early work directed to substantiating (or disproving) the Watson-Crick model concentrated on the predicted semi-conservative mode of replication. Perhaps it was felt that a convincing demonstration of semi-conservative replication *in vivo* would be the most effective way to gain the attention of classically trained biologists. There was, of course, not much that was really "debatable" about the chemical and physical structure of the DNA molecule *per se*; to the extent that the model was formulated in purely physical and chemical terms, it didn't leave much for the classical biologists to talk about at all, for few of them at the time were trained rigorously in either physics or chemistry.

The best-known demonstration of semi-conservative replication was carried out by M. Meselson and F. Stahl (1958)[4] using *E. coli*. To follow their work, the technique of **density gradient centrifugation** needs to be understood.

DENSITY GRADIENT CENTRIFUGATION

It is often important in biological experiments to separate molecules that have very similar chemical and physical properties. If the two molecules have (or can be made to have) slightly differing densities (mass divided by volume), a technique devised by J. Vinograd can often be used to effect a separation. A 6 molar solution of the heavy (and expensive) salt cesium chloride (CsCl) or cesium sulphate (Cs_2SO_4) is spun for many hours in an ultracentrifuge. As a result of the high centrifugal forces acting on the heavy cesium ions, they tend to settle toward the bottom of the centrifuge tube. This tendency, on the other hand, is continuously counteracted by the randomization tendencies of Brownian movement. The result of these two opposing forces is the establishment of a smooth concentration, and hence density, gradient from the top to the bottom of the test tube. The gradient is not steep, as the top of the gradient might be about 1.60 gm/cc and the bottom about 1.80 gm/cc. It is, however, precisely the fact that the top-bottom density difference is not great, and the fact that nucleic acids have densities averaging around 1.70 gm/cc,

that make such gradients so useful. Molecules whose densities lie within this range and differ by as little as the third decimal place can often be satisfactorily separated.

In practice, the molecules to be separated are added to the CsCl solution and the mixture spun so as to establish the required density gradient. Each molecule will, in the course of the centrifugation, migrate to the point in the centrifuge tube where the density of the CsCl exactly equals its own. At the end of the centrifuge run, all the molecules of identical density will form one band and those of a different density will be "floating" in a band at some other level within the centrifuge tube. To preserve these separations—they will quickly disappear under the influence of Brownian movement once the centrifugal forces have been removed—one punches a hole in the bottom of the (plastic) centrifuge tube and collects individually each drop as it runs out. The drops are numbered in the order of their collection and each can be analyzed chemically or physically (e.g., for radioactivity, if one of the molecules to be separated was radioactive) at one's leisure. The data are usually presented in the form of a plot or graph; the bands of material separated in the density gradient then become the peaks of the graph.

There are several means of creating artificial density difference between molecules that otherwise would be identical. The commonest is by the use of heavy atomic isotopes. These atoms have an additional neutron in their nuclei, and consequently their mass is increased without significant increase in their volume. (The nucleus of an atom contributes most of its mass but little of its volume.) The most familiar heavy isotope is H^2 (deuterium); its greater mass is acknowledged in the name of its well-known oxide, "heavy water". In biology, the most useful heavy isotopes are C^{13}, N^{15}, and H^2, none of which is radioactive.

REFERENCES

1. Anyone even remotely interested in the history of molecular biology must read H.F. Judson's *The Eighth Day of Creation* (Simon and Schuster, New York, 1979). The book presents an incomparable account of the personalities and circumstances involved in many of the subjects discussed in these lectures. This quotation is from p. 228.

2. F. Crick, 1954. The structure of the hereditary material. Sci. Amer. (Oct.).

3. Felsenfield, G., 1985. DNA. Sci. Amer. 253: 58-66.

4. Meselson, M., and F. Stahl, 1958. Proc. Nat. Acad. Sci. 44: 671.

LECTURE 8

SEMI-CONSERVATIVE REPLICATION OF DNA

INTRODUCTION

In the previous lecture we noted how very slight differences in density can be exploited to separate molecules that are very much alike in all other properties. The first significant application of this methodology in biological research came with the work of M. Meselson and F. Stahl in 1958. This experiment is considered a classic not just for the introduction of this technique into biology but also for its unambiguous demonstration that DNA replicates *in vivo* in the semi-conservative manner postulated by Watson and Crick. Some have called it "the most beautiful experiment in biology".

THE MESELSON-STAHL EXPERIMENT[1]

The Meselson and Stahl experiment makes use of both density gradient centrifugation and the heavy isotope N^{15}. *E. coli* were grown for many generations on a medium that contained N^{15} as the predominant nitrogen source. In consequence, all of the bacterial DNA substituted N^{15} wherever N^{14} was normally found. The heavy DNA was extracted and subjected to density gradient centrifugation in CsCl. It was found to band at a distinct location lower in the centrifuge tube than that of a band of DNA from ordinary ("light") *E. coli*. These two band locations, that for fully heavy DNA and that for fully light, establish the reference points for the further work.

An initial small volume of *E. coli* was taken and grown for many generations on N^{15} medium. A sample of these cells was then washed and transferred to a medium containing only N^{14}. Half of these cells were subsequently removed from the light medium, and their DNA was extracted and banded in CsCl; the remainder was allowed a second round of DNA replication in N^{14} before the DNA from these cells was also extracted and banded in CsCl. In each case the location of the band or bands of DNA obtained on density gradient centrifugation was compared with those of the standards mentioned earlier.

The bacteria whose heavy DNA had replicated only once in light medium yielded a single band of DNA; its density was exactly intermediate between that of fully heavy and that of fully light DNA. This is what one would expect on a semi-conservative mode of replication, since a double helix composed of a parental heavy strand and a daughter light one would, of course, be of intermediate density. However, it is also consistent with a dispersive mode of replication since a double helix composed of 50% old (heavy) and 50% new (light) nucleotides would be of intermediate density no matter how these nucleotides were distributed throughout the helix. The occurrence of a single band of intermediate density after one round of replication rules out only a fully conservative mode of replication, which would have yielded two bands (i.e., one fully heavy parental band and another fully light progeny band).

The question of semi-conservative vs. dispersive replication is settled by examining the density of the heavy DNA which had replicated *twice* in light medium. On a dispersive model, we would predict the appearance of a single band of DNA with a density intermediate between that of hybrid and that of light DNA; i.e., after two rounds of replication in light medium, the previously heavy DNA would be three-quarters light. On a semi-conservative model, we would predict the occurrence of two bands, one hybrid in density and one fully light—and this is what in fact was found. DNA does replicate semiconservatively.[*]

DENATURATION AND RENATURATION OF DNA

If DNA in solution is heated to near 100°C, the hydrogen bonds holding the two complementary strands together break, and the double helix in consequence unwinds into its two separate polynucleotide chains. The temperature range over which this **denaturation** occurs is usually fairly narrow; its mid-point is called the **melting temperature (Tm)** of that particular DNA. When DNA melts, a number of its physical properties are noticeably altered. The viscosity of the solution decreases, for example, and the quantity of U.V. light (2600Å wavelength) absorbed increases by about 40%. Either of these changes—the latter with greater ease—can be used to determine the Tm of a given DNA sample.

Since G-C base-pairs are held together with three hydrogen bonds and A-T base-pairs with only two, it would require more energy (in this case heat) to melt DNA that is relatively rich in G-C base-pairs. Indeed, each source of DNA has a unique Tm, which in turn directly correlates with the amount of G-C base-pairing in the sample. Looking at it another way, one sees that the Tm of a particular sample becomes a very accurate measure of its G-C content. In addition to heat, a very high or very low pH, urea and alcohols will often cause denaturation of DNA.

---

[*] If one reads the actual scientific paper by Meselson and Stahl and thinks carefully about the data that are presented, one sees that there is an obvious anomaly that was left unexplained by the authors. It takes somewhere between 30 and 45 minutes for fast-growing *E. coli* to replicate all their DNA (the exact time depends upon the precise culture conditions). Hence, if these *E. coli* begin the experiment with fully "heavy" (dense) DNA, the appearance of fully hybrid DNA would not be expected before the first round of replication had been completed, that is, not until at least about 30 minutes. The data and reproduced photographs of the density gradient centrifugation runs, however, clearly show fully hybrid DNA in evidence long before 30 minutes. In fact, fully hybrid DNA is present in the very first extractions taken shortly after the heavy bacteria began to replicate in the light medium. How can this be? Meselson and Stahl made no mention if this puzzling fact in the text of their paper, although it cries out for explanation. Following this "tradition", I shall not explain it here either, except to say that there is indeed a reasonable explanation for this observation and that the astute student may find clues to it buried in the text of subsequent lectures.

Denatured DNA lacks all of the rigidity of the double helix and becomes in solution a random coil. Parts of the individual polynucleotides which are by chance complementary to other short sections of the same or of another adjacent polynucleotide chain may become partly helical again, but these are chance occurrences in an otherwise mainly disorganized mess of single strands. Needless to say, denatured DNA loses all its biological activity. Melted transforming DNA, for example, is wholly incapable of transforming bacteria.

It came as somewhat of a surprise, therefore, when in 1959 J. Marmur showed that denatured DNA could sometimes be **renatured**. If the melted DNA, instead of being rapidly cooled, is only slowly brought back to room temperature, some of the complementary strands will have had the opportunity of aligning themselves in the proper complementary register and reforming a partial or complete double helix. The likelihood of this happening is obviously related to the length of the interacting strands: shorter strands will have greater success in encountering their complements in an orientation suitable for renaturation than will longer ones, which are probably more tightly knotted up.

Renatured DNA re-acquires all the chemical and physical properties of native DNA, including increased viscosity and lessened absorbance at 2600Å. Even its previous biological properties are restored, at least in part. Melted transforming DNA which is then renatured re-acquires about 15% of its former transforming ability, the remainder being lost through incomplete or inexact renaturation.

It is also possible for strands of DNA to anneal (another term for "renature") even if they lack exact complementarity for a few of their bases. The absolute number of infelicitous pairings that is still compatible with formation of a structure whose overall character is that of a double helix will depend upon the number and position of the "correct" pairings. Long polynucleotides, with thousands of correct pairings to hold the two chains together, are better able to withstand the structural distortions of a few mispaired (i.e. non-complementary) bases then are small pieces with fewer hydrogen bonds keeping them together. There is a whole field of inquiry devoted to determining the minimum amount of sequence complementarity needed for annealing.

One practical application of these studies is in investigations of evolutionary kinship. If DNA is the hereditary material, evolution must proceed by the accumulation of changes in the DNA sequence. Organisms that are closely related in evolution should, therefore, have greater similarity in their DNA sequences than organisms more distantly related. The readiness with which the DNA from different organisms forms hybrid molecules when denatured and then renatured together can be construed—with many caveats—as a measure of their phylogenetic relatedness. In a later lecture we shall see that it is now possible to sequence large stretches of DNA and hence to compare directly, base by base, related sequences in two different organisms.

We shall also see later in these lectures that complementary sequences of DNA and RNA molecules will also pair to form DNA-RNA hybrid structures. This observation has become the basis of many experimental procedures in molecular biology.

STRUCTURE OF RNA

If the ratios of the bases A, U, C and G are determined for RNA from a variety of viral, animal and plant sources, it will be only fortuitous if either of Chargaff's Laws are followed for any of the samples. Nor does RNA generally give any physical measurements (e.g., as in X-ray diffraction studies) indicative of extensive helical structure. By and large, RNA is single- stranded. Like the individual strands of the DNA molecule, it is basically a polymer of nucleotides (ribonucleotides, in this case) linked by 3'-5' phosphodiester bonds. Since it is the double helical character of DNA that gives this molecule its rigidity, single-stranded RNA in solution takes on the appearance of a seemingly formless random coil or "tangle", in everyday speech. If, however, it is complexed with protein, as when, for example it forms part of a TMV virion, it can nonetheless achieve a high degree of order. Moreover, as we learn more in about the function of particular RNA molecules (e.g., messenger RNA, transfer RNA, etc. discussed later in these lectures) we see that the "randomness" of an RNA "tangle" is more illusory than real. The three dimensional shape assumed by RNA in solution is a consequence of complementarity that exists (by design or by chance) between parts of the molecule that may be close together or far apart along the chain. The functioning of the RNA often depends on the precise three dimensional shape assumed by RNA, a shape that on first inspection appears random and unpredictable, but in fact is neither. As we introduce other forms of RNA we shall have occasion to refer to what is known of their structures.

REFERENCES

1. Meselson, M., and F. Stahl., 1858. Proc. Nat. Acad. Sci. 44: 671-682.

LECTURE 9

IN VITRO SYNTHESIS OF DNA

INTRODUCTION

The time-honored chemists' demonstration that the true structure of an organic compound has been defined is the *in vitro* synthesis of the molecule on the basis of the postulated structure. This synthesized material must, of course, have all the chemical and biological properties of the naturally occurring substance. Hence, shortly after the publication of the Watson-Crick theory, chemists began to consider how the synthesis of DNA could be accomplished in the laboratory.

As it turned out, the chemical synthesis of DNA did not immediately tell us much of significance about natural DNA, but it did serve to sweep away the last vestiges of mysticism that tended to surround a molecule of such over-riding importance to life. The first successful synthesis of DNA was accomplished in 1956 by A. Kornberg[1] and his associates.

Table 9.1 Kornberg's results from various cell-free fractions derived from *E. coli* used to synthesize DNA *in vitro*.

System	Deoxynucleotide incorporated, $m\mu$moles
Complete system	0.50
Omit dCTP, dGTP, dATP	<0.01
Omit dCTP	<0.01
Omit dATP	<0.01
Omit Mg^{++}	<0.01
Omit DNA	<0.01
DNA pretreated with DNAase	<0.01

KORNBERG'S SYNTHESIS OF DNA[2,3]

Since the end product in the synthesis of both purines and pyrimidines *in vivo* appears to be not the free bases but rather the nucleosides (base plus deoxyribose sugar), Kornberg predicted that these latter compounds must be the true precursors (building blocks) of DNA. He also predicted that the assembling of these nucleoside precursors into DNA would not occur spontaneously but like every other metabolic process in a cell must be mediated by some **enzyme**—a form of biological catalyst about which we shall learn a great deal in later lectures. If this were so, then the key to successful *in vitro* synthesis of DNA becomes the isolating of the appropriate enzyme from a cell and getting it to work in a test-tube. To find such an enzyme, one would naturally look where one would expect it to be most abundant in nature, as, for instance, in *E. coli* cells that are rapidly synthesizing DNA. It is from this source, then, that Kornberg tried to extract the enzyme responsible for DNA synthesis.

In a carefully monitored experiment in which highly labelled thymidine (the nucleoside) was added to extracts from rapidly growing *E. coli*, Kornberg noted the consecutive conversion of this nucleoside first into the 5'-nucleoside monophosphate and then into the nucleoside-5'-triphosphate. Subsequently the compound was converted into material which, unlike the free nucleoside triphosphate, was insoluble in a dilute acid solution. On analysis this radioactive acid-insoluble material proved to be DNA.

The actual amount of newly synthesized DNA isolated during these initial experiments was exceedingly small. Even though the precursor material was radioactively labelled to the extent of about 10^6 counts per minute (cpm), the amount of acid-insoluble activity recovered was only about 50 cpm. This low yield of DNA is readily explained by the fact that the *E. coli* extract Kornberg used not only contained those enzymes responsible for the observed polymerization of the nucleotides into DNA but contained as well a number of contaminating enzymes (nucleases) known to degrade DNA to the level of the free nucleotides. Hence, the synthetic activity of the DNA polymerizing enzyme (now called **DNA polymerase I**, or **pol I**) was masked by the destructive activity of the nucleases; were it not for the extreme sensitivity of the radio-tracer technique, the newly formed DNA would have escaped detection altogether.

Using Kornberg's observation as their starting point, workers then set about to isolate the DNA polymerase free of contaminating nucleases and other extraneous matter—or, in Kornberg's own metaphor, "Through this tiny crack we tried to drive a wedge, and the hammer was enzyme purification".

THE REACTION CONDITIONS AND CHEMICAL DETAILS

As the DNA polymerase extract was further purified, the necessary conditions for its optimal activity could be investigated. It was soon recognized that four conditions must be met:

1. DNA polymerase must be present; the nucleotides will not polymerize spontaneously in the absence of this enzyme.

2. All four deoxyribonucleoside-5'-triphosphates must be present in the reaction mixture. The absence of any one of these nucleotides, the substitution of the corresponding nucleotide mono- or di-phosphate for the triphosphate, the substitution of the corresponding ribonucleotide for the deoxyribonucleotide or the substitution of the nucleoside-3'-triphosphate for the 5' ones virtually eliminated the yield of DNA polymer.

3. A specific concentration of the divalent magnesium ion (Mg^{+2}) must be maintained in the reaction mixture. The exact function of the magnesium ion is still somewhat in doubt, although a role in the binding of the enzyme to the DNA is postulated for it.

4. The final and, at first, most unexpected requirement is for a "primer molecule" of DNA. Specifically, this means that DNA must be added to the reaction mixture before any appreciable synthesis of DNA can be observed*. In retrospect, however, the necessity for primer DNA seems not at all surprising. If the *in vitro* reaction is viewed as one of *replication* rather than simply of *synthesis*, the need for a pre-existing molecule to be replicated is obvious. Moreover, if the reaction is replicative, the role of the added DNA is not just that of a primer—an initiator of the reaction—but in addition must be that of a **template** to specify the exact nucleotide composition of the product DNA. The true relationships between the primer DNA and the product will be discussed in a later section.

The accompanying table indicates the effects on the yield of DNA when any of the essential substrates for the reaction is omitted or varied. It can be seen that a significant yield of DNA is obtained only when all of the conditions discussed above are satisfied.

The enzyme-mediated polymerization can be described by the structural representation in Figure 9.1. The polynucleotide chain is lengthened by the addition of an incoming nucleotide through a new phosphodiester bond; this necessitates the release of a pyrophosphate molecule for each nucleotide added. Note that the new strand is growing only at its 3' end. The 3'-OH of the added nucleotide now itself becomes available for the addition of still another nucleotide. The reaction is fully reversible. The yield of DNA can at times exceed twenty times the amount of primer added.

The DNA product of the *in vitro* reaction demonstrates many of the usual properties of native DNA. It is always a double-stranded molecule and has the same viscosity and ultraviolet adsorption spectrum as is found for double-stranded DNA. Both the primer and product are

---

*The DNA was added by Kornberg to the reaction mixtures in the expectation that it might serve as a primer in a manner analogous to the way in which added carbohydrate provides a starter molecule to be further lengthened during the in vitro synthesis of starch-like molecules. The added DNA was also thought useful in "soaking up" the activity of the contaminating nucleases so that some of the hoped-for newly synthesized DNA might survive to be detected and isolated. As we shall see, the added DNA is indeed serving as a primer, although the mechanism of this priming is only distantly related to starch model. The added DNA had an additional role not immediately obvious to Kornberg at the time: it was the source of the three other nucleotides needed for the reaction to proceed. In the initial experiments, Kornberg had established that thymidine triphosphate was a building block. The triphosphates of the remaining three nucleotides had not yet been discovered! However, nucleases in the impure enzyme preparation degraded some of the added DNA and still other unrecognized enzymes lurking in the crude extract converted the liberated nucleotides to the necessary triphosphate forms that were essential for their use as substrates for DNA synthesis. Sometimes being too careful in scientific work is inimical to success. Max Delbruck has termed this phenomenon "the principle of limited sloppiness", crediting it with a major but largely unsung role in scientific progress.

degraded by DNAse (an enzyme that specifically degrades DNA and nothing else) and both are digested to the same extent. These observations tend to confirm that *E. coli* polymerase does assemble a "genuine" molecule of DNA.

DNA polymerase is not found just in *E. coli*. Although its existence was first demonstrated in this organism, enzymes with the same function were soon found in many other species as well. For example, a common source of another DNA polymerase enzyme is the thymus gland of calves. Although it has the same function as the *E. coli* enzyme, calf thymus DNA polymerase is serologically distinct. (Antibodies prepared to *E. coli* pol l do not bind to the DNA polymerase isolated from calf thymus.) Another easily obtained DNA polymerase is found in T2 or T4 phage-infected *E. coli*. It seems that the phage prefers to make its own DNA polymerase instead of using that already found in the host[**]. All these DNA polymerases, besides differing in their structure, also differ in subtle ways with respect to the kind of DNA primer-template they prefer (e.g. single or double stranded); a fuller discussion of these differences lies outside the demands of this account.

OTHER ACTIVITIES OF DNA POLYMERASE l

Besides being an enzyme capable of synthesis, *E. coli* pol l turns out to possess specific DNA degradative properties, as well. In particular, pol l can chew away, one nucleotide at a time, hydrogen bonded nucleotides from the 5' ends of a double stranded DNA molecule. This ability is known as the **5' to 3' exonuclease** activity of pol l. It will also chew away in the same manner nucleotides

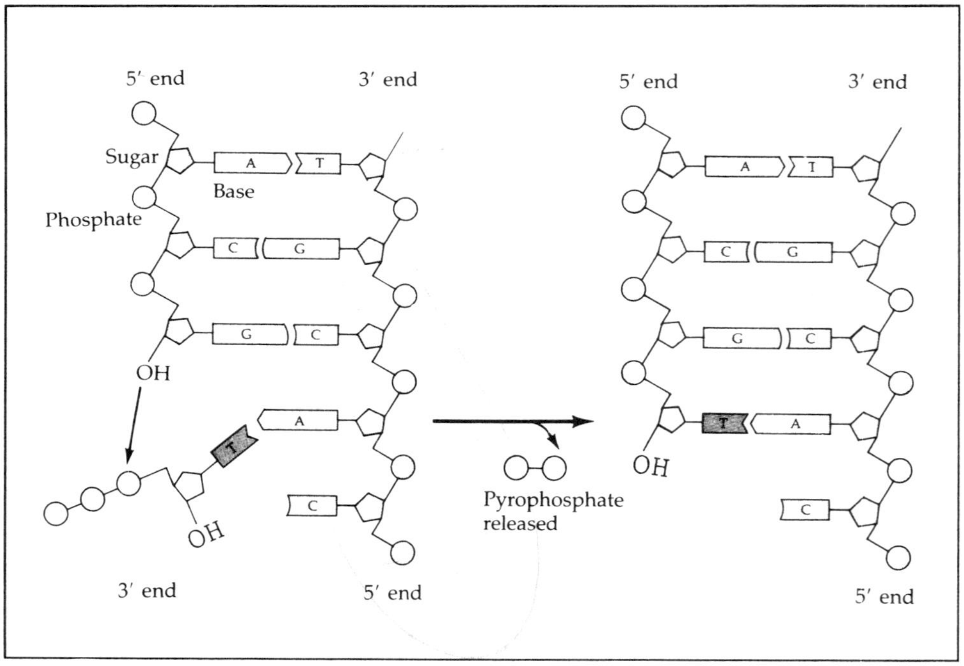

Figure 9.1 A new nucleotide is added to a chain of DNA. From Biology, Second Edition, by Neil Campbell (Redwood City, CA: Benjamin/Cummings Publishing Company, 1990), p. 317. Reprinted by permission.

[**]Probably because T-even phage have hydroxymethylcytosine in place of cytosine in their DNA.

that are part of a DNA/RNA hybrid double helix. Under certain circumstances, the 5' to 3' synthetic activity and the 5' to 3' exonuclease activity proceed simultaneously on the same primer-template. This might happen, for example, when the pol I is given as a primer-template a double-stranded DNA molecule with a "nick" in one of the two strands. By a nick is meant a single break in the phosphodiester bonds linking the nucleotides, but where none of the nucleotides is actually missing from the strand. This nick creates a 5'-P end on one side of it and a 3'-OH end on the other. Pol I can then delete nucleotides from this 5' end at the same time as it adds nucleotides to the 3' end, thus first creating a space and then filling it with a new nucleotide of the same type. The effect of this is to move the nick along the DNA molecule. A break in the backbone remains, because pol I cannot seal the break by linking two adjacent nucleotides *already in place*, but the break is progressively displaced towards the 3' end of the broken strand. The biological significance of this process, called **nick translation**, will become apparent when we examine DNA replication *in vivo*.

PRIMER-PRODUCT RELATIONSHIPS

For some purposes—more apparent later in these lectures than now—it is important to distinguish between the priming and the template functions of the initiating DNA molecules. As a template, this DNA serves, presumably, to specify the sequence of nucleotides polymerized into product DNA. For this purpose it must provide an unpaired single strand whose sequence can be copied in complementary fashion. (It is worth noting at this point that this template is *copied* 3' to 5', whereas the complementary strand being assembled along this template is being *synthesized* 5' to 3'. We shall have occasion to return to this point in a later lecture.) As a primer, the initiating DNA offers a free 3'-OH end to which the first incoming nucleotide can attach. A direct consequence of this distinction is the fact that not all DNA molecules can initiate *in vitro* DNA synthesis, being defective either as primers or templates. For example, a closed, single-stranded circle of DNA would be an excellent template, but since it has no 3'-OH end, it cannot prime DNA synthesis. On the other hand, a double-stranded linear molecule whose two ends are exactly even has, of course, two free 3'-OH ends (one at either end of the molecule), but it cannot serve as a template for DNA synthesis because it lacks unpaired strands. In any discussion of the detailed mechanism of DNA synthesis these considerations loom large.

REFERENCES

1. Kornberg, A., 1960. Science <u>131</u>: 1503.

2. The best general references for DNA synthesis and replication are the two books on the subject by A. Kornberg: *DNA Replication* (Freeman, San Francisco, 1980) and *Supplement to DNA Replication* (Freeman, San Francisco, 1982).

3. Kornberg, A., 1968. The synthesis of DNA. Sci. Amer. (Oct.).

LECTURE 10

FURTHER STUDIES ON THE PRIMER-PRODUCT RELATIONSHIP *IN VITRO*

UNWINDING DNA *IN VIVO*

The moment one begins to investigate the truth of the simplest facts which one has accepted as true, it is as though one had stepped off a firm, narrow path into a bog of quicksand—every step one takes one sinks deeper into the bog of uncertainty.

L. Woolf

Although primer DNA and product DNA possess much the same physical properties and the same over-all base composition, one would not be justified in calling the former the template for the latter unless the base sequence of the product were identical to that of the primer. The straightforward way to settle the question, a complete base pair sequence analysis of both molecules, was technically impossible at the time of Kornberg's original discovery but could now be done with relative ease. As a poor alternative to a sequence analysis, J. Josse, A. Kaiser and A. Kornberg developed a technique, known as **nearest neighbor analysis**, which permitted them to determine the frequency with which each nucleotide is adjacent to every other nucleotide in product DNA. If this synthetic DNA is then used to prime a second synthesis, the 16 dinucleotide frequencies (4X4) in the second product can be compared with those found for the first product (i.e., the primer for the second reaction).

The results of many such analyses carried out by Kornberg's group showed conclusively that, for primers of many different base compositions, the dinucleotide frequencies of the particular primers and their products are always identical. Since the likelihood of this occurring by chance is vanishingly small, it seems fully justified to refer to the primer DNA as a primer-template[*]. Thus DNA not only initiates the reaction but also provides the model base sequence to be copied into product molecules.

---

[*]This work also showed experimentally for the first time what has been taken more or less on faith since Watson and Crick first proposed it: namely, that the DNA molecule is antiparallel in structure. The X-ray diffraction work and other physical studies on which their theory was based were insufficient to discriminate between a parallel and an antiparallel construction, although Watson and Crick strongly favored the latter notion.

DISQUIETING ANOMALIES ARE FOUND

If by this point you have become convinced that DNA polymerase I makes *in vitro* a product DNA that is identical to its primer-template—you will now have to change your mind. It doesn't. The first piece of evidence that primer and product DNA have different properties came from studies that tested the ability of product DNA to carry out transformation in bacteria. DNA, which has the capacity to transform bacteria, should, if used as a primer-template in a DNA synthesis reaction, yield product DNA with the same capacity. Not only is the product DNA from such a synthesis quite unable to transform, but the primer-template, which initially had that ability, loses significant amounts of it during the course of the synthesis. Product DNA, therefore, does not possess the biological properties of its primer-template, however similar they appear to be by chemical and physical criteria.

A second difference between primer-template DNA and product DNA concerns their denaturability. Natural (primer-template) DNA will, as discussed earlier, denature at a characteristic Tm and will remain single-stranded if the melted DNA is quickly cooled to room temperature. The product DNA synthesized by the Kornberg enzyme melts at the same Tm as does the particular primer-template used in the reaction, but it reanneals to a double helix even if the subsequent cooling to room temperature takes place quickly. In other words, under conditions where the primer-template remains single-stranded the product of the polymerase I reaction reforms into a double helix. It is as if something holds the two complementary strands of the synthetic DNA together so that they cannot drift apart when the hydrogen bonds are broken; as soon as the source of heat is removed, the two strands then readily snap back into a double helix.[1]

The reason for the lack of biological activity and the non-denaturability of synthetic DNA became somewhat clearer when the primer-template and its product molecules were both examined under the electron microscope. The natural DNA had the expected thread-like appearance of linear strands; the product DNA, by contrast, appeared as a multi-branched structure. Further investigation by S. Mitra and Kornberg led to the belief that the product of the DNA polymerase I reaction was not a molecule composed of two separate strands but one composed of a lengthy self-complementary, single strand! Therefore, instead of having four ends, it has only two; at the ends of the branches and at one end of the stem, the DNA is simply doubled back over on itself in hair-pin fashion. For this reason, when it melts, the two complementary sequences cannot drift apart and are ready to reform a helix as soon as the temperature is lowered.[2]

The reason DNA polymerase I cannot make two complementary strands has to do with the chemical directionality of the DNA strands. The polymerase makes a new strand of DNA by following a template DNA strand in the 3' to 5' direction. The new strand being synthesized is in consequence assembled in the 5' to 3' direction; i.e., the polymerase cannot assemble a new strand by synthesizing it in the 3' to 5' direction. Moreover, all it can do is add new nucleotides to the 3'-OH end of an already synthesized chain; it cannot initiate on its own a new DNA strand. If these data derived from *in vitro* studies reflect the true specificity of DNA polymerase I as it functions *in vivo*, they would appear to rule out the possibility that the polymerase functions alone in synthesizing DNA in the living cell.

In 1969, P. deLucia and J. Cairns made a discovery that at first appeared to rule out *any* critical role for polymerase I in the normal synthesis of DNA.[3] In contrast to its ramifications, their finding was indeed a simple one: a mutant strain of *E. coli* which contained almost no detectable synthetic activity for DNA polymerase I, yet which made its DNA in a completely normal manner. Subsequent studies confirmed that polymerase I was not involved *in vivo* in the synthesis of DNA as the principal synthesizing enzyme, but was necessary for the repair of already synthesized DNA**. (A discussion of DNA repair, a system of continuous monitoring and mending of DNA, is found in later lectures.) If this repair enzyme is removed from its *in situ* environment and given specific conditions *in vitro*, then it will synthesize long-chain polynucleotides, but, despite a number of tantalizing similarities with natural DNA, this DNA is anything but normal.

Exactly how DNA synthesis occurs *in vivo* is still a matter of some conjecture, although it is known that a system of cooperating enzymes is involved, the key one being DNA **polymerase III** (discovered, incidentally, by T. Kornberg, son of A. Kornberg). But instead of getting enmeshed in the detailed enzymology of this synthesis—before we have even discussed what enzymes are and how they function!—we shall first examine some other puzzles related to DNA synthesis *in vivo*.

UNWINDING THE MOLECULE DURING REPLICATION

The two single strands of the DNA helix are wound around each other in the manner of two strands of a rope. Such an arrangement, technically known as a **plectonemic coil**, means that the two complementary DNA chains must unwind if the coil is to separate into its component strands. (For the curious, the other type of double helix, a **paranemic coil**, would separate into its components without unwinding; think of the double helix that results from pushing two bed springs together.) Replication of DNA, therefore, should require either that the two unravelling strands continuously twist around each other or that the intact part of the parental helix rotate about its own axis. In either case, this would present a formidable mechanical problem. Consider the following information.

> 1. Each base-pair is off-set by an angle of 36° relative to its neighbors; consequently there are 10 base-pairs per complete turn of the DNA helix.

**The mutant deLucia and Cairns found was unable to undergo normal DNA repair. Later studies showed that the enzyme molecule itself was not missing from the mutant cells, but the enzyme molecule was simply lacking most—but not quite all—of its synthetic activity. The exonuclease activities remained intact.

2. The single DNA molecule that comprises the entire genetic apparatus of *E. coli* is 1100 microns long—which is 500 times the length of the *E. coli* cell.

3. Each complete turn (10 base-pairs) has an axial length of 34Å (= 3.4 x 10^{-3} microns).

4. The total genetic apparatus of *E. coli* must, therefore, comprise 1100 x 10/(3.4 x 10^{-3}) or 3.2 x 10^6 nucleotide pairs.

5. *E. coli* can replicate its entire DNA molecule in 40 minutes, or 2400 seconds.

6. Therefore 3.2 x 10^6/2400, or 1400 nucleotide pairs must be added per second.

7. The DNA must rotate, therefore, at a speed of 1400/10 or 140 turns a second or 140 x 60, i.e., 8400 rpms.

How does one get a slender molecule which is 600,000 times as long as it is wide and which is packed into a container (along with a whole lot of other submicroscopic *tchotchkes*) 500 times smaller than its length to rotate along its axis at the speed of a modern centrifuge? One doesn't. It seems that there is a special enzyme or enzyme system that continuously "snips" the phospho-diester bonds of the parental helix in advance of the replicating fork. The diminishing portion of parental helix is then permitted to rotate until the torque is relieved. The enzyme, appropriately called **gyrase** in *E. coli*, then reseals the break and the backbone of the strand is made whole again. As the replicating fork passes through that site, another cut is made lower down. The process just described keeps repeating itself until the entire molecule has replicated. Hence at no time is it necessary either for the unravelling strands to move around each other or the parental helix to rotate along its entire length.

Although the net effect of gyrase activity—the release of the torque occasioned by the unwinding of a plectonemically coiled double helix—is adequately described by the foregoing account, the actual manner in which this is brought about is somewhat more complex than that account re-veals. The advancing replicating fork causes the double helix to "supertwist" and in consequence to loop back over itself. The gyrase then puts a double strand cut in one section of the helix and draws the intact superimposed double helix through the cut to the other side. The cut is then re-sealed. Perhaps the easiest way to visualize this is to imagine a telephone hand-set cord, a single helix, which often becomes supertwisted from frequent use. This causes the helix to loop around over itself in an effort to accommodate the torque. One can take out these loops by twisting the handset in the direction opposite to that which created them in the first place, but one could also accomplish the same thing—although it is not recommended you do so—by cutting the cord (helix) at every place the helix crosses over on itself and pulling the top helix through the cut to the other side. The cut ends would then have to be spliced back together again.

Gyrase, then, plays a key role in DNA replication in bacteria; analogous enzymes accomplish the same thing in higher organisms, although their activities are not yet so well characterized.

Recently there has come onto the market a class of antibiotics known as "quinolones" that cause the death of bacteria by specifically interfering with the action of gyrase and thus preventing DNA and cell replication. The equivalent enzyme in the host organism (usually humans) is not affected by the quinolone and hence escapes injury.

SEPARATING THE STRANDS AND KEEPING THEM SEPARATE

In many organisms specific **unwinding proteins** (called **helicases**) have been isolated and characterized. When mixed with DNA *in vitro*, these proteins will denature the duplex DNA into its separate strands at a temperature about 40° C below the Tm characteristic of that DNA. These proteins appear to play an obligatory role in unwinding the DNA during replication *in vivo*. In *E. coli* two unwinding proteins, **rep protein** and **helicase II**, act cooperatively to unwind the helix. It seems likely that they bind to different strands and utilize the energy derived from the hydrolysis of two molecules of ATP for each base-pair separated.

The separated DNA strands, being fully complementary, have a strong tendency to reanneal into a double helix and would quickly do so if something did not prevent them. This "something" is **single-stranded DNA binding proteins (SSB proteins)**, which bind to the newly separated strands and keep them apart.

REFERENCES

1. Richardson, C., Schildkraut, C., and Kornberg, A., 1963. Cold Spring Harbor Symp. Quant. Biol. 28: 9

2. Mitra, S. and A. Kornberg, 1966. J. Gen. Physiol. 49 (pt. 2): 59

3. DeLucia, P., and J. Cairns, 1969. Nature 224: 1164-66.

LECTURE 11

DISCONTINUOUS SYNTHESIS OF DNA *IN VIVO*

SOLVING THE DIRECTIONALITY PROBLEM

The next item under DNA replication brings us back to the problem of directionality in DNA replication. In order to replicate in the manner originally set forth by Watson and Crick, nucleotides would be polymerizing in the 5' to 3' direction along one of the parental strands and 3' to 5' along the other. Not only does DNA polymerase I lack the specificity needed to polymerize nucleotides 3' to 5', but no enzyme with such specificity has yet been discovered. In fact, none is likely to exist. It is consequently not possible for both nascent polynucleotide chains to lengthen in the same geometric (as opposed to chemical) direction, i.e., in the same direction as the travelling fork is moving.

The key to the puzzle was the finding by R. Okazaki that the very first products of DNA synthesis *in vivo* were often short DNA single strands about 1000 to 2000 nucleotides in length. These single-stranded fragments, called eponymously **Okazaki fragments**, can be synthesized in the required 5' to 3' direction along one of the DNA parental strands (as the replicating fork proceeding in the opposite direction, leaves a gap to be filled in by the synthesis of this fragment) and then the separate Okazaki fragments that lay along this parental strand can be stitched together (by an enzyme called **ligase**) to form a continuous strand complementary to the parental strand. Of course, while a particular Okazaki fragment is being synthesized, the fork continues to move ahead in the opposite direction, producing a new gap that must be filled by the initiation and synthesis of another fragment, and so on. Okazaki fragments, therefore, are continuously being initiated, synthesized and attached together along *one* of the two parental strands.

Along the other of the two parental strands there is no need for Okazaki fragments at all, since 5' to 3' synthesis proceeds in the same direction as the replicating fork is headed. Synthesis along this new strand is **continuous**, since as the moving fork produces a single-stranded gap along the parental strand, its complement can be synthesized right behind the fork in the required 5' to 3' direction. Hence, new DNA synthesis is continuous along one parental strand (often called the **leading strand**) and discontinuous along the other (the so-called **lagging strand**).

SOLVING THE PRIMING PROBLEM

A final point concerns the primer for DNA synthesis *in vivo*. It was noted earlier that DNA polymerase I must add 5'-nucleotide phosphates to pre-existing 3'-OH ends. Another way of putting this is to say that polymerase I cannot initiate a new 5'-triphosphate end. This same inability attaches to all known DNA polymerases, specifically including pol III, the enzyme that is synthesizing both the continuous and the discontinuous new strands of DNA. Since DNA chains grow 5' to 3', one is thus left without a mechanism to account for the initiation of any independent new strands of DNA.

Subsequent studies have pointed to a role for RNA as an initiator of DNA synthesis. An enzyme, called **primase**, synthesizes along the lagging strand a short, single-stranded piece of RNA less than a dozen or so nucleotides in length. (In the case of T7 phage DNA synthesis, the size and sequence of these RNA primers are known: A-C-C-A or A-C-C-C.) The direction of their synthesis is, as always, in the 5' to 3' direction. Primase does not itself need a primer, but can initiate RNA synthesis *de novo*. This short piece of RNA (which is presumably complementary to its DNA template) can, however, serve as a primer for DNA synthesis, since it would have a free 3'-OH end to which DNA polymerase III can now add on deoxyribonucleotides to make the Okazaki fragment. Another enzyme (probably pol I, using its 5' to 3' exonuclease activity) later removes the RNA primer and fills in the gap by extension synthesis from the 3'-OH end of the adjacent Okazaki fragment. The removal of the priming RNA and the filling in of the gap could be carried on simultaneously by "nick translation" (see Lecture 9). Finally the ligase enzyme would covalently join the individual fragments. *In vivo*, therefore, pre-existing DNA is the template for DNA replication, but small molecules of RNA are the actual primers.

The primase enzyme appears to function within a larger aggregate of proteins that constitutes what is known as a **primosome**. Among the activities associated with the primosome, apart from priming, is the removal of the SSB proteins so that the primase—and subsequently, the pol III—can have direct access to the template DNA. One particular protein of the aggregate, the **n' protein** has been identified with the removal of the SSB protein, but the aggregate as a whole may contain as many as 20 proteins. The primosome appears to follow behind the advancing replicating fork, moving in the same geometrical direction as the fork. The primer RNA, however, is made, like DNA, in the 5' to 3' direction. Since for lagging strand synthesis this is in the opposite geometrical from which the primosome is moving, there is an obvious problem. The topological problem was compounded when mounting evidence suggested that there was likely only one location where DNA was being synthesized and not, as might have been expected, two (i.e.,one at or near the fork synthesizing the leading strand and another a little way back—and progressively moving further back as synthesis proceeded—synthesizing the lagging strand).

The answer to these dilemmas rests with knowing the particular configuration of the DNA at the replicating fork. First, it now appears that the primosome and the various proteins associated with the pol III are part of a larger structure called a **replisome**, which follows closely the advancing replicating fork along the DNA molecule. The replisome contains two molecules of pol III, one for each of the strands being replicated. The parental strand being replicated continuously passes through the replisome in a straightforward fashion, since the synthesis of the complementary strand presents no problem of directionality: this strand is being synthesized in the same direction in which the fork is moving. The parental template strand for lagging strand synthesis, by contrast, is looped 180° before it passes through the replisome. The effect of this looping is that the portion of the parental strand which is being replicated (or on which the priming RNA sequence is being laid down) is running in the same chemical direction as the opposite parental strand. Synthesis along both parental strands can now proceed in the same geometrical direction as that in which the fork is travelling. At intervals, the newly synthesized Okazaki fragment will hit up against the one previously laid down. At this point, the loop must be released from the replisome and reform further down the strand, close to the fork.

Most of our understanding of DNA replication has come from studies on *E. coli* and phage, but it now appears that the enzymological steps in the replication of eukaryotic chromosomes are largely similar, *mutatis mutandis*, to those in prokaryotes. In particular, eukaryotes replicate one strand continuously and one discontinuously, employ helicase to unwind the strands, SSB proteins to keep them apart, primases to initiate the synthesis, and so on. One difference is that prokaryotes use a single enzyme, pol lll, to replicate both the leading and lagging strands; eukaryotes, by contrast, employ two distinct polymerases, termed **polymerase alpha** and **polymerase delta**. Polymerase alpha makes the DNA on the lagging (discontinuous) strand and polymerase delta makes it on the leading (continuous) strand.[1] There appears to be a separate protein that acts as a primase very tightly associated with polymerase alpha.

REPLICATING THE E. COLI "CHROMOSOME"

Up to now we have been tacitly assuming that the replicating DNA molecules are linear in structure. In the case of most bacteria and several phages, the single DNA molecule that carries the genetic information is circular. There are, therefore, no ends to the molecule and no physically obvious starting point for DNA replication. In such cases, DNA replication begins by the localized limited unwinding of the helix; this "bubble" creates in effect two forks in the parental molecule. A priming RNA molecule is then laid down along one of the single strands within the bubble and this primer is lengthened by pol III. This growing DNA becomes the leading strand. Meanwhile, as the leading strand grows and the size of the bubble increases in consequence, the complementary strand has also been initiated and is growing by the expected discontinuous synthesis involving repeated priming and Okazaki fragment formation. In all these particulars, the growth of the two chains is what one expects of linear DNA, except for the fact that the parental template strands are hydrogen bonded together on the other side of the increasingly enlarging bubble—now more properly referred to as an "eye", since the strands delineating it have become double. Circular DNA with such growing eyes are often called **theta structures** after their resemblance to the Greek letter of that name (Θ).

Theta structures, however, have of necessity *two* forks, which leads us directly to the obvious question: Do both of them serve as replicating forks? Put another way, is the replication of bacterial DNA unidirectional (only one fork moves) or bidirectional (both forks move)? If the latter situation occurred, in order to preserve the necessary chemical directionality, the parental strand that is being replicated continuously at one fork would have to be replicated discontinuously at the other, and *vice versa*.

It is not easy to settle the question of whether DNA synthesis in, say, *E. coli* is uni- or bidirectional, since, although one can observe with an electron microscope the fact that the eye grows larger as replication proceeds, *where* this growth is taking place is not apparent.

Increasing size is compatible with either model, and DNA provides no visual markers to serve as reference points. All parts of the circular molecule look the same.

The answer to this question came quickly when means were devised to introduce into the *E. coli* DNA visual markers at specific locations along the molecule. The technique makes use of the fact that regions of the DNA that are rich in AT base pairs will have a melting temperature somewhat lower than that of the molecule as a whole. When isolated intact DNA is heated to just below its characteristic melting temperature, these AT-rich areas open up (denature) first, creating small bubbles wherever they occur around the molecule. The addition of formaldehyde causes them to remain open after the DNA is cooled. Because all *E. coli* would have the same base sequence along their DNA, these bubbles are at reproducible locations from one cell to the next. When theta structures from bacteria in various stages of DNA replication are examined, it was readily determined that both ends of the replicating eye are moving around the parental circular DNA. DNA replication in *E. coli* is bidirectional, the two forks eventually meeting about half-way around the circle.

Even organisms whose DNA is linear usually begin DNA replication not at the physical end of the molecule but instead at one or more interior locations. The phages known as T7, for example, have linear DNA and a single origin for DNA synthesis. But this site is about 17% of the way in from one end. A replication bubble forms there and becomes an eye as growth proceeds bidirectionally towards both ends of the molecule. Since one end will be reached long before the other, the replicating structure will then become transformed into a Y-shaped structure and replication is completed unidirectionally.

The origin of replication is at a fixed point, and there is only one origin site per genome. The place on the *E. coli* chromosome where replication is initiated (called **oriC**) is known with considerable precision. It consists of 245 base pairs. DNA replication is initiated by the binding of several different proteins to this site; some of these proteins are identical to ones that play a role in the replisome itself. The base sequence of this site is remarkably similar across a wide range of bacteria separated by many millions of years of evolution.

DNA REPLICATION IN EUKARYOTES

In organisms that have very long DNA, a single origin site would not give sufficient time for replication of the complete molecule within the time allotted, even allowing for bidirectional replication. In higher eukaryotes, for example, it is believed that a single DNA molecule runs from one end of the chromosome to the other (in a highly coiled-up way, of course, since the DNA would be thousands of times longer than the chromosome—more about this in a later lecture). In eukaryotes it is not uncommon for there to be many hundreds of replication origins along the molecule, each growing bidirectionally and eventually joining up to form the fully separated daughter molecules.

In yeast (*Saccharomyces cerevisiae*) the origins of replication along the DNA are known to specific segments comprising at least 11 bases . To these initiation signals attach a complex of proteins known as the **Origin Recognition Complex (ORC)**, which, in ways yet to be unravelled, initiates the replication of the DNA.[2]

REMOVING PRIMING RNA

There is one left-over problem in DNA replication that can now be addressed. The use of RNA primers in DNA synthesis obviously necessitates a mechanism for their removal in order to ensure the chemical homogeneity of the product. In the foregoing account, the 5' to 3' elongation of the Okazaki fragment closest to the fork fills in the gap left by the removal of the RNA from the adjacent fragment. In linearly replicating DNA, this mechanism would leave a gap at one end of each daughter molecule unfilled, as there would be no adjacent Okazaki fragment to lengthen into the gap. Specifically, the removal of the primer from the 5' end of the leading strand and from the 5' end of the final Okazaki fragment leaves projecting unreplicated single-stranded 3' ends. Unless some means existed to fill in these gaps, the DNA molecule would grow shorter at every round of replication.

A number of different strategies have been adopted to solve this problem. The simplest merely eliminates it as a problem at all. In organisms like *E. coli* that have circular DNA molecules as their genomes, there are no ends to the molecule and therefore there will always be DNA on the 3' side of a gap. This DNA can be lengthened so as to fill in any gaps created by the removal of primers.

It is in linearly replicating DNA that the problem becomes acute and the solutions exceedingly complex and varied. It is beyond the needs of this account to examine them in any detail, but something should be said about the probable way the matter is handled in eukaryotes. In some protozoa it has been noted that the ends of the chromosomes (chromosome ends are called **telomeres**[*]) are composed of a large but variable number of DNA segments composed of six nucleotide pairs. These repeated units—each telomere has about fifty of them—have the following composition:

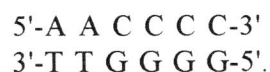

5'-A A C C C C-3'
3'-T T G G G G-5'.

The GGGGTT segments (the "G-rich segments") are synthesized *de novo* in variable numbers and added enzymatically to the 3' end of other GGGGTT segments at the 3' end of the chromosomal DNA. (After several of these G-rich segments have been added, the C-rich

---

[*]The term "telomere" long predates the advent of molecular biology. It has always been understood that the ends of chromosomes, the classical telomeres, are somehow different from the rest of the chromosome. For example, chromosomes that become fragmented (and lack, therefore, a telomere at the broken end) have a propensity to fuse with other chromosome fragments that lack a telomere, whereas normally-ended chromosomes display no such tendency to join end to end with other chromosomes.

complementary strand is believed to be synthesized in a conventional manner by primase and pol III, analogous to the way in which Okazaki fragments are synthesized. This means that the final 3' end of the molecule will still have an overhang of single-stranded DNA.) The enzyme that does the telomere synthesis, **telomerase**, is really a ribonucleoprotein containing a small piece of RNA. This RNA piece has been isolated and sequenced.[3] Not surprisingly, it contains the sequence which is complementary to the GGGGTT* it adds.[4] Telomerase, therefore, is a polymerase that contains its own template. The DNA complements to the added segments are synthesized conventionally in the course of DNA replication. The continual replacement of these segments prevents the DNA molecule from growing shorter at each replication.

The initial work on telomeres was mostly done with *Paramecia*, a single-celled organism with hundreds of chromosomes (and hence a good supply of telomeres). This protozoan is essentially immortal, since each cell divides indefinitely. In consequence, the enzyme that synthesizes the telomeres, telomerase, must also be always active. When telomerases were discovered in mammalian cells, it was natural to assume that they too would be always active, at least in those cells of an animal that were still dividing. This turns out not to be the case: in almost no cell of an adult mammal has evidence of telomerase activity been found. Moreover, it has also been observed that mammalian telomeres *do* become somewhat shorter (by about 50 nucleotides) at each cell division. Humans, for example, have between 1000 and 3000 repeats of a TTAGGG telomere sequence at the ends of each chromosome, but the actual number varies depending upon the age of the person (or, more precisely, on how many cell divisions have preceded the sampling of the cells for telomere length). Unlike the cells of *Paramecia*, mammalian cells are not immortal; after about 70 to 100 rounds of cell division, most cell lines die. The intriguing question currently under investigation is whether the shortening of the telomeres at each cell division is causally related to this cell death. Is telomere length a type of biological clock that limits our cells' (and therefore our) life span?**

---

*The sequence of the telomeres has been examined in a large number of vertebrates (including humans) and found to be GGGAATT instead of the GGGGTT sequence found in the ciliate, *Tetrahymena*. In another ciliate, *Euplotes*, it has been found to be GGGGTTTT. In all cases, he sequences appear to be synthesized by a telomerase enzyme that incorporates the appropriate RNA template.

**Equally intriguing is the relationship between telomerase activity and cancer. Cancer, of course, is the uncontrolled division of cells; but cancerous cells should also be limited in the number of divisions they can undertake, if this limit is established by diminishing telomere length. It has indeed been observed that cancers in later stages have shorter telomeres than those in earlier stages (and therefore telomere length might provide a means of measuring how advanced a cancer is), but it has also been observed that in very late cancers, telomerase becomes active, conferring potential immortality on these cells. Specific inhibitors of telomerase, if any can be found, should not harm normal cells (which have no telomerase activity), but should be able to limit the growth of cancer cells.

The fact that telomerase contains an RNA template necessary for the synthesis of a DNA polymer should not be skipped over lightly. It has been for several decades an item of dogma in biology that DNA can only arise *in vivo* by the replication of pre-existing DNA. Later in this account the activities of certain viruses, known as retrovirus, will be discussed. The defining characteristic of retroviruses is their ability to make a DNA copy of their RNA genome, using an enzyme known as reverse transcriptase. The discovery of retroviruses necessitated a revision of this "dogma" concerning the origin of DNA, but still they were viewed only as an unusual and special exception to the general rule. Since telomerases are also capable of making DNA from an RNA template and, moreover, appear to be ubiquitous in eukaryotic cells....well, it shows how foolhardy it is to be dogmatic about anything in biology.

REAL ODDITIES

Just to emphasize that there are exceptions to almost every rule in Biology, it should be noted in passing that there are some viruses (e.g., reovirus) that have double-stranded RNA as their hereditary material, and others (e.g., ϕX 174) that employ single-stranded DNA. The single-stranded phages like ϕX 174 maintain their DNA molecule as a closed circle.[5] When they infect a cell and initiate the replication of their DNA, the first step is the synthesis of a complementary, circular strand hydrogen-bonded to the entering one. At this point, therefore, the phage DNA is in the form of a closed double-stranded circle. A nick is then put in the original entering strand, thus creating a 3' and a 5' end. DNA synthesis now begins at the 3' end (the usual end at which DNA molecules are lengthened), and as this occurs, the 5' end is displaced from the circle as a single strand. By the time the advancing 3' end has gone all around the complementary intact circle, an entire linear phage genome would have been freed from the circle, but because the liberation of the single strand has occurred as a result of the concomitant synthesis of another strand in its place on the circle, the circle remains double stranded. Attached to the circle, however, will be a single-stranded "tail" one genome in length. By a rather complicated mechanism, this tail is cleaved from the circle and can become encapsulated in phage protein as a progeny phage particle. Another trip around the double circle will liberate a second single-stranded genome, and so on. This mode of replication is called **rolling circle replication**, but the more colorful name, "the toilet paper model", serves to fix the details of this mode of replication more clearly in our minds. Later in this narrative other examples of rolling circle replication will be encountered.

REFERENCES

1. Tsurimoto, T, T. Melendy, and B. Stillman, 1990. Nature <u>346</u>: 534-539.

2. Bell, S., and B. Stillman, 1992. Nature <u>357</u>: 128-134.

3. Shippen-Lentz, D. and B. Blackburn, 1990. Science <u>247</u>:546-552.

4. Greider, C. and E. Blackburn, 1989. Nature 337:331-337.
   Blackburn, E.H., 1991. Nature 350:569-573.

5. Sinsheimer, R., 1962. Single-stranded DNA. Sci. Amer. (July).

LECTURE 12

MANIPULATING DNA

SYNTHESIS & REPLICATION OF RNA

SEQUENCING OF DNA

It was not until 1977 that a method to sequence lengthy segments of DNA became available. The most commonly used method was devised that year by A. Maxam and W. Gilbert; it is applicable to pieces of DNA up to about 300 to 500 nucleotides in length. Usually it is used in conjunction with the procedures for "cloning" DNA discussed in the lectures on Recombinant DNA, since the method requires beginning with a very large number of DNA pieces that are of identical length and sequence. By a rather simple enzymatic procedure, a radioactive phosphorus atom is put on the two 5' ends of each of the (identical) DNA molecules in the collection. The now-radioactive DNA is then denatured and the strands separated by a procedure that isolates all of the "Watson" strands free from all the "Crick" strands. Either the isolated "Watson" or the "Crick" strands (both of which carry a 5' radioactive phosphorus atom) are used in the procedures that follow.

The single stranded pieces are divided into four lots, which are treated separately. One group of molecules is treated in such a way that the two purines (Adenine and Guanine) are eliminated from the DNA and the backbone is broken at that point. Another group is treated so that only Guanine is eliminated and the backbone broken where just that base resided in the molecule. The third sample is treated so that the pyrimidines (Cytosine and Thymine) are tagged with a chemical marker so that breaks in the backbone can be placed on the 5' side of the pyrimidines. Finally, only Cytosine is tagged in the fourth sample, and the backbone then broken on the 5' side.

All these treatments are of a chemical nature, the details of which need not concern us here. What is important to note, however, is that the treatments are adjusted so that an average of only one break is put in each DNA molecule. Consequently, although there may be, for instance, 20 Guanines in a particular strand, only two fragments result from cutting this strand by the treatment that eliminates Guanine alone. One of these fragments will be radioactive (because it carries the radioactive phosphorus attached at the 5' end), and the other will not. Note that if this guanine was, for example, the *fifth* base from the 5' end, the radioactive fragment would be *four* nucleotides long. A similar argument holds for the other three samples: if the eliminated or tagged base was n nucleotides from the 5' end, the *radioactive* fragment obtained will be n-1 nucleotides in length. (The non-radioactive fragments do not interest us in this procedure.)

The four treated samples are then separately spotted on to a polyacrylamide gel and subjected to an electric current running through the gel. This procedure, called **gel electrophoresis**, will separate the fragments according to their length, since shorter fragments will migrate in the electric current further along the gel than will larger ones. Fragments differing in length by only one nucleotide can be resolved by this technique. The location of the fragments (which are invisible on the gels) can be manifested by applying a sensitive photographic film to the gels and noting the parts of the film that become exposed by the radioactive phosphorus. Noting, for example, that the sample in which Guanine was eliminated yielded radioactive fragments that were four, five, and ten nucleotides long means that there had been a Guanine at positions five, six and eleven of the original strand, counting from the 5' end. Examining all four samples yields the complete nucleotide sequence of the original strand.*

It is now possible to sequence stretches of DNA 300 to 500 base pairs in length in a single run and to establish the complete DNA sequence of many viruses. To sequence these larger molecules, the intact genome is first cut into several smaller pieces at specific locations, and these fragments are sequenced individually. The first viruses whose sequence became known were ϕX174 and simian virus-40 (SV-40), each of which has a genome of about 5000 base pairs (or bases in the case of ϕX174, which consists of single-stranded DNA). Within about five years of this feat, the ten-fold larger genomes of the T7 and lambda bacteriophages were sequenced. Now scientists have sequenced the genome of a herpesvirus, cytomegalovirus, which consists of almost a quarter million base pairs, and even one whole chromosome of a eukaryote, the common backers' yeast.** They are also beginning a controversial 15-year project to sequence the entire three billion base pairs of the human genome. The cost of this latter extravaganza—and part of the reason for the controversy—is estimated between $1.00 and $2.00 per base pair. The variously ascribed ancient Greek dictum, "Know thyself", has become a somewhat costly proposition of late.

As mentioned in an earlier lecture, the comparison of DNA sequences from different organisms gives clues to their evolutionary relatedness. As an example, large stretches of DNA from higher

---

*Almost. It is not possible to learn the identity of the nucleotide **at** the 5' end, since no fragment will contain a radioactive phosphorus when it is eliminated. For complicated technical reasons, it is also not possible to learn the identity of the penultimate nucleotide, either. For this reason, both isolated single strands are usually sequenced. A strand with an undetectable Thymine, for example, at the 5' end will have its complement (Adenine) at the 3' end of the other strand.

**The sequenced chromosome is number 3 of the 16 chromosomes this yeast possesses. The DNA from this chromosome is some 315,000 bases in length and comprises about 180 genes. Only about 40 of these genes were previously known from conventional genetic analysis. An insight into the complexity of this effort can be gained by the fact that the scientific paper describing the achievement has 147 authors working in 35 laboratories in 17 countries! (Oliver, S., et al., 1992. Nature 357: 38-46)

primates have been sequenced and compared in an effort to sort out who begat whom. In comparable regions of the DNA (the various genes coding for the haemoglobins), it was found that chimpanzees and gorillas differed by about 2.1% whereas chimpanzees and humans differed by only 1.6%. This suggests that chimpanzees are our closer, and likely closest, relatives. These findings are consistent with the results based on mitochondrial rather than nuclear DNA. (In later lectures we shall note that chloroplasts and mitochondria have their own DNA.) Comparable mitochondrial sequences in chimpanzees, humans and gorillas show a 9.6% divergence between chimpanzees and humans and a 13.1% divergences between chimps and gorillas.

"AMPLIFYING" DNA

In the lectures on Recombinant DNA technology, various methods will be described that enable investigators to **clone** or amplify DNA so that almost limitless copies of a given stretch can be obtained. More recently a method has been devised that does not rely on living vectors such as viruses or bacteria; this newer *in vitro* method uses DNA polymerase I. The only requirement in addition is that the base sequence of short segments of DNA to the right and left of the target sequence being amplified be known so that short **priming sequences** of single-stranded DNA complementary to them can be synthesized. The method works in the following way. The double-stranded DNA containing the target sequence is melted into its single strands and the single stranded priming sequences added, along with pol I. The primers will attach to their complementary sequences on the longer single-stranded molecules, making part of these longer molecules double-stranded. The polymerase will then elongate these primers in the 5' to 3' direction; this synthesis will include the regions complementary to the target sequence and all the rest of the DNA to the 5' end of the template molecule. The mixture is then heated to render all the DNA single stranded again. When it is cooled, more of the same single-stranded priming sequences are added, along with more polymerase (the earlier polymerase would have been inactivated by the heat treatment used to melt the DNA). When the priming sequences have been extended in this second cycle, the heating, cooling, and addition of more polymerase and priming sequences is repeated—and so on, over and over again. With increasing rounds of synthesis, the DNA in the target sequence increases much more rapidly than does the extraneous DNA and hence the required amplification of the target DNA is achieved. This technique is known as the **polymerase chain reaction** or **PCR**.

Because the technique as outlined above requires fresh pol I to be added at each cycle, it could become very costly to carry it through very many cycles. However, a way around this limitation has been devised with the use of DNA polymerase from bacteria that normally grow in hot springs. These organisms have, of necessity, evolved proteins, including polymerases, that are not inactivated at the temperatures that are required in this procedure; their use in the polymerase chain reaction makes it possible to carry out the procedure without having to replenish the polymerase at each cycle of the reaction.

The polymerase chain reaction is finding increasing use in forensic studies to detect minute amounts of DNA at crime scenes or in medical diagnosis to detect, for example, the presence of viral DNA present in quantities well below that which is required for conventional techniques to discover. In the former case, the DNA amplification starting, say, from the amount present in a few cells shed with a single hair, is combined with "restriction fragment length polymorphisms" (RFLPs)—discussed elsewhere in these Lectures—to permit the unambiguous identification of an individual with an assurance equal to or greater than that achieved with fingerprints.

In paleobiology, the polymerase chain reaction can be used to clone minute amounts of DNA that might have survived intact from prehistoric times. This amplified DNA can then be compared with DNA from modern species to determine evolutionary relationships and rates of divergent evolution. A recent astonishing example of these techniques occurred when paleobiologists discovered intact DNA in a fossil magnolia leaf that had lain in sediments of an ancient lake bed for an estimated 17 to 20 million years. These researchers were able to identify at least one actual gene from this cloned DNA, a gene involved in photosynthesis that turns out to be not that much different from the one found in modern magnolias. It can be estimated that this gene underwent only one mutation every million years.

Insects trapped in the sap of an extinct tropical legume called *Hymenaea* become pemanently embedded as the sap polymerizes, hardens, and turns into the glass-like substance we call amber. In amber, destructive oxygen is excluded and, moreover, the sap contains antibiotic substances that preserve the insect from bacterial decay. Recently one such insect, an now-extinct bee trapped 40 million years ago, was extracted from the amber and found to still contain fragments of its DNA, which have been amplified by the polymerase chain reaction and analyzed. Because the fragments were much smaller than an entire gene (one ten-thousandth of one percent), the identity of the gene cannot be established. It is hoped that further studies if this sort will eventually uncover an insect that had just drawn blood from a dinosaur before being trapped in amber. This happy circumstance could enable scientists to obtain and analyze dinosaur DNA from the insect's gut, with all the attendant opportunities for creative science and perfervid science fiction.[*]

SYNTHESIZING AND REPLICATING RNA

Once the critical role played by DNA in the heredity of cells was introduced into these lectures, RNA has been largely ignored, except as it was involved in the replication of DNA. However, before RNA all but disappeared from this account, it was noted that some viruses do use RNA, and not DNA, as their hereditary material. This alone would be sufficient justification to examine more carefully the synthesis and replication of RNA; later in these lectures we shall uncover an even more cogent reason when the central role RNA plays in protein synthesis is discussed.

---

[*]Jumping the gun somewhat, Michael Crichton's novel, *Jurassic Park*, envisioned just such a scenario.

POLYRIBONUCLEOTIDE PHOSPHORYLASE

The very first *in vitro* synthesis of nucleic acid was reported in 1956 by M. Grunberg-Manago and S. Ochoa, a year before Kornberg's discovery of DNA polymerase. Given the ribonucleoside-5'-*di*phosphates, **polyribonucleotide phosphorylase**, an enzyme isolated from *Azotobacter vinelandii*, will assemble a polymer of single stranded RNA. No primer molecule is required; in fact, not even all four nucleoside diphosphates are needed. The phosphorylase will take whatever such ribonucleotides are present and assemble them into a polymer—whose over-all composition seems simply to reflect the concentration of the various nucleotides in the reaction mixture. For instance, if only uridine 5'-diphosphate is present, a polymer (poly-U) containing only uridine is produced; if all four diphosphates are present in, say, a ratio of 2U:IC:lA:lG, then the polymer will contain in random order 40% U, 20% C, 20% A, and 20% G.

Although the RNA synthesized by the phosphorylase has all the chemical and physical properties of some types of naturally occurring RNA, it soon became apparent that this reaction had little to do with the way RNA is synthesized *in vivo*. For one thing, the equilibrium for the reaction is decidedly in favor of the free nucleotides. In order to promote appreciable RNA synthesis, the liberated phosphate must be continuously removed from the reaction mixture. A more serious objection is the seeming unlikelihood that the cell would or could depend upon such an aleatory method of establishing the composition of its RNA. If the base composition (not to mention base sequence) of its RNA has any meaning for the cell—and surely it has— then sole dependence on the fluctuating internal concentration of the free nucleotides to establish this composition would be out of the question. Moreover, further investigation has shown that polyribonucleotide phosphorylase is never a synthetic enzyme in the cell. Its function in the intact cell is likely that of a **nuclease** to degrade RNA to the free mononucleotides. Only outside the cell and in the presence of atypically high concentrations of free nucleotides can it be persuaded, grudgingly, to synthesize a polymer.

RNA POLYMERASE

The bulk of the RNA in a cell is made by an enzyme, **RNA polymerase**, which uses a primer-template of double-stranded DNA and synthesizes a single-stranded molecule of RNA as a product. Except for the rather special case of RNA viral replication, RNA is never the template for its own synthesis. When discussing the mechanism of protein synthesis in later lectures, we shall examine RNA synthesis by RNA polymerase in some detail. For the present, we shall merely note that RNA polymerase uses the ribonucleoside-5'-triphosphates as substrates and assembles a product RNA in the 5' to 3' direction, while copying the DNA template 3' to 5'.

REPLICATION OF VIRAL RNA[1]

Since some viruses have only RNA as their hereditary material, it is obvious that a means must exist for this RNA to self-replicate. The fact that viral RNA is single-stranded introduces obvious difficulties in making facile comparisons with DNA replication. Indeed, the two mechanisms are quite different.

RNA synthetase
= RNA replicase

When a cell becomes infected with an RNA virus, a new enzyme is induced in the cell. This enzyme, **RNA synthetase** (also known as **RNA replicase**), is never found in uninfected cells; the information for its synthesis is encoded only by the entering viral RNA and not by the host DNA. The first task of this enzyme is to convert the single-stranded viral RNA into a double-stranded molecule called a **replicative form**, or **RF**. This it accomplishes by using the entering RNA as a template and assembling along its length a complementary strand.[*] Beyond this point, events in the replication of RNA become somewhat obscure. The RF does not, despite its similarity with the DNA double helix, replicate semi-conservatively. Instead it seems to continue peeling off strands which are complementary to the entering viral strand, and in so doing the RF becomes converted to a structure known as a **replicative intermediate**, or **RI**. The exact structure of the RI is still disputed, but it is agreed that at some point it is composed predominantly of single strands that are complementary to viral RNA. These in turn serve as templates for further synthesis of strands complementary to themselves. These new strands, being complementary to the complement of viral RNA, are obviously identical to the viral RNA strand that entered the cell in the first place.

All of these conversions are mediated by the one enzyme, RNA synthetase, and all of them can be carried out by the isolated enzyme *in vitro*. The enzyme that is induced by a particular RNA virus is remarkably specific in its primer-template requirements in that it will only replicate RNA from that particular virus and from none other. The substrates for the reaction are the four ribonucleoside-5'-triphosphates found in RNA. The final products of the *in vitro* synthesis have all the properties of the primer, including its biological activity.

ONCOGENIC RNA VIRUSES[2]

It has been well-established that certain RNA viruses induce specific cancers in animals. For example, certain leukemias in chickens and breast cancers in mice are caused by, respectively, "C-type" and "B-type" viruses that can readily be isolated and characterized. The capsids of these oncogenic viruses all contain a peculiar DNA polymerase known as **reverse transcriptase**, which employs a template of single-stranded RNA—the genetic material of the virus itself—to make a double-stranded DNA product. An RNA virus which makes a DNA copy of itself as a means of replicating is called a **retrovirus**. One much-talked-about retrovirus is HIV (human immunodeficiency virus), the virus that causes **AIDS** (acquired immune deficiency syndrome) in humans.

But why would an RNA virus ever want to synthesize DNA? It appears that these viruses replicate their RNA by transcribing it from a DNA template which they first synthesize by means

---

[*]To distinguish these two strands apart, the strand found in the mature virion and any strands identical to it will be called **plus strands**; their complementary strands are called **minus strands**. (To be precise, the technically correct definition of plus and minus strands has to do with which one—the plus strand—can also serve as a messenger RNA, but this detail need nor concern us here.)

└→ read up on mRNA

of their reverse transcriptase. This DNA template, moreover, need not be transcribed into RNA right away. Instead, it can be stably integrated into the DNA of the host cell and remain there, replicating in synchrony with the host DNA for many cell generations. The integration of this "RNA viral DNA" (i.e., the DNA transcript of the viral RNA) may effect no detectable change in the host cell or its descendants, but they all remain capable of producing RNA viruses in response to various stimuli, some known, most not. The presence of these oncogenic virus-related DNA sequences has been detected in both apparently normal and tumorous cells of many animals, including humans.[3]

Besides its importance to investigations into the origin of cancer, reverse transcriptase has also become an immensely useful enzyme in the laboratory. In later lectures we shall note its utility in the field of recombinant DNA technology ("gene splicing").

REFERENCES

1. Spector, D. and D. Baltimore, 1975. The molecular biology of poliovirus. Sci. Amer. 232: 24 (May).

2. Rafferty, K., 1973. Herpes viruses and cancer. Sci. Amer. 229: 26 (Apr.).

3. Temin, H., 1972. RNA-directed DNA synthesis. Sci. Amer. 226: 24 (Jan.).

retrovirus — RNA virus which makes a DNA copy of itself

Some such oncogenic viruses contain reverse transcriptase → employs a template of single stranded R.N.A. to make double stranded DNA product.

65

LECTURE 13

THE CELL NUCLEUS

THE CELL DOCTRINE

In the earlier lectures, I have been somewhat cavalier in discussing cells as if the reader already knew a great deal about them. In this section some of the lacunae will be filled in.

It is common knowledge that all indisputably living organisms are aggregates of the fundamental living unit, the cell. R. Hooke, in 1665, is usually credited with discovering cells while examining a slice of cork under his microscope. Purists might and do argue that Hooke was observing only vacuous cell walls and not living cells, cork being a dead and desiccated tissue. No matter. The "cell doctrine", the idea that all organisms are just cells and cell products, is usually attributed to M. Schleiden (a botanist) and T. Schwann (a zoologist), both of whom were active in the mid-nineteenth century. But in fairness to many other workers who predated these two, it should be recognized that the names of Schleiden and Schwann are associated with the concept more as a matter of alliterative convenience than of historical truth.

Before launching into a discussion of cells one ought to issue at least one *caveat*, which thereafter the present writer shall proceed to ignore. The noun "cell", like the noun "soup", must be preceded by an adjective (or an adjectival noun) to have much meaning. There is onion soup, turtle soup, chicken gumbo soup; and there is a liver cell, a blood cell, a bacterial cell; but there is neither quintessential soup nor cell anywhere. The "typical cell", therefore, is just abstraction of the particular things I want to talk about, gathered from various cell types and, figuratively, have drawn a membrane around with pen and ink.

PROKARYOTES AND EUKARYOTES

One of the great biological divides is drawn on the basis of whether a cell possesses a nucleus or not. Primitive organisms (if one may be finger-pointing for a minute) such as bacteria and blue-green algae do not have a nucleus; they are called **prokaryotes**, as if to suggest they are headed towards acquiring a nucleus, but haven't reached that enviable state yet. The genetic material (DNA) of these organisms is free within the cell membrane, although it is attached to it at one place, which is believed to be where the replication of the DNA takes place. Electron micrographs of bacteria often reveal one or two diffuse regions within the cell where the DNA is bunched; these regions are often called **nucleoids**. The **genome** of bacteria—that is, their total amount of genetic material—is the naked DNA molecule itself, which structurally is a closed circle. Sometimes this DNA molecule, which technically is known as a **genophore**, is ecumenically or jocularly called a **chromosome**, after the analogous organelle in the nucleus of a higher organism. It is, of course, no such thing.

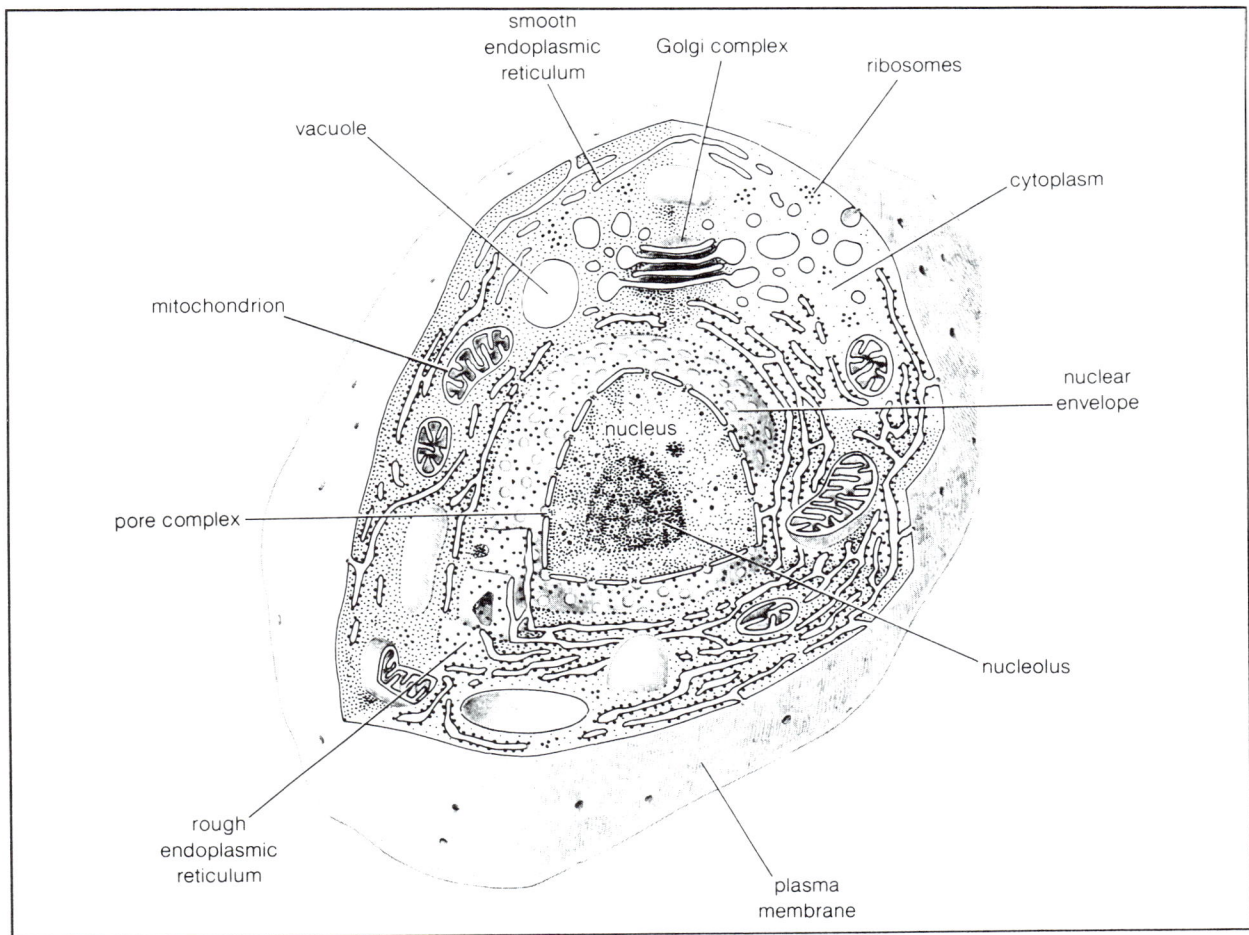

Figure 13.1 Structures commonly found in animal cells. From BIOLOGY OF THE CELL, 2ND EDITION by Stephen L. Wolfe. Copyright © 1981 by Wadsworth, Inc. Reprinted by permission of the publisher.

The cells of higher organisms possess a **nucleus**; organisms with nuclei are called **eukaryotes**. The possession of this membrane-bound nucleus means the cell is divided spatially into two inter-dependent areas, nucleus and **cytoplasm**. Because the double-layered membrane that surrounds the nucleus, besides being semi-permeable in the manner of all biological membranes, is also perforated by physical holes visible in the electron microscope, there is considerable opportunity for the exchange of materials between the nucleus and cytoplasm, and a great deal of evidence exists that this does in fact take place. For the time being, we shall focus our attention on this nucleus, for it is here that the hereditary apparatus of the eukaryotic cell lies.

THE NUCLEUS AND ITS CHROMATIN

Unless a cell happens to be in the process of dividing, one is actually examining an **interphase** cell at a time when there is not much to see in the nucleus either with a light or an electron microscope. There may be one or more large, dark-staining bodies, the **nucleoli**, which are rich

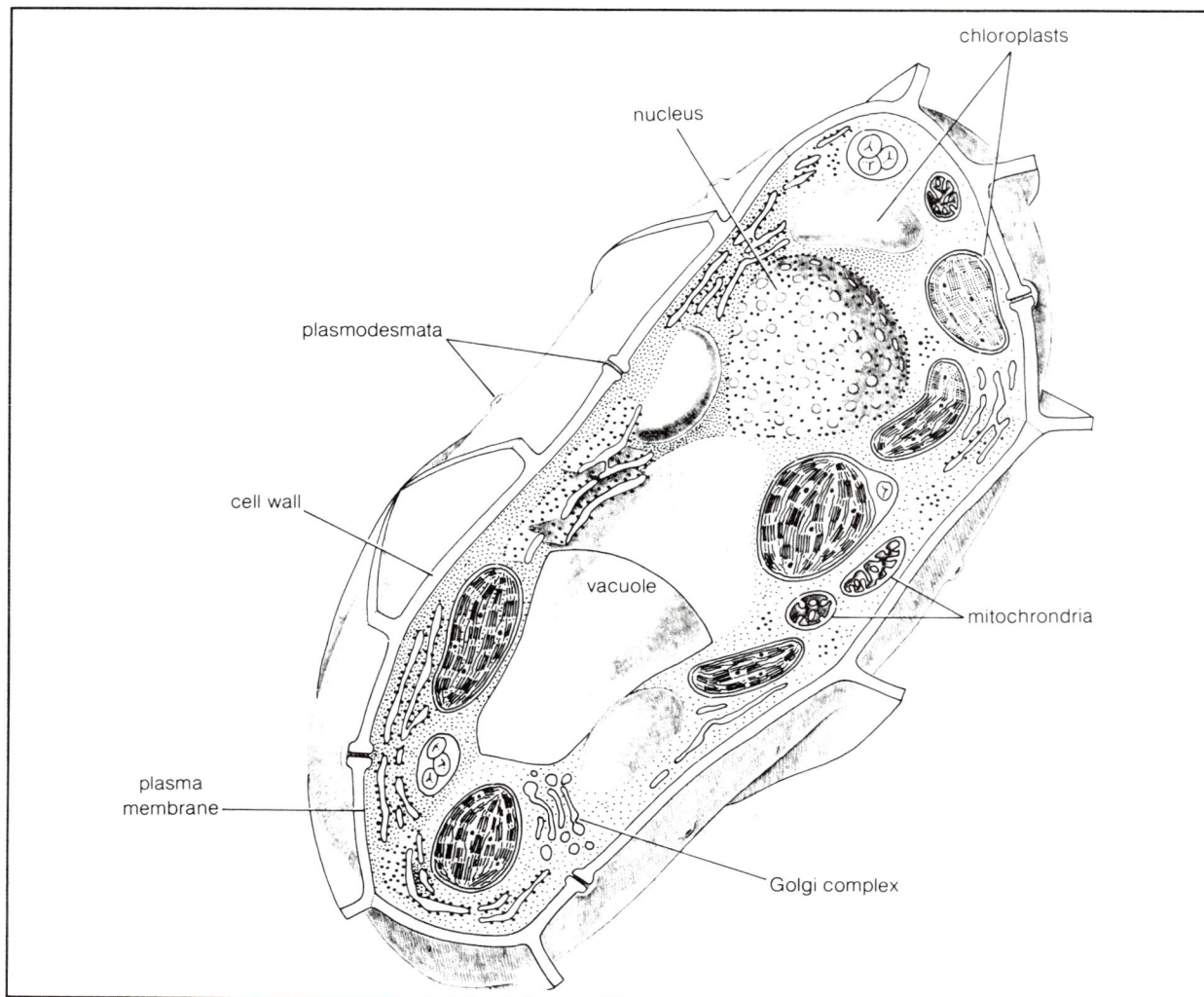

Figure 13.2 Structures commonly found in plant cells. From BIOLOGY OF THE CELL, 2ND EDITION by Stephen L. Wolfe. Copyright © 1981 by Wadsworth, Inc. Reprinted by permission of the publisher.

in RNA and turn out to be where most of the cellular RNA is manufactured. Beyond that there is an unresolvable network of DNA-rich ground material called **chromatin**, which until recently has resisted our most determined efforts to learn anything worthwhile about its structure and organization. Chemically, it is composed of DNA, some RNA, and a number of different specialized proteins called **histones**. These histones are intimately bound to the DNA and clearly play a role in packaging the DNA to fit into the nucleus and are also thought important in regulating the activity of individual genes on the DNA. This packaging function, incidentally, ought not to be taken lightly; if all the DNA in a single human chromosome of average length were stretched out in a single linear molecule, it would be about 4.0 centimetres long. (This is far from a boastable quantity: a salamander has about 2.7 meters of DNA in each of its 24 chromosomes, and even a single-celled organism such as the *Amoeba* has many times more DNA

68

histones → proteins attached to DNA...

than a human.) A cell, remember, isn't that big ("typically" 0.5 to 40 microns in diameter)* and its nucleus is even tinier; stuffing all that DNA inside a nucleus without hopelessly tangling it becomes a formidable engineering problem.

By a combination of X-ray diffraction, electron microscopic, and biochemical studies investigators have recently discerned what appears to be a fundamental subunit, the **nucleosome**, in mammalian interphase chromatin. The nucleosome is a spherical particle about 125Å in diameter. Its structure is determined by the four different histones (two molecules of each) it comprises and by the piece of DNA about 200 base-pairs long that is associated with it. Since this DNA is compressed about five-fold in length, it has been suggested that the DNA segment is wrapped around the histone particle. Individual nucleosomes are spaced about 25Å apart and connected by a thread of DNA. This connecting DNA may itself be complexed with a molecule of histone. The actual compaction of the chromatin DNA is, of course, much greater that can be accounted for by this fundamental structure; consequently the string of nucleosomes must itself be additionally folded in ways that are still matters of conjecture.[1]

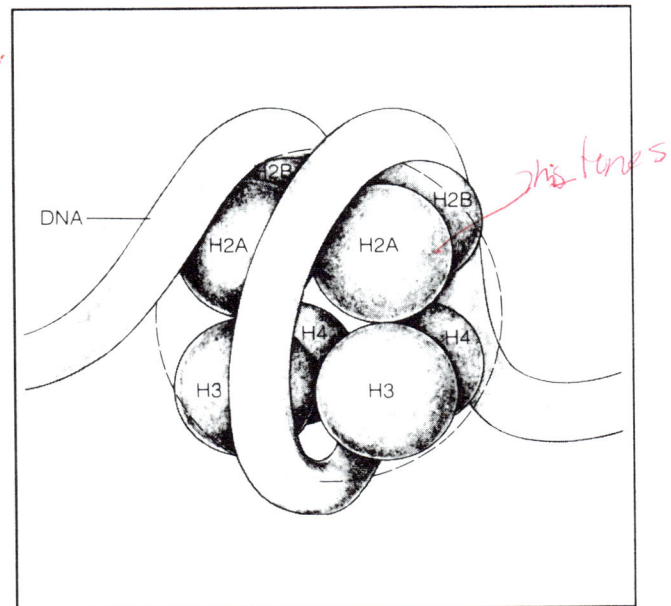

Figure 13.3 The nucleosome. (An Hl histone binds to DNA between nucleosomes.) From BIOLOGY OF THE CELL, 2ND EDITION by Stephen L. Wolfe. Copyright © 1981 by Wadsworth, Inc. Reprinted by permission of the publisher.

CHROMOSOMES

When a cell sets out to divide, events in the nucleus become more entertaining for the microscopist. Condensing out of the blurry network of chromatin, dark-staining shapes, called **chromosomes**, begin to take form. The nucleoli, which now can be seen to be attached to specific chromosomes (at a location called the **nucleolar organizer**), begin to shrivel and, as the condensation of the chromosomes continues, to disappear altogether. Later too, the nuclear membrane will dissolve, and the chromosomes burst out into the surrounding cytoplasm. The stage is now set for the division of the cell to get under way, but before describing what takes place next, let us focus for awhile on the chromosomes themselves. The chromosomes carry, of course, the DNA and hence determine the genetic fate of the cell. The DNA, in turn, is functionally subdivided into a large number of the fundamental units of inheritance, called

---

*There is, of course great variation in the size of cells. The axons of nerve cells, for example, can be many feet in length, and the largest cell—the yolk of an ostrich egg—is almost a meal in itself.

genes. Exactly how this DNA is organized within the chromosome is again a mystery, although now that we seem to be gaining an understanding of chromatin structure, one that may soon be unravelled. The largest human chromosome is five microns long, yet if all the DNA it carried was in a single continuous piece, it would be 16 centimetres long, i.e., 80,000 times longer. Much of the available evidence suggests that the DNA of the chromosome is in one piece and that replication of this unwieldy molecule during interphase proceeds at many different sites along its length simultaneously.*

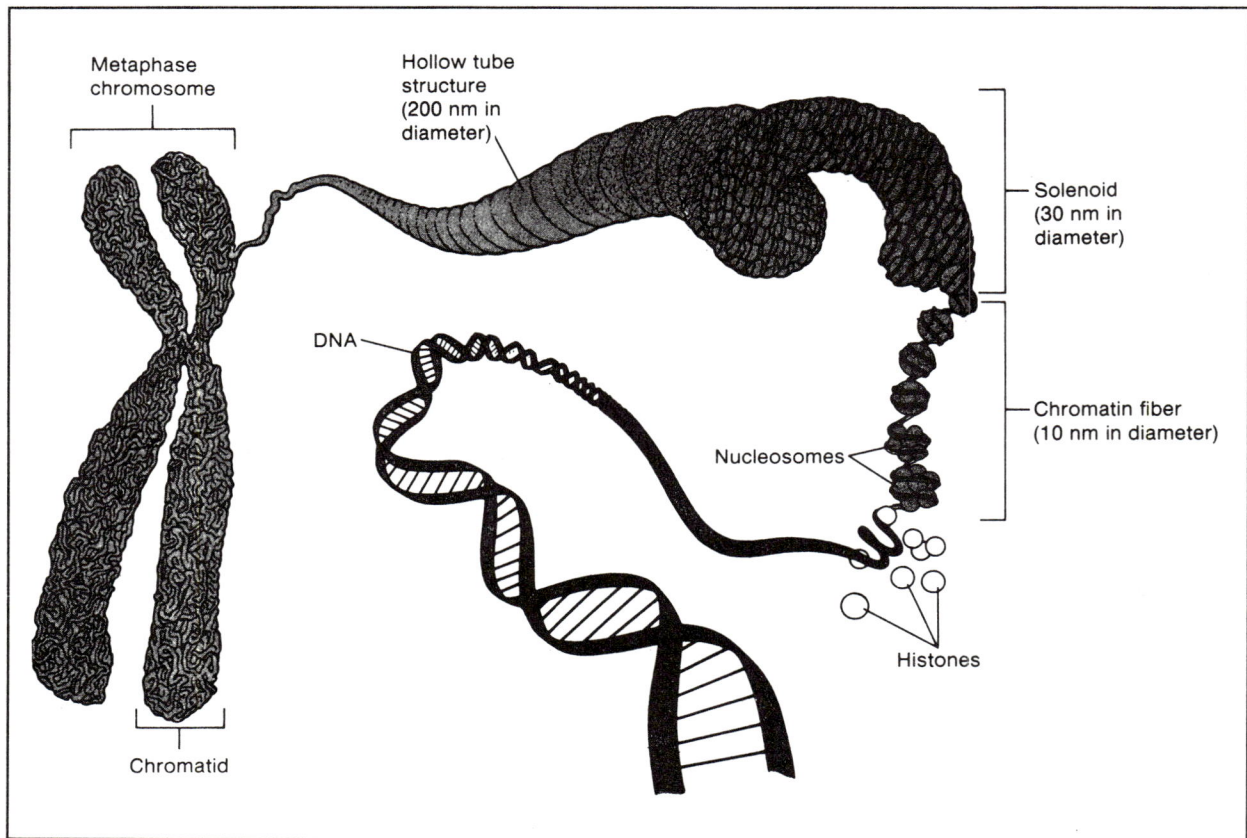

Figure 13.4 The relationship between chromosomes and DNA. From Leland G. Johnson, Biology, 2nd ed. Copyright © 1987 Wm. C. Brown Communications, Inc., Dubuque, Iowa. All Rights Reserved. Reprinted by permission.

Morphologically, chromosomes may be of different sizes, but all have somewhere along their length a constriction called a **centromere**, which plays, as we shall see, a very important role in

---

*DNA synthesis proceeds at about a rate of 3000 nucleotide pairs added per second. If there were only one replication fork per chromosome, it would take about 30 days to replicate the DNA in an average-sized human chromosome.

the movements of the chromosome during cell division. The position of the centromere also helps to identify individual chromosomes. If it is midway along the length, the chromosome is called **metacentric**; an off-center position for the centromere labels the chromosome as **submetacentric**; if it is close to the end, the chromosome is an **acrocentric** one. The lengths of chromosome material lying to either side of the centromere are known as the **chromosomal arms**. Unless the chromosome is metacentric, these arms will be of different lengths; in this case, the shorter arm is called "p" and the longer one "q".

All higher organisms are **diploid**, and a diploid organism has, by definition, two sets of chromosomes in every cell. One of these sets is inherited from the female parent and the other from the male. Since there are many more genes in a genome than there are chromosome pairs to accommodate them one-for-one, it is obvious that each chromosome must carry many genes. Since there is a remarkable constancy in the size and shape of the chromosome complement found in all normal individuals of the same species—your two chromosome number 7's, for example, look exactly like the next person's—it stands to reason that a particular set of genes is always carried by a particular chromosome and that none of these genes is ever found on any other chromosome. Hence genes, like the chromosomes on which they are carried, will also exist in pairs in diploid organisms. A pair of chromosomes is known as a **homologous pair**; a pair of genes is known as an **allelic pair**. The individual chromosomes of the pair are called **homologues**; the individual genes of the pair are called **alleles**.

The only exception to this rule has to do with the two so-called **sex chromosomes** or **allosomes**, named X and Y. Female humans and fruit flies (and most *male* birds, by the way) have two X chromosomes, and so do not differ from the pattern established for the non-sex chromosomes, or **autosomes**. Male humans and fruit flies (and *female* birds), by contrast, have non-identical allosomes, an X and an obviously dissimilar Y partner.[*] The Y chromosome is the smallest chromosome in the human genome, and being much smaller than the X, cannot possibly contain copies of all the genes that are on the X chromosome, which is our eighth largest. In fact, there are few genes of singular importance on the Y (unless, that is, one considers gender singularly important, since the Y chromosome does contain a gene responsible for the development of testes and hence for male differentiation), and although the total absence of an X chromosome is a

---

[*] The X-chromosome was first noticed late in the nineteenth century as being present in half of the spermatozoa of a particular insect; the Y-chromosome, being much smaller, was not noticed at the time. Because the single X-chromosome was present in males but seemingly absent in females, C. McClung suggested around the turn of the century that the X determined maleness. However, the apparent lack of a partner for the X spawned considerable doubt about whether it was really a chromosome at all—hence the designation "X"-chromosome and its original description as an "accessory" chromosome. The Y-chromosome partner was first discovered by N. Stevens a few years later; she correctly recognized that it was the presence of the Y, not the X, that determined maleness in these insects, and in many other organisms, besides.

71

lethal condition, a goodly number of people manage to live seemingly quite normal lives without any Y chromosomes at all*. (Technically such individuals are known as **females**.)**

The number of chromosomes per nucleus varies widely from species to species and does not correlate in any way with what we consider to be the complexity of the species. Humans, as has been mentioned, have 46 chromosomes; chickens and dogs have 78; turtles have 56; fruit flies have 8. This seemingly aleatoric distribution of chromosomes also extends to the total DNA content of cells. It might seem sensible to some that turtles and dogs have about 20% less DNA per nucleus than humans have, but then how does one explain the fact that salamanders have *30 times* more than humans, and lilies and lungfish have *16 times* more?*** It's a strange world.

HUMAN CHROMOSOMES

For a long time the human chromosome number was taken to be 48 (i.e. 24 pairs), but in 1956 J. Tjio and A. Levan, on a recount, discovered that the true number was 46. Since the chromosome complement of the human species is a matter of considerable selfish interest, the fact that this important piece of information eluded investigators so long points out the difficulties of recognizing and differentiating the individual chromosome pairs. Tjio and Levan succeeded, first, by treating large numbers of dividing human cells with a poison called colchicine, and then by immersing the cells in a hypotonic salt solution. The colchicine has the effect of arresting the division of the cells uniformly at a stage of division (metaphase) when the chromosomes are maximally condensed and hence visible, and the hypotonic salt causes the cells to swell and hence permits the greatest amount of chromosome separation. The cells are then gently squashed (an interesting oxymoron!) and photographed. The best photographs will be those where the chromosomes are all distinct and non-over lapping. This permits one to cut the individual chromosomes out of the photograph, match the homologous pairs, and arrange the whole set in a paste-up in descending order of size. Such an orderly display of chromosomes is called a **karyotype**.

---

*For what it is worth, it has been noted that certain men among the Amish have suffered a lengthy deletion of the long arm of the Y chromosome. These men are otherwise normal in every respect except one: they live significantly longer than other Amish men with normal Y chromosomes.

**Some species rely on a seemingly more aleatoric method of determining sex. Crocodile eggs, for example, yield either male or female crocodiles depending upon the ambient temperature at which development takes place. Higher temperatures produce males and lower temperatures females.

***The total amount of DNA in a *haploid* chromosome set is known as the **C-value** of that species. (See the following Lecture for the definition of "haploid".) The apparent lack of correspondence between the C-value and the presumed complexity of the organism is termed the *C-value paradox*.

In 1960 a group of human cytogeneticists meeting in Denver, Colorado, established what has become known as the **Denver classification** of human chromosomes. The chromosomes are fitted in descending order of size into seven groups lettered A to G (see illustration). Originally it was often difficult or impossible to make a judgment about whether a particular chromosome was, for example, no. 17 or no. 18, but nevertheless it was usually possible to place it reliably in its particular group (in this case, E.). Recently, sophisticated staining techniques reveal unique staining patterns (bands) for each chromosome. This enables skilled cytologists to identify any human chromosome with assurance. In fact, the whole process can now be computerized so that a complete human karyotype can be generated from good cytological preparations with little human intervention.

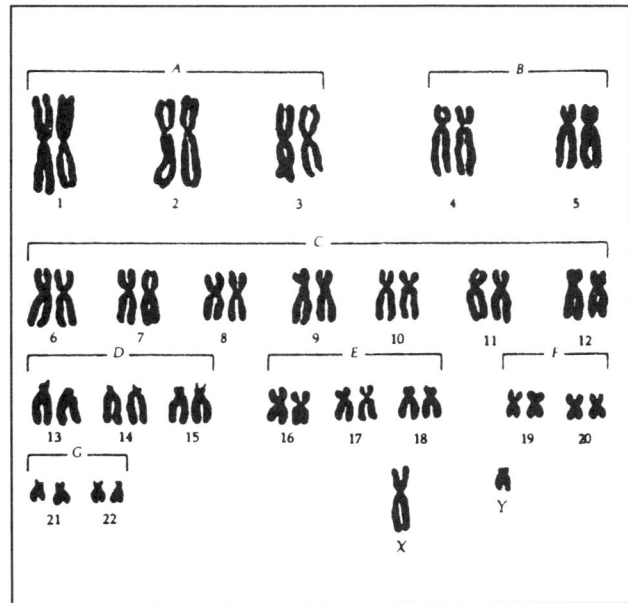

Figure 13.5 A normal human male karyotype—the chromosome complement arranged according to the Denver classification.

PRENATAL CHROMOSOME DIAG-NOSIS[2]

The development of techniques for the ready assessment of the human chromosome constitution has rapidly lead to many practical applications, especially in the area of prenatal diagnosis. For a variety of reasons, both legitimate and pernicious, women may, for example, want to know the sex of their developing fetus. In most cases the legitimate reason has to do with the possibility of sex-linked inherited diseases that would afflict a male child, and in those cases where the presence of the disease itself cannot be ascertained before birth, any male fetus will be aborted. Two methods for prenatal sex determination are currently in general use: **amniocentesis** and **chorionic villus sampling**. In the former case, a needle is inserted through the woman's abdomen and a small amount of the fluid (the amniotic fluid) surrounding the developing fetus is withdrawn. Fetal cells that had been sloughed off into the amniotic fluid can be cultured and later assayed for their chromosomal constitution as well as for a large and growing number of fetal metabolic defects. It is not practical to do an amniocentesis before about 16 weeks of pregnancy, a time approaching that at which a safe abortion becomes less so*. The procedure itself carries a risk of spontaneous

---

* It also takes about two weeks to culture the cells extracted by amniocentesis, thus adding to the delay before the results become known. Recent technical developments may result in this two-week delay being reduced to 24 hours. It appears possible to create specific "DNA probes" that attach by complementary base-pairing only to designated chromosomes. These probes also

(continued...)

abortion of about one-half of one percent. Chorionic villus sampling can be done much earlier in pregnancy, at about 10 or 11 weeks. It involves removing and culturing cells from part of the fetal membranes. This is accomplished by means of a catheter threaded into the uterus, a procedure that carries about a one percent risk of abortion.

Both amniocentesis and chorionic villus sampling require that the cells removed be grown in culture in order that a sufficient number of cells in the appropriate stage of division (metaphase) can be derived for analysis. This means that it is often two to three weeks after the procedure that the results are obtained. Investigators are currently experimenting with methods that will shorten that waiting period to two days. These newer methods involve DNA "probes"—in this instance, pieces of single-stranded DNA that are complementary to specific chromosomes and to which a florescent dye has been attached. These probes will seek out and bind to their specific chromosomes even if the cell is not in the process of division. The florescent dye can readily be visualized under the microscope, and hence any abnormality in the number of these chromosomes is apparent. Probes specific for the X and Y chromosomes have been developed, and for several other chromosomes commonly involved in human abnormalities, as well. (Some of the various human diseases that result from abnormal chromosome numbers are discussed in the subsequent lecture.)

Recently another technique for determining the sex and possibly many other attributes of the developing fetus has been devised. Because it originates as a simple blood test of the mother, it is totally harmless to both her and the fetus. It has long been a matter of controversy whether fetal cells circulated in the blood of the mother, but investigators using highly sensitive biochemical techniques (primarily the polymerase chain reaction, discussed elsewhere in this text) have now determined that such fetal cells do exist in the maternal circulation and can be identified as such. If the fetus is a male, these cells will carry identifiable genes that are found only on the Y-chromosome; if no such genes are found in the blood sample, the fetus must be a female. If this straightforward technique proves to be as accurate as the initial reports suggest, it could open the door to widespread prenatal sex determination, even where there is no suspicion of fetal abnormality. The fear is that gender itself—that is to say, a particular gender that differs from the one the mother wished for (or bought clothes for, or....)—could begin to be considered an "abnormality" to be avoided by abortion.

---

*(...continued)
carry a molecule of dye that fluoresces under ultra-violet light. When a particular probe, for example the one specific for chromosome 21, is added to the fetal cells and the cells then examined under UV, the number of chromosome 21s in a cell can be directly counted simply as the number of fluorescent spots per cell. Any number other than two spells trouble.

There has also been a recent report that amniocentesis can actually be safely done as early as the 11th week of pregnancy. If this is confirmed, the results from an amniocentesis should be routinely available during the first trimester of pregnancy, when abortions are safer and less controversial.

Fetal cells in the maternal circulation are extremely rare; approximately 1 in 20 million cells is of fetal origin. Because these fetal cells have surface characteristics that differ from those of the mother's, it is possible to identify and select for them in an automated cell sorter. Current techniques have resulted in preparations derived from the mother's blood in which fetal cells represent about 1% of all cells present. This is probably sufficient concentration to permit the use of florescent DNA probes to identify chromosomal abnormalities in the fetal cells and therefore permit totally harmless and very early fetal chromosome analysis.

REFERENCES

1. Kornberg, R, and A. Klug, 1981. The nucleosome. Sci. Amer. 244: 52-64

2. Friedman, T., 1971. Prenatal diagnosis of disease. Sci. Amer. (Nov.)
   Fuchs, F., 1980. Genetic amniocentesis. Sci. Amer. (June)

LECTURE 14

CELL DIVISION

MITOSIS

The elegant shuffle of chromosomes that enables the cell to divide into two daughter cells, each with exactly the same number of chromosomes as the parent cell, is called **mitosis**. Strictly speaking, mitosis refers only to events that involve the chromosomes and not to the division of the cell cytoplasm, which is a separate but related process called **cytokinesis**. The events during mitosis are somewhat arbitrarily divided, like the calendar year, into four stages purely for purposes of reference; but just as summer fades by degrees into fall, so too it must be borne in mind that there are generally no abrupt transitions between one stage of mitosis and the next.

The first stage of mitosis, **prophase**, has actually already been described. It occurs when the chromosomes condense out of the indistinct chromatin network, the nucleoli (when present) start to fade away, and the nuclear membrane dissolves to give the chromosomes some elbow room (chromosomes do, after all, have arms). As the chromosomes become more distinct, it is possible to discern that each is actually composed of two parts held together by the centromere. These parts are known as **chromatids**. When the chromosome is maximally condensed, these chromatids may even partially separate, but since they are still bound together at the centromere, a metacentric chromosome will often look like a figure

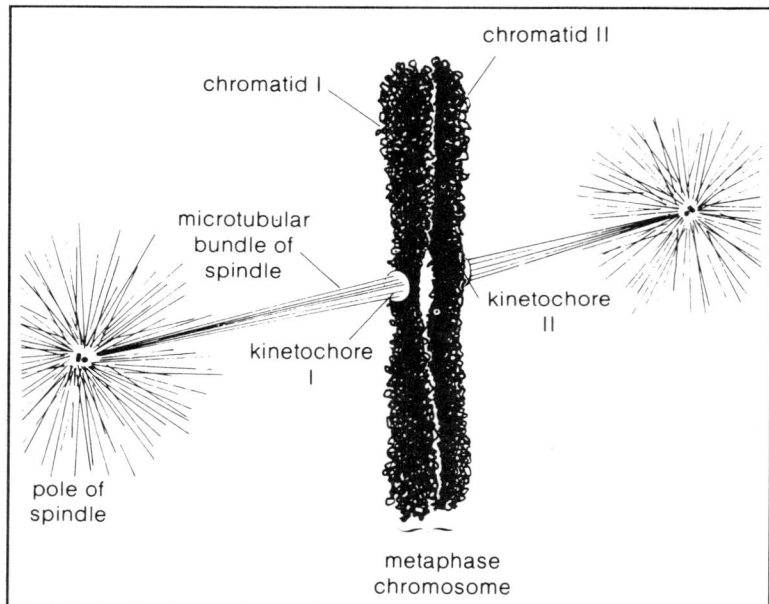

Figure 14.1 **A metaphase chromosome with the spindle apparatus attached. From BIOLOGY OF THE CELL, 2ND EDITION by Stephen L. Wolfe. Copyright © 1981 by Wadsworth, Inc. Reprinted by permission of the publisher.**

"X". These chromatids must not be mistaken for a pair of homologous chromosomes, nor should they be mistaken for the individual strands of a DNA double helix. Each chromatid has a complete double-helical molecule of DNA. The two chromatids of a chromosome, usually called **sister chromatids**, are the result of the replication of the chromosome sometime just prior to prophase; these two sister chromatids are therefore genetically identical in the same way that the two

DNA helices formed after one round of DNA replication are genetically identical. At this point, remember, they are physically joined by their common centromere and constitute a single chromosome.

As prophase fades into the next stage, **metaphase**, the chromosomes begin to line up in a single-file row across the center (or **equator**) of the cell. (Some scientists call the period during which the chromosomes are moving towards the equator a separate stage: **prometaphase**. But most do not.) The actual line is best discerned by examining the position of the centromeres, since the arms of the chromosome tend to be sprawled higgledy-piggledy all over the place. Here it is important to note that there is no relation between one chromosome and its homologue with respect to their relative location in this line-up. They could, for example, be at opposite ends of the equatorial line, right next to each other, or anywhere else. Each is free to crowd into the line wherever it happens to find itself.

The movements of the chromosomes are controlled by a network of fine fibers called the **spindle apparatus**. These fibers, which appear in the cell only during cell division, run at right angles to the equator of the cell. Spindle fibers emanate from a body known as the **centrosome**, which had divided into two, each part migrating to an opposite side of the cell. In animals and some of the lower plants centrosomes are inhabited by a fascinating cell organelle called a **centriole**. Centrioles are inherited cytoplasmically and arise only by division of another centriole. They have recently been shown to contain a small amount of DNA[1], a fact that begins to explain their autonomous character. Higher plants lack centrioles altogether.

Because the spindle fibers converge at their two ends, these convergences establish what are often called the two **poles** of the dividing cell. As metaphase draws to a close, it becomes clear that the centromere of each chromosome has become attached to a separate spindle fibre. The attachment site of the spindle fibre to the centromere is a protein complex called a **kinetochore**.

The one abrupt transition during mitosis signals the end of metaphase and the beginning of the following stage, **anaphase**. Suddenly and almost simultaneously the centromeres of all the chromosomes split into two parts and the two chromatids that once composed each chromosome become fully separate. Since each of these newly-freed chromatids has its own centromere, they can now be called chromosomes in their own right. At this moment, therefore, the cell has twice the (diploid) number of chromosomes it should have.

Each of the newly separated centromeres still remains with the same spindle fibre the parental centromere was attached to. The spindle fibers then seemingly contract with the result that the new chromosomes (which were formerly sister chromatids) are reeled in toward the poles, each to a pole opposite to the one its former sister chromatid is pulled to.* At the end of anaphase,

---

*The word "seemingly" in this account disguises a great deal of ignorance about what actually motivates chromosome movements during cell division. It does *appear* that the spindle fibers
(continued...)

77

the obvious result is a cluster of chromosomes around each cell pole. The more significant point, however, is that each of these clusters contains a physically and genetically identical diploid set of chromosomes.

The succeeding stage, **telophase**, sees the gradual reverse of the events that occurred in prophase: the chromosomes begin to lose their distinct shapes, the nucleoli begin to regrow on the particular chromosomes that previously anchored them, the nuclear membrane reforms and the spindle fibers melt away. Meanwhile the cytoplasm has constricted around the cell equator so that by the end of telophase two distinct nucleated cells are present where before there was only one. These cells are now said to be in **interphase** during which time active metabolic processes, largely suspended during mitosis, are resumed. It is a bit ironic that this inter-division stage, which is when the cell is really its busiest, is sometimes called the **resting stage**, as if it were just the entr'acte of a continuous performance of dancing chromosomes staged for the chance observer's entertainment.

One very important task to accomplish during interphase is the duplication of the DNA[*] in preparation for the subsequent mitosis (assuming, of course, that there is to be a subsequent one). Synthesis of DNA begins not immediately after mitosis, but only after an interval called "the first gap" phase (G1). The period of synthesis (the S phase) is followed by another interval, "the second gap" (G2), and then by mitosis (the M phase). The duration of these phases varies considerably depending on the species in question and the cell type being observed. For example, in the human white blood cell, these times would be approximately 2 hours for the M phase, 11 hours for G1, 7 hours for S, and 4 hours for G2. This sequence of events goes under the name of the **cell cycle**.

---

[*](...continued)
are contracting and pulling the chromosomes apart, but this is probably an illusion. Investigators have succeeded in cutting the spindle fibers at a point close to the pole, and still the chromosomes move apart. It is difficult to see how fibers could be pulling on chromosomes if these fibers were not attached to something relatively immobile at the other ends. Current speculations on chromosome movements suggest that the kinetochore is, or is part of, the "motor" and that the spindle fibers are more like pathways or tracks along which the chromosomes are moved. Despite being intensively investigated for almost a century, the movement of chromosomes is still a mysterious process.

[*]Students sometimes equate the two strands of a prophase/metaphase chromosome (i.e., the sister chromatids of a single chromosome) with the two strands of the DNA molecule replicated during interphase. Don't. Each chromatid has a complete (i.e., double-stranded) DNA molecule running—in a very convoluted manner—its entire length. The replication of the DNA during interphase results in two complete DNA molecules; these two sister (or daughter, depending on your perspective) DNA molecules wind up in the sister chromatids.

What controls the initiation and the highly coordinated subsequent events of cell division has long intrigued biologists, but until very recently little progress has been made towards finding out. Now, however, it is known that two proteins are the key to the answer. One of these, **CDC protein** (which stands for **Cell Division Cycle protein**[*]), is found as an essentially identical protein in all eukaryotic cells from yeasts to humans. By itself, CDC has no effect on cell division, but when it combines with the other protein, **cyclin**, the combined protein triggers a cascade of activities that are essential for the initiation and orderly completion of cell division. The principal activity of the combined protein is the activation of other pre-existing proteins by the addition of phosphate groups to them, and it is these secondarily activated molecules, of which there can be many, that orchestrate cell division first-hand. Whereas CDC is a single species of protein, cyclin is actually a family of different proteins that are capable of combining with CDC to form, therefore, different combinations with reactivities that vary according to the particular cyclin that is present. The cyclins are produced sequentially during cell division, and each is degraded as the subsequent one is produced. In this way, a continuously variable master molecule, the CDC/cyclin compound, appears throughout cell division, each variant capable of activating a different series of secondary proteins that coordinate cellular for awhile until the next series is activated.[2]

In summary, mitosis is the means by which the cell ensures that its chromosomes and hence its hereditary characteristics are passed down exactly and in their entirety to the two daughter cells to which the parental cytoplasm is also apportioned more or less equally.

Most cells of the body, by the way, don't divide at all any more. In many cases, once a cell has become highly specialized (e.g. muscle cells, brain cells), it loses altogether the capacity to divide. If more cells of a specialized type are required, as would happen, for example, when bone or skin is regenerating, these new cells are replenished by divisions triggered in un-specialized "stem cells" that have remained unspecialized for just this turn of events. Once the daughter cells differentiate into actual bone or skin cells, they cease dividing forever. Mature human red blood cells don't even have a nucleus.

MEIOSIS

The precise chromosome movements that constitute mitosis neatly explain how the full genetic complement is passed from one parental cell to two daughter cells. What it cannot explain is the constancy of chromosome numbers from one generation to the next, for if gametic cells had the same number of chromosome characteristic of all the rest of the body cells, the fusion of these eggs and sperm would produce an individual with twice the parental chromosome numbers. Clearly this doesn't happen, and the mechanism that prevents it from happening is termed **meiosis**. Meiosis is a special type of cell division that occurs only in cells giving rise to eggs and sperm (or their equivalent in other phylogenetic kingdoms) and which reduces the chromosome number in the gametes produced to exactly half that of the parental cells. In

---

[*] It is also called MPF (Maturation Promoting Factor).

parcelling out to a gamete just one representative of each homologous pair of chromosomes, meiosis also results in a random shuffle of the chromosomes present in the parental cell; consequently the gamete inherits a new, unique combination of the chromosomes which the parental cell had previously inherited from both its male and female parents.

Meiosis is actually a double cell division, one succeeding the other, and each is further divided into the same four stages that characterize mitosis. There are, however, some very striking differences between the stages of mitosis and the identically named ones of meiosis. This is especially true during the critical first set of meiotic divisions.

Prophase I of meiosis proceeds seemingly like a mitotic prophase with respect to the condensation of the chromosomes, the disappearance of the nucleoli and eventually the nuclear membrane, and the laying down of the spindle apparatus. The significant difference rests with the behavior of the homologous chromosomes. In mitosis it is only a matter of happenstance should homologous chromosomes come to rest anywhere near each other, but in meiosis the homologous chromosomes actively seek each other out and come to rest parallel to each other side by side. Since each chromosome is composed of two chromatids by this time, each homologous pair will comprise altogether four chromatids. For this reason, it is often called a **tetrad**.

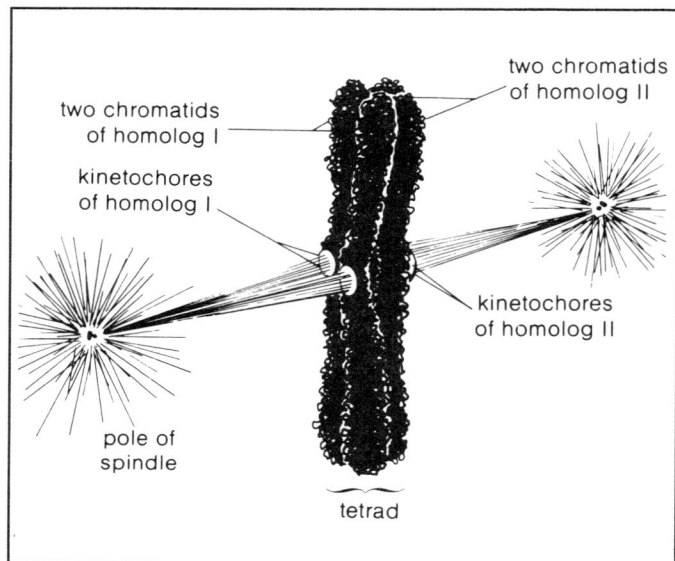

Figure 14.2 **Meiotic metaphase pair of homologous chromosomes.** From THE BIOLOGY OF THE CELL, 2ND EDITION by Stephen L. Wolfe. Copyright © 1981 by Wadsworth, Inc. Reprinted by permission of the publisher.

Prophase I of meiosis is somewhat more protracted than prophase of mitosis (more will be said about this later), but eventually it fades into metaphase I as the homologous pairs line up at the equator. In a human *mitosis*, the centromeres are arranged in a single file 46 in length; in a human *meiosis*, by contrast, the centromeres are arranged in a double file only 23 pairs in length.

There is no division of centromeres at the end of metaphase I, but instead the spindle fibers pull the homologues apart, and during anaphase I they are dragged to opposite poles, a phenomenon universally known as **segregation**. Each of these chromosomes is still composed of two chromatids, but since the two new nuclei regenerated in telophase I contain only 23 chromosomes the chromosome number of the parental cell is effectively halved. Cells with only one representative of each homologous pair of chromosomes are called **haploid**.

80

But there is still the problem of converting these chromosomes composed of two chromatids in the two haploid daughter cells to ones composed of single chromatids. This is accomplished by having each of the daughter cells undergo a second series of division stages—prophase II, metaphase II, etc.—which resembles the stages of a conventional mitosis. This time the centromeres do split at the end of metaphase II reducing each chromosome to two. These are then pulled to opposite poles, and the cells at telophase then differentiate into functional gametes, eggs or sperm as the case may be. This second division, which creates chromosomes having only one chromatid, is known as the **equational division** of meiosis. The first division, which actually accomplishes the reduction of the chromosome number by half, is called the **reduction division**.

Over-all, the process of meiosis results in, first, two haploid cells with chromosomes composed of two chromatids, and then a total of four haploid cells with chromosomes composed of a single chromatid. The time interval between meiosis I and II may be minutes or years, and the number of functional gametes differentiated as a consequence of each complete meiosis may vary from one to four. Each species seeks to put its own imprimatur on the process, but the underlying chromosome movements as described here are fairly standard throughout.

THE FORMATION OF FUNCTIONAL GAMETES (GAMETOGENESIS)

After meiosis II the resultant cells undergo a process of differentiation that transform them into functional egg or sperm cells. In mammalian males, each of the four cells resulting from the two meiotic divisions differentiate into a functional **spermatozoon**. The different stages in the formation of spermatozoa have individual names: **spermatogonia** (the diploid mitotically dividing cells in the testis that eventually undergo meiosis), **primary spermatocytes** (the still-diploid cells that have reached the end of their mitotic divisions and are about to enter meiosis), **secondary spermatocytes** (the haploid products of the reduction division of meiosis), **spermatids** (the final haploid products of the equational division), and finally spermatozoa (the differentiated products of spermiogenesis, the process whereby the spermatids lose much of their cytoplasm, grow tails, etc. and become functional sperm. This differentiation process requires a temperature slightly lower than that of the body cavity. The solution is obvious.

One peculiarity of gametogenesis in mammalian males is that the spermatogonia often do not fully separate when they divide, but the daughter cells remain attached to each other by slender cytoplasmic bridges. The same thing occurs during the meiotic divisions up to and including the spermatid stage. This results in a raft-like structure containing many spermatids attached together by these cytoplasmic bridges. During spermiogenesis the spermatozoa are pinched off from the individual cells of this "raft", which are left with most of the cytoplasm but no nucleus. These **residual cells** disintegrate.

Human males produce about 300 million spermatozoa in each ejaculate, but of these only about 200 actually reach the ovum. All the factors at work in this winnowing process are not understood. Recent studies suggest that the egg (or possibly the surrounding cells) produces a chemical attractant for the sperm. This substance, as yet unidentified, may also play a role in activating sperm or selecting those that are in an appropriate physiological state for fertilization.

81

In mammalian females, instead of four functional products of meiosis, there is usually only one—a single **ovum**. At the end of the reduction division, there is a vastly unequal sharing of the cytoplasm at cytokinesis. One of the two cells gets almost all of it, whereas the other, called a **polar body**, gets almost none. The same thing happens again at the end of the equational division. Hence, there results only one functional ovum and two polar bodies (or three, if the first polar body divided in an equational division of its own). These polar bodies are not functional and eventually disintegrate. The various stages in the formation of the functional ovum have names similar to those of the equivalent stages of spermatogenesis: **oogonia, primary** and **secondary oocytes**, and mature ovum. In humans, females are born with their primary oocytes arrested in the meiotic prophase stage. Beginning at puberty, about 1000 of these begin to develop further each month, but usually only one completes the process to become an ovum, and the rest die. During a women's reproductive lifetime, only about 400 to 500 ova are released, although at puberty there were about 300,000 primary oocytes waiting at the starting line. Most of these have disintegrated by menopause.

The ova that do mature in women over 40 years of age are considerably less capable of initiating and maintaining a successful pregnancy than those of younger women. It used to be thought that older women had difficulty conceiving because their uterus is less capable of sustaining a pregnancy, but now it has been shown that older women have about the same success rate for pregnancy as younger women (normally about 12% using current menthols) if eggs from younger women donors are fertilized *in vitro* and transplanted into their uterus. Even menopause is no barrier to a successful pregnancy if appropriate hormone therapy is used to replace hormones such as progesterone that are no longer produced naturally. Through the use of donated eggs and *in vitro* fertilization it appears likely that women in their 40s and 50s will become increasingly common in maternity hospitals of the future.

CHROMOSOMAL SEX DETERMINATION

There is one important exception to the pairing of homologous chromosomes during prophase I and metaphase I of meiosis. It has already been pointed out that the X and Y are so dissimilar in appearance that they could in no sense be considered homologues. The human X, for example, is a relatively large Group C chromosome and the Y much smaller Group G chromosome. Nevertheless the two allosomes do pair during meiosis in the male because of homologies between sections of the short arm of the Y and the X chromosome. This meiotic pairing will result in half the sperm produced carrying an X chromosome and half carrying a Y. All eggs produced by a female will, of course, have an X chromosome, since the two X chromosomes in a female cell will pair and then segregate in the same manner as any other homologous pair. The sex of the diploid conceptus will depend, therefore, only on the type of allosome carried by the sperm that succeeds in fertilizing the chromosomally uniform eggs.

The male:female ratio of human embryos at conception is termed the **primary sex ratio**. Until very recently it was thought that it was heavily skewed in favor of males, one explanation being that Y-bearing spermatozoa had greater success in racing up the track to fertilization than did their slower X-bearing "sisters". This was seemingly borne out by the much greater number of male fetuses that aborted spontaneously ("miscarried") during pregnancy, especially during the first trimester. It now turns out that early female fetuses may have been misidentified as males. If there is any skewing in the primary sex ratio, it may actually favor females; the **secondary** (birth) **sex ratio** is slightly biased towards males. Males drop off at a faster rate thereafter, so that equality in numbers is only achieved between the ages of 10 and 40. Thereafter the proportion of females steadily increases to reach 75 to 80 percent by the tenth decade of life.

REFERENCES

1. Hall, J., et al., 1989. Cell 99: 121-132.

2. Murray, A., and M. Kirschner, 1991. What controls the cell cycle. Sci. Amer. 264: 56 (March)

LECTURE 15

CHROMOSOMAL ABNORMALITIES IN HUMANS

AUTOSOMAL NON-DISJUNCTION

The easiest way to fix in one's mind the chromosomal events that take place during meiosis is to examine the consequences of aberrant segregations in humans. Historically the first example of a human chromosome abnormality reported in the literature caused the congenital disorder known as mongolism or **Down's Syndrome**. Such individuals, who comprise about one in six hundred live births, are severely mentally retarded, and constitute about 10% of all institutional-ized mentally handicapped patients. Besides the mental disability, there is also an accompanying spectrum of physical abnormalities such as an open mouth with a protruding furrowed tongue; misshapen, small, low-set ears; epicanthal folds about the eyes (hence the earlier name, "Mongoloid Idiocy"); and distinctive dermatoglyphic patterns. People with Down's syndrome have about a 1% chance of getting leukaemia (i.e., 20 times the usual risk), and by the time they are 35 to 40 years old, they usually get Alzheimer's syndrome. Because their thymus glands are abnormal, there are associated abnormalities of the immune system that make such individuals much more prone to infections.

Although such individuals have been known for centuries, it was not until 1959 that the etiology of the abnormality was discovered. In a karyotype of several Down's individuals, J. Lejeune discovered that they typically possessed 47 chromosomes, instead of the normal number of 46. The extra chromosome, it was found, resulted from the triplication of one of the smallest chromosomes of the genome, number 21. This chromosome accounts for only about 1000 to 1500 of the estimated 100,000 genes carried on all the human chromosomes. Although most Down's individuals have the entire chromosome 21 in triplicate, it is now thought that the crucial genes responsible for most of the characteristics of this disorder constitute only about 5 to 10% of the chromosome and are concentrated near the end of the long arm.

Theoretically, a **trisomy** for any particular pair could arise as a result of an improper segregation of homologous chromosome pairs during anaphase I of meiosis or as a result of an improper separation of the new sister chromosomes during anaphase II. (Both of these anomalous events—improper segregation of homologues and improper separation of sister chromosomes derived from one chromosome—are covered by the general term, **chromosomal non-disjunction**.) Non-disjunction could occur, of course, in either sex with the same result: an otherwise haploid gamete with just one of its chromosomes present in double dose. The fusion of this **aneuploid** gamete with one carrying the standard haploid set of chromosomes establishes a diploid individual which is trisomic for one chromosome.

In the case of Trisomy-21 (i.e., Down's Syndrome) it was long thought that most such individuals arise as a consequence of non-disjunction at anaphase I or II in the female parent. This origin was inferred because of a striking correlation between the age of the mother and the frequency of mongolism among her offspring. The older the mother, the greater the fetus is at risk, although there is no similar correlation with the age of the father, taken as an independent variable. Young mothers (under 29 years old) yield offspring with Down's Syndrome at a frequency of about one in 3000; for old mothers (over 45) the rate is one in 40. The favored interpretation of this phenomenon at present is that younger mothers are more efficient at spontaneously aborting (miscarrying) the defective fetus than are older mothers. This miscarriage occurs at a very early stage in development, often before the woman knows she is pregnant. At variance with this interpretation is the very recent observation that about 95% of all Down's syndrome individuals have two chromosome 21s from the mother, rather than from the father. This information was obtained from molecular comparisons of the DNA banding patterns from the two parents and the affected child.[*] (Such identification techniques are described in Lecture 40, "Genetic Screening in Humans".) It has also been ascertained that the non-disjunctional event takes place 75% of the time in meiosis I. This seems especially odd, since meiosis I in human females takes place before the mother is even born! Meiosis II in each reproductive cell does not take place until much later, just prior to the mensual production of a mature egg during the reproductive years of that female .

Since Lejeune's discovery of the origin of Down's Syndrome, several other human autosomal trisomies have been discovered. All of them (e.g. Trisomy-18, Edward's Syndrome; and Trisomy-13, Patau's Syndrome) exhibit severe mental retardation and a constellation of physical malformations so severe that such infants almost never survive beyond a few months.

SEXUAL DIFFERENTIATION AND ALLOSOMAL NON-DISJUNCTION

The sex chromosomes, X and Y, are frequently involved in non-disjunctions that affect the sexual characteristics of the individual who inherits the abnormalities. Before discussing these, however, we need to see how mammals such as humans differentiate into males and females in the first place.

For the first six to eight weeks after conception the developing embryo is neither male nor female. The undifferentiated gonad began as a bulge of tissue near the embryonic kidney and consists of two parts: an inner core called the **medulla** surrounded by a surface layer of tissue called the **cortex**. At this point it is capable of developing into either an ovary or a testis,

---

[*]Also at variance with this interpretation is the recent (1991) observation that Down's syndrome is about the only one of 43 common birth defects that appear correlated in frequency with increasing maternal birth age. In this survey, which examined records of the nearly 27,000 babies born with birth defects in British Columbia between 1966 and 1981—about 5% of all babies born there during those years—a few common birth defects (e.g., pyloric stenosis and a heart defect called patent ductus arteriosis) actually declined in frequency in older mothers.

depending upon what additional stimulation it is given. This "additional stimulation" arrives in the form of an invasion by the so-called **primordial germ cells** that had been forming in the yolk sac, outside the embryo proper. These germ cells migrate at about the seventh week to the still undifferentiated gonad, where they take up permanent residence. If the primordial germ cells are chromosomally XX, they populate the cortex and the medulla degenerates. Hence, the ovary is derived from the gonadal cortex. On the other hand, if the invaders are chromosomally XY, they penetrate to the medulla, where they settle in and proliferate. This causes the cortex to degenerate, leaving the medulla to develop into a testis. On rare occasion for genetic or developmental reasons the invaders fail to arrive, and when this happens the primitive gonad degenerates and no functional gonads develop in the embryo.

Besides having undifferentiated gonads, the early embryo also has two sets of ducts known as

Table 15.1 Some common human disorders caused by improper chromosome numbers.

Syndrome	Abnormal chromosomal	Frequency
	Trisomics	
Patau's	13, 13, 13	1 in 500 births
Edwards'	18, 18, 18	1 in 3,000 births
Down's	21, 21, 21	1 in 600 births
Triplo-X	X, X, X	1 in 1,200 female births
Klinefelter's	X, X, Y	1 in 400 male births
Jacobs'	X, Y, Y	1 in 300 male births
Turner's	X, O	1 in 2500 female births

the Mullerian and the Wolffian ducts. If the early gonad differentiates into an ovary, the Mullerian ducts in due course develop into structures such as oviducts, uterus and parts of the vagina. If the gonad develops as a testis, however, it secretes a hormone, **Mullerian Duct Inhibitor (MDI)**, which prevents development of the Mullerian ducts. Then the testis secretes a second hormone, testosterone, which stimulates the Wolffian ducts to differentiate eventually into the epididymis and sperm ducts. Note, therefore, that in the absence of particular stimuli (MDI and testosterone) secreted by the testis, development will proceed in the general direction of femaleness. An ovary does not have to be present for this to happen by default. A similar situation exists for the various embryonic structures that develop into clitoris, labia, penis, etc.: in the absence of testosterone, the development is in the female direction; in the presence of testosterone male structures develop.

The key element in mammalian sex determination is the Y-chromosome. If it is absent *in toto*, the embryo will develop strongly in the female direction; if it is present, the development will be primarily male-directed, regardless of how many X-chromosomes are also present. Normally, course, there would be only one X chromosome present with the Y, but exceptions do exist. The first exception to be discovered in humans are individuals suffering from **Klinefelter's Syndrome.**[1] These individuals appear male but have small, non-functional testes. At puberty they do not develop much body hair and may develop prominent breasts. As a group, they are decidedly mentally subnormal, although many individuals fall well into the normal range and function well in society.

The origin of Klinefelter's Syndrome rests with a non-disjunction of the sex chromosomes in one of the parents, an event which yields a double-X egg (later fertilized by Y-bearing spermatozoon) or an XY spermatozoon (that later fertilized an X-bearing egg). In either event, the result is a zygote (i.e., a fertilized egg) that is XXY in sex chromosome constitution and with, therefore, a total of 47 chromosomes. (About two-thirds of individuals with the XXY condition derive both X-chromosomes from the mother; in the remaining cases the two Xs come from the different parents.) Occasionally non-disjunction in both the first and second divisions of meiosis yields individuals with even more allosomes--for example, XXXXY, XYY, etc. In all cases where at least one Y chromosome is present, the general appearance of the individual is male, although the abnormalities associated with Klinefelter's Syndrome are usually accentuated with increasing numbers of X chromosomes, relative to Y chromosomes.

A similar condition is known in cats. Because the genes for the coat colors of black and orange are alleles (see later Lecture on Genes and Alleles) carried on the X chromosome, in order to have both genes, two X chromosomes must be present[*]. Such black and orange cats (known as "tortoise shell" cats) are, in consequence, usually female. Now and then male tortoise shell cats are reported. Like Klinefelter's individuals in humans, such cats are XXY, and likewise are sterile.

If non-disjunctions can result in XX or XY eggs and sperm, it can also result in eggs and sperm entirely lacking an allosome. When, for example a "nullo-X" egg (lacking an X chromosome) is fertilized by an X-bearing spermatozoon, an "XO" zygote having a single allosome is produced. In humans these people are females suffering from **Turner's Syndrome**. They are short in stature and because their ovaries fail to develop, they never menstruate and their breasts do not develop. Swellings in the skin of the neck give the person a broad or "webbed" appearance. Often there are heart defects, as well[**].

---

[*]As for why the orange and black colors appear as discrete patches, see the section later in this Lecture on X-chromosome inactivation and consequent mosaicism in females.

[**]About 20% of all spontaneous abortions are XO in chromosome constitution; 98% of all human XO conceptions are eliminated during pregnancy. Still, about 1 in 2500 live-born females

(continued...)

Individuals with XYY and XXX chromosome constitutions are also fairly common. The former tend to be taller than normal males and the latter are females who sometimes experience early cessation of menstruation or fertility problems, but in general neither condition causes extensive enough or frequent enough abnormalities to be considered a medical syndrome.

Because the X-chromosome carries many genes that are indispensable for viability (not just these relating to sexual differentiation), cases such as "YO" where there is not even one X chromosome are never found. Presumably they are aborted spontaneously very early in the pregnancy. The small Y chromosome, by contrast, seems to carry mainly genes concerned with sexual differentiation, although there have been some reports that one human condition (hairy ear pinna) is caused by a gene for a somatic condition carried on the Y.

The incidence of chromosomal abnormalities (including significant structural abnormalities as well as the non-disjunctional aneuploidies) is surprisingly high in humans. It is estimated that about 5% of all human gametes carry a significant abnormality, meaning that about 10% of all fertilizations initiate a chromosomally defective embryo. The vast majority of these embryos are spontaneously aborted during pregnancy, usually at a stage so early that the woman was unaware she was pregnant. At birth, about 1.0% of human still-born and 0.6% of live-born babies have a significant chromosomal defect.

SEX REVERSAL

More curious than the obvious chromosomal anomalies are the cases of individuals who appear to be sex-reversed—that is, people who appear clearly male but are XX or clearly female but are XY. Given that the Y chromosome is male-determining and the lack thereof is female-determining, these cases are clearly puzzling. Some of the latter (XY females) suffer from **Congenital Insensitivity to Androgen Syndrome** ("CIAS") whereby through some genetic defect tissues in the developing embryo have lost their ability to be stimulated by the androgen (mainly testosterone) that the developing testis produces. In consequence, it is as if androgen were absent altogether and development proceeds in the female direction. Usually these individuals appear as normal (but usually taller and thinner) women, but having no ovaries they neither menstruate nor develop breasts. This usually leads them to seek medical advice at puberty, when their true condition is revealed and their internal testes discovered. (They should be surgically removed, since they have a tendency to become cancerous if left *in situ*.)

[**](...continued)
have Turner's syndrome. This suggests that Turner's syndrome is the most frequent chromosome aneuploid known. It also suggests that, given the very pronounced intra-uterine selection against XO fetuses, most or perhaps all live-born XO individuals are not wholly XO in chromosome constitution, but are to some degree mosaics, having normal as well as XO cells. (See the discussion of mosaicism later in this chapter.)

There is, however, another explanation for many XY females (especially when no evidence of internal testes is found), one that is also relevant to the origin of XX males. Very detailed study of the Y chromosome in such cases often reveals that the Y chromosome is missing a minute piece of the small arm, a piece so small (constituting about 0.2% of the chromosome) that the Y chromosome appears entirely normal under the microscope. Similar studies on the XX males, in contrast, often reveal that one of the X chromosomes has attached to it that same minute piece of the Y chromosome. These studies suggest that the crucial signal that initiates development in a male direction may be a single gene on the Y-chromosome located on the small arm.[2] If a 14,000-base segment of the mouse Y-chromosome is injected into XX mouse fertilized eggs, some of these pieces insert themselves into the DNA of the host, and in these cases, the mouse develops as a seemingly normal male.[3] (This technique for transferring individual genes is discussed in Chapter 39.) This result strongly suggests that the injected segment of the Y contains the so-called sexual switch that initiates male development. This gene is called **SRY** (for "sex-determining region of the Y") in humans, and **Sry** in mice. The gene is activated during the seventh week of pregnancy in humans, that is, between the 43rd and 49th day.[*] (Other genes besides SRY are needed for the complete development of functional males and females, since XX males and XY [and XO] females do not have functional testes or ovaries.)

CHROMOSOMAL MOSAICISM

Many individuals with chromosome abnormalities are **mosaics**: that is, composed of cells of differing genetic constitutions. This happens when non-disjunction occurs during the mitoses that follow fertilization of the ovum. In general, the earlier the non-disjunction during growth of the embryo, the greater the proportion of cells with abnormal chromosome numbers in the adult, but a great deal, too, will depend upon the type of cell line in which the event took place and the severity of the handicap this abnormality poses to the cell and its descendants.

It turns out that all normal human females are, in a sense, mosaics with respect to their two X chromosomes. It had long been a puzzle to biochemists why the female, with two X chromosomes, was not over-supplied with the products of all genes on the X-chromosome when compared with the quantities produced by a male, with only one X. In 1961, M. Lyon and L. Russell proposed that early during the development of a female embryo (before the 16th day) one of the X chromosomes in all cells outside the germ line is inactivated and that this particular X chromosome remains inactive in all the descendants of these cells. (Cells in the germ line, i.e., those destined to become oocytes, are exempt from X-inactivation). The X chromosome chosen to become inactive, whether the maternal- or paternal-derived X, would be a matter of chance in each cell, but once one of the two is "Lyonized", the same one remains inactive in all the de-

---

[*] Students of the Talmud may note certain resonances between this finding an the advice given by the ancient rabbis on the utility of prayers for the birth of a son. According to the tractate B'rachot (p. 60a), the rabbis counselled that such prayers by husbands are worthwhile up to the 40th day of pregnancy, but thereafter are in vain, since the sex has already been determined.

scendants of that cell. Inactivation is initiated from a region on the long arm near the centromere. It is known as the **X-inactivation center (IIC)**.

A manifestation of this inactive X-chromosome, it turns out, can be detected in interphase female cells as a dark-staining, small "blob" closely associated with the inner surface of the nuclear membrane. It was first noticed in 1949 by M. Barr in stressed cats, but he subsequently discovered that it was due not to their being stressed but to their being female. The existence of this **Barr body** makes it a relatively easy task to determine the genetic sex of a mammal without doing a complete karyotype. If there is more than one X chromosome present, all extras are Lyonized and each Lyonized X becomes a separate Barr body. A Triplo-X female, therefore, will have two Barr bodies; a Klinefelter's male, one; and a normal male will have none at all.

This cannot be the whole story, however, for if all X chromosomes in excess of one were totally inactivated, there would be no reason why XXY (Klinefelter's) individuals should be in any way different from XY males. Indeed, it is the case that not all of the "inactive" X is inactivated. Small parts of the chromosome near the tip of the short arm remain active. The products of several of the genes located in these regions are produced in proportion to the number of X chromosomes in the cell. In the case of Turner's syndrome, the X-chromosome location of the relevant gene (or genes) has been tentatively identified; early evidence suggests that it specifies one of the proteins of the small ribosome subunit. Hence, when this region is present in single dose (as in XO individuals), Turner's syndrome results.

REFERENCES

1. Jacobs, P., 1959. Nature 183:202-3.

2. Sinclair, A., et al., 1990. Nature 346: 240-44.

3. Koopman, P., et al., 1991. Nature 351: 119-121.

LECTURE 16

GENES AND CHROMOSOMES

Heredity is the last of the fates, and the most terrible.

O. Wilde

PAIRS OF CHROMOSOMES AND GENES

It was mentioned earlier that the ultimate units of genetic function, the **genes**, were contained within the chromosomes and that a single chromosome will carry hundreds of genes along its length. Each chromosome (homologue) of a particular autosomal pair will carry the same genes as its twin, but the two genes (alleles) composing any given gene-pair may be in different physiological states. In an extreme case, one member of the pair may be fully functional, while its mate may be damaged and totally inactive. If the two alleles are identical (both normal or both damaged), we say the individual is **homozygous** for those genes; if they differ significantly in activity, we say the individual is **heterozygous** for them. A human male, of course, cannot be either homozygous or heterozygous for any genes on his X chromosome; he is said instead to be **hemizygous** for these genes.

In summary, because chromosomes exist in pairs, the genes they carry must also exist in pairs, at least for genes on the autosomes. When meiosis takes place, the homologous chromosomes segregate to opposite poles of the dividing cells. Consequently, egg and sperm cells will contain but one copy of each chromosome pair and hence but one allele of each allelic pair.

DOMINANCE AND RECESSIVENESS

The appearance of an organism is called its **phenotype**—what it looks like, in other words.[*] Its underlying genetic constitution is called its **genotype**. A gene that has not been damaged (changed, or **mutated**) is said to be **wild-type**. Similarly, an individual who seems to have the standard or usual characteristics common to that species is said to have a wild-type phenotype.

---

[*] The meaning of the term phenotype extends considerably beyond the simple visual appearance of an organism. It can connote "appearances" that are only obvious with sophisticated biochemical tests (e.g., whether a particular enzyme is present in the organism or not) or even such things as behaviors or talents (e.g., the recent suggestion that the human phenotype "having absolute musical pitch" is the result of possessing the particular autosomal dominant gene that confers it).

It is quite possible, however, for an individual to appear normal (to possess a wild-type phenotype) and still have a genotype made up of many mutant genes which could later be passed on to his or her progeny. The reason for this rests with properties of genes called **dominance** and **recessiveness**.

If an individual is heterozygous for a particular pair of alleles, the phenotype of the organism will, in general, be identical to the phenotype of another individual homozygous either for the wild-type gene of the pair or else for the mutant gene of the pair. If the phenotype of the heterozygote is wild-type, then the mutant allele of the pair is said to be recessive. If the phenotype of the heterozygote is mutant, then the mutant allele is said to be dominant. Note that dominance or recessiveness is a property ascribed only to the mutant allele and not to the wild-type one. One does not say that the wild-type allele is dominant, but instead says that the mutant allele is recessive; one does not say that the wild-type allele is recessive, but instead says that the mutant allele is dominant. This convention (which may be at odds with what you have learnt in the past) is extremely important to keep in mind when naming genes, since genes are usually named after the mutant allele and not the wild-type one. The following rules apply for most organisms:

1. A mutant gene is given a name and symbol which, as far as possible, is sensibly related to the phenotype of the organism when it is homozygous (if the mutant is recessive) or heterozygous (if the mutant is dominant) for the mutant allele in question. Hence, a gene in fruit flies (*Drosophila*) that causes white eyes when homozygous will be called white and symbolized w, and one that causes the fly's wings to be extended at a peculiar angle can be named held-out and symbolized ho. One tries to hold the maximum number of letters in a genetic symbol to three.

2. If the mutant is recessive, the first letter of the name and symbol are in lower case; if the mutant is dominant, the first letter of the gene name and symbol are capitalized. For example, white, dumpy and scarlet (w, dp and st, respectively) are all obviously recessive mutants (of *Drosophila*); Plum, Bar and Bithorax (Pm, B and Bx, respectively) are all dominant mutants (of Drosophila).

3. Names of genes and symbols for genes are always underlined, as was done in the examples given above.

4. Wild-type (i.e., unmutated) alleles of mutant genes are named after their mutant alleles simply by placing a "+" as a superscript after the symbol for the mutant. Therefore w+, dp+ Cy+ and Bx+ are all wildtype alleles of the white. dumpy, Curly and Bithorax genes, respectively.

5. Oftentimes when it is obvious what allelic pair is being referred to, one omits the gene symbol altogether when designating the wild-type allele and just keeps the "+" sign. For example, +/vg is heterozygous for <u>vestigial</u> wings and +/<u>Cy</u> is heterozygous for <u>Curly</u> wings. What is the phenotype of these two flies?

[handwritten margin note: +/vg will have vestigial wings; Cy will have curly wing]

6. A phenotype designation is also derived from the name of the mutant by omitting the underlining and putting the name or symbol in parentheses. For example (vg), (Cy) and (st) flies are (vestigial), (Curly) and (scarlet) in appearance, respectively. Flies that are (vg^+), (Cy^+), (st^+) are all wild-type in appearance, with respect to the genes in question. A (+) fly is wild-type in phenotype for all genes (although it may, of course, be heterozygous in genotype for one or more recessive mutants).

7. Genes or groups of genes on separate chromosomes are separated by a semi-colon; genes on the same chromosome are usually separated by commas and connected by the underscoring line. For example: +/<u>a</u>; <u>b</u>$^+$/<u>b</u> vs. <u>a</u>$^+$, <u>b</u>$^+$/<u>a</u>, <u>b</u>. In the first case the <u>a</u> and <u>b</u> genes are on different chromosomes; in the second case they are linked on the same one.

[handwritten margin note: important later on for recombination]

The names of genes in bacteria and phage follow other conventions, however.

MENDELISM

Because genes are carried on chromosomes, they will be concomitantly dragged along as the chromosomes are apportioned to the daughter cells during meiosis. At fertilization, these chromosomes acquire new homologues with which they will remain partners for a whole generation, only to be divorced and forever separated when that new zygote itself later undergoes meiosis and separates its homologues. New homologous pairs mean new allelic combinations for the allelic pairs; genes that were homozygous in one parent may be heterozygous in the offspring, and *vice versa*. Every fertilization starts a whole new genetic ball game, one that has never been played before and will never be played again.

The pattern traced by the distribution of genes (and hence genetic characteristics) from parents through their gametes to their offspring is called **mendelism**. An understanding of mendelism requires an understanding of just two simple phenomena, both of which are by now very familiar to you:

1. chromosome movements during meiosis, and
2. dominance or recessiveness of mutant genes.

An organism heterozygous for a mutant gene $\underline{a}$ (i.e., $\pm/\underline{a}$ in genotype) will produce two types of gametes: those with the chromosome that carries $\underline{a}^+$ and those with its homologue that carries $\underline{a}$. If this individual mates with another of the same genotype, the random fertilizations will produce the following zygotes in the proportions given: $1\ \pm/\pm$: $2\ \pm/\underline{a}$: $1\ \underline{a}/\underline{a}$. This will yield a phenotypic ratio of 3 $(+)$: 1 (a). Technically this is known as a **monohybrid cross**.

Needless to say, this 3:1 ratio and the 1:2:1 ratio that underlies it are both statistical

Table 16.1 These are the actual results Mendel obtained. His ratios can be seen to be very close to the theoretical 3:1 expected in each case.

Dominant form	No. in F_2 generation	Recessive form	No. in F_2 generation	Total examined	Ratio
Round seeds	5,474	Wrinkled seeds	1,850	7.324	2.96:1
Yellow seeds	6,022	Green seeds	2,001	8.023	3.0.:1
Gray seed coats	705	White seed coats	224	929	3.15:1
Green pods	428	Yellow pods	152	580	2.82:1
Inflated pods	882	Constricted pods	299	1.181	2.95:1
Long stems	787	Short stems	277	1.064	2.84:1
Axial flowers	651	Terminal flowers	207	858	3.14:1

concepts which admit of some fluctuation in the real world. For example, when G. Mendel, the eponymous discoverer of mendelism, first carried out the experiments in pea plants that led to his formulating these "laws of heredity", his data yielded ratios in different experiments of 5474:1850 (2.96:1), 6022:2001 (3.01:1) and 705:224 (3.15:1). But these ratios seem "too perfect", and there have been intermittently heard dark suggestions that Mendel fudged his results or, putting the best face possible on it, "improved on Nature" a bit, perhaps.*

When crossing two organisms which are heterozygous for two sets of alleles located on different chromosomes, one is making a **dihybrid cross** of the following sort: $\underline{a}/\pm$; $\underline{b}/\pm$ X $\underline{a}/\pm$; $\underline{b}/\pm$. At metaphase I of meiosis the homologous chromosomes will be paired and all homologous pairs will be lined up along the equator. The orientations of the $\underline{a}$- and $\underline{a}^+$-carrying chromosomes rel-

---

*Mendel published an account of his work in 1866 in a scientific journal that enjoyed a fairly wide distribution (Rumor has it that after the rediscovery of Mendel's work a copy of this journal was found in the library of Columbia University—its pages still uncut). Mendel's failure to gain recognition in his own lifetime must be attributed more to his own shortcomings than to any fault of the general public. When C. Nageli, the most distinguished botanist of his day, criticized his work, Mendel seems meekly to have accepted these pronouncements as the last word on the subject and abandoned his research in favor of other interests. If he would discard his own ideas so readily, why should anyone else be expected to take them more seriously than he?

Mendel's principles were rediscovered independently by three investigators in 1900: C. Correns, H. DeVries, and E. von Tschermak.

ative to the chromosomes carrying the b alleles can be a/+; b/+ or +/a; b/+. This means that ana-phase I in some cells will result in a segregation that produces (eventually) pairs of haploid gametes that are a; b and ±; ± while in other cells the pairs of gametes produced will be ±; b and a; ±. Altogether, then, four different gametes are possible from each parent; the sixteen possible pairwise fusions that may follow produce a phenotypic ratio of 9(+): 3(a): 3(b): 1(a;b).

The principle underlying this ratio is the simple fact that one pair of homologues aligning themselves at the metaphase equator does not exert any influence on the relative orientations of other homologous pairs. Geneticists call this principle **independent assortment**.

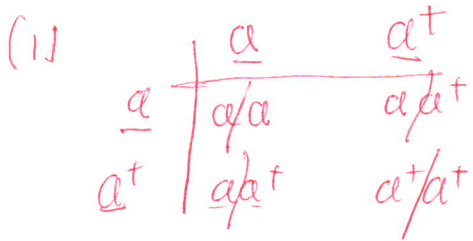

(1)

	$\underline{a}$	$\underline{a}^+$
$\underline{a}$	a/a	a/a^+
a^+	a/a^+	a^+/a^+

as can be seen, 3(+): 1. This was demonstrated by means of a Punett square....

(2) a/+; b/+ yields gametes that are a/a+; b/b+
a/+; b/+ yields same gametes.

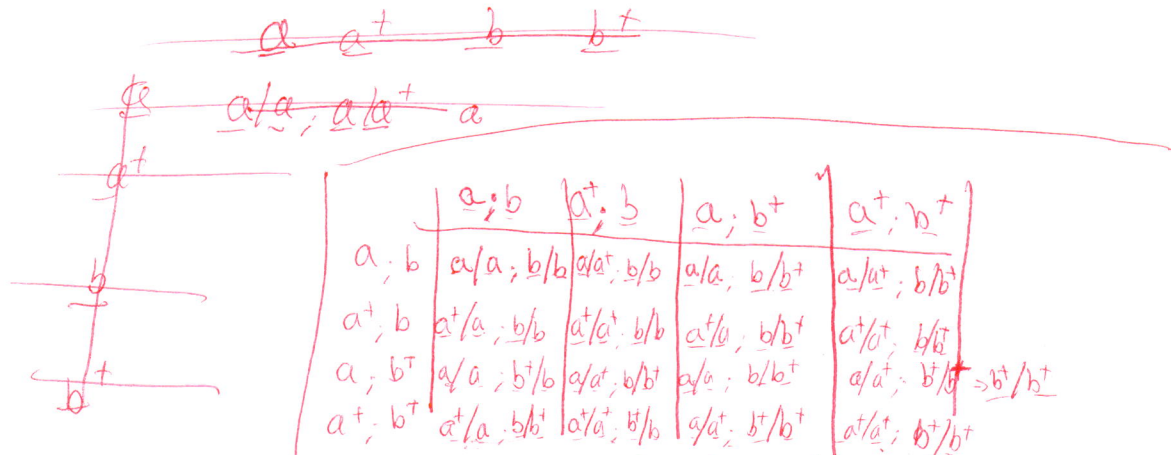

	$a; b$	$a^+; b$	$a; b^+$	$a^+; b^+$
$a; b$	$a/a; b/b$	$a/a^+; b/b$	$a/a, b/b^+$	$a/a^+; b/b^+$
$a^+; b$	$a^+/a; b/b$	$a^+/a^+; b/b$	$a^+/a, b/b^+$	$a^+/a^+; b/b^+$
$a; b^+$	$a/a; b^+/b$	$a/a^+; b/b^+$	$a/a; b/b^+$	$a/a^+; b^+/b^+ \Rightarrow b^+/b^+$
$a^+; b^+$	$a^+/a; b/b^+$	$a/a^+; b^+/b$	$a/a^+; b^+/b^+$	$a^+/a^+; b^+/b^+$

Phenotypic Ratio
9 (+) : 3 (a) : 3 (b) : 1 (a; b)

This means   Translation:

9 (a^+, b^+) : 3 (a, b^+) : 3 (a^+, b) : 1 (a, b)
9 all wild type   3 wild type   3 wild type : 1 total
in phenotype   for only a gene   for only b gene   mutant meaning homozygous for both mutant gene's

95

LECTURE 17

SEX LINKAGE

GENE RECOMBINATION

SEX LINKAGE

When genes are located on the X chromosome (and because the X chromosome is a reasonably large one, there are many such genes), a different pattern of inheritance is noted than is observed with autosomal genes. The reason for this is the lack of homology between the X and Y chromosomes; the Y chromosome carries almost no alleles of any of the genes on the X. Hence, an XY male carries only one copy of all genes on his X chromosome, whereas an XX female has two. An XY male cannot, in consequence, be either heterozygous or homozygous for genes on his single X chromosome: he is said instead to be **hemizygous** for all such genes. An important practical consequence of hemizygosity is that a male expresses the phenotype of *all* his X-linked genes, regardless of whether such genes are recessive or dominant when functioning in an XX female. This fact, in turn, causes an anomaly in the pattern of inheritance of X-linked recessive genes in comparison with autosomally linked recessive genes. A female which is heterozygous for such a gene (and therefore wild-type in phenotype) will, when mated to a wild-type male, produce no phenotypically mutant *female* offspring (half of whom being heterozygous for the mutant gene and half being homozygous wild-type), but half her *male* offspring will be mutant in phenotype (being hemizygous for the mutant-bearing X-chromosome inherited from the mother) and the other half will be phenotypically wild-type (being hemizygous for the wild-type X-chromosome inherited from the father) This phenomenon is known as **sex linkage**.

A number of well-known human genetic diseases are sex linked, including the common variety of color blindness and "bleeder's disease" or hemophilia. In both these diseases it is common knowledge that the number of male sufferers greatly exceeds the number of affected females. The reason, as explained above, is that the disorders are caused by recessive X-linked genes. In order for a female to be affected, she must inherit a defective allele from both parents (i.e., be homozygous for the mutant gene), whereas a male has only to inherit one copy of the mutant gene to manifest the disorder (i.e., be hemizygous for the mutant gene). — Guys get a bummer deal!

"PARENTAL IMPRINTING" OF GENES[1]

It has been a basic tenet of genetics since the laws of Mendel were first rediscovered that it makes no difference to the expression of an autosomal gene whether that gene has been inherited from the male or the female parent. Now, about ninety years later, that view is being seriously questioned, at least in mammals. Almost all the classical laws of inheritance were worked out using insects, particularly the fruit fly, or various plants. In such organisms it is indeed the case that the sex of the parent contributing a gene is irrelevant to how it will function in the offspring. This is probably true also for the vast majority of genes in mammals and other classes. However, there has been recent highly suggestive evidence that some genes in mammals do behave differently depending upon whether they were inherited from the male or female parent.[*] In humans, for example, some genetic diseases are more or less severe depending solely upon whether the defective gene came from the mother or the father, and a similar phenomenon has been observed in the mouse. It has also been demonstrated that the presence of a pair of homologous chromosomes inherited both from the same parent results in distinctly abnormal growth and development in mice. It does not matter whether both homologues derived from the male or female parent.[2] It appears that genes can be modified or "imprinted" differently in the two sexes and that these modifications affect how that gene will function in the offspring.[**]

A particularly striking example of imprinting in humans is seen in the case of two rare genetic diseases, Prader-Willi Syndrome and Angelman Syndrome. The former are characterized by mental retardation, obesity because of compulsive eating, small hand, feet and penis (if male); the latter are also mentally retarded, but have a large mouth, red cheeks, a jerky gate and are prone to unprovoked laughter. Because the two genetic diseases are so dissimilar, it had been assumed they were caused by different mutant genes. In fact, they are caused by the same gene mutant in the same way (a deletion on chromosome 15). If the mutant gene is inherited from the father, the child gets Prader-Willi Syndrome; if from the mother, Angelman Syndrome.

It is assumed at present that modification of the DNA is achieved by differential methylation of chromosomal DNA in the two sexes. When appropriate follow-up studies on these observation are completed, a major re-writing of classical genetic theory may be required.[***]

---

[*]Nonetheless, there has always been noted the curious difference in the offspring when horses are bred to donkeys. When the horse is the female parent, the hybrid offspring is known as a mule and resembles more the donkey father than it does the horse mother. (It has its father's ears, tail, hooves, legs, and a bit of a mane, but it can be as big as its mother.) But when the female parent is a donkey, the hybrid, known as a hinny, still resembles more its father than its mother.

[**] Other studies have also been carried out in mice whereby the entire diploid chromosome set was derived either wholly from the father or wholly from the mother. In both such cases, the embryos did not survive at all, despite the fact that they carried a seemingly complete genome. It would appear, therefore, that parthenogenic (i.e., uni-parental) development is not possible in mammals, although it occurs naturally in many lower animals, is a well-established laboratory phenomenon in higher forms such as frogs, and figures prominently in the religious mythology of many human tribes.

[***] A particularly puzzling instance of genomic imprinting is that involved in the so-called "fragile X syndrome", the most frequent cause of mental retardation in humans. The syndrome

(continued...)

97

RECOMBINATION

An older name for recombination is **crossing-over**. The latter term is no longer used much.

We have already seen one type of "recombination": that occasioned by the formation of novel allelic pairs on fertilization. This occurs simply as a consequence of segregation and independent assortment of chromosomes during meiosis. The technical term **recombination**, however, is reserved for new combinations of genes located on the same chromosome.

Up to now we have been assuming that genes on the same chromosome must stay together forever in the same combinations in which they were found. For example, if $a^+ b$ denotes two genes carried by the same chromosome (i.e., are **linked genes**), then these two genes would seem destined to stay together forever, since independent assortment results in new combinations only of genes located on different chromosome pairs.

Nevertheless, there is a mechanism which allows recombination of genes on the same chromosome. During prophase I of meiosis the paired homologous chromosomes sometimes exchange segments. When the breakpoint for the exchange occurs between genes, recombination occurs. For example, if before recombination an individual had the gene configuration $a^+ b$ / $a b^+$, after recombination it would have the configuration $a^+ b^+$ / $a b$. It is still heterozygous for both a and b, but these genes are now arranged differently on the two homologous chromosomes. After segregation occurs, one gamete will acquire the a^+ and b^+ genes and another their a and b alleles. Before recombination took place, all gametes would have acquired the a^+ and b or the a and b^+ combination.

If one homologue has only mutant genes and the other has only their non-mutant alleles (e.g., the configuration $a^+ b^+$ / $a b$), the genes are said to be in the **cis** arrangement; if each homologue has both a wild-type and a mutant allele (i.e., the configuration $a^+ b$ / $a b^+$), the genes are said to be in the **trans** arrangement. Recombination between genes will convert them from the cis

$a^+ b / a^+ b$ or $a^+ b / a b^+$

***(...continued)
derives its name from the fact that affected individuals, usually males, have the end of their X chromosome only loosely attached to the remainder. This situation makes the chromosome liable to breakage at that point. Although all individuals who have the characteristic retardation manifest the fragile X, up to half of the males who do have a fragile X are asymptomatic. Moreover, some females who have only one fragile X are also symptomatic. Asymptomatic males produce asymptomatic daughters, as expected, but these daughters produce symptomatic males, and often symptomatic females, too. Apparently "passage through a female" is necessary to activate the chromosomal defect. Methylation of the fragile X region of the chromosome has been proposed as a mechanism for this activation, but this interpretation is complicated by the additional observation that the fragile X region has considerably more DNA (20 times or more) in symptomatic people than in asymptomatic ones. As I said, the inheritance and causation of the fragile X syndrome are very puzzling.

to the trans arrangement or from the trans to the **cis** arrangement. The offspring that inherit the $a^+ b$ and $a b^+$ chromosomes in the example given are said to be the **recombinants**.[*]

Whether recombination occurs or not is a chance phenomenon along any particular homologous pair of chromosomes undergoing meiosis. The likelihood that it will occur between any two linked genes is directly related to the distance separating them (horizontally) on the chromosome. Hence genes close together (**closely linked**) will rarely be recombined; genes far apart (**distantly linked**) will almost always be recombined. The frequency with which linked genes can recombine provides a method of **mapping** genes on the chromosome. If genes recombine 1% of the time, they are said to be one **map unit** apart on the chromosome; 5% recombination means 5 map units, and so on. It is increasingly common to call a map unit a **centi-Morgan (cM)** after T. Morgan, the American *Drosophila* geneticist who did much of the early research on recombination and gene mapping.

The molecular mechanisms that result in recombination are still a matter of some debate. The process is known to involve a physical breakage and joining of originally distinct DNA molecules. That two homologous DNA molecules (i.e., the DNA in two homologous chromosomes) could be cut in exactly the same location and joined together without error boggles the imagination. A way around this difficulty has been proposed by R. Holliday, whose molecular explanation of recombination enjoys considerable support and acceptance at present. The details of the **Holliday model** are outside the needs of this account.

---

[*] Although grammarians beg to differ, as this writer to *Nature* is at pains to point out.

> The Latin ending —ans, which became —ant in French and English, characterizes the present participle (active voice) and means —ing in English. Thus "recombinant" means "recombining".
>
> The use of this word is correct, for example, in "recombinant technology". But it is applied wrongly when speaking about the product of recombination, for example a gene, after it has been recombined. Then it is not a "recombinant gene" (recombining gene) but a "recombined gene" (past participle, passive voice)....[I]f "recombined" does not look learned enough, I would propose "recombinated": "recombinated gene" instead of "recombinant gene" does not sound bad and would be logically and linguistically correct. (Katscher, F., 1991. Nature 351: 179.)

(His references to "recombinant technology" are relevant to Lectures 38 and 39; "recombinant genes"—as opposed to the "recombinant *chromosomes*" discussed here—are discussed in Chapter 30.)

WHY BOTHER WITH SEX?

The answer to this question may appear obvious to some, but in fact sex is a puzzle to biologists. The predominance of sexual reproduction in the organisms alive to-day suggests that very powerful selective pressures in favor of it have long existed, despite its seeming disadvantages. Chief among the disadvantages is the restriction in the number of offspring possible when only one parent produces them. The conventional explanation for sex has emphasized the offsetting advantage of sexual reproduction conferred by recombination. This frees beneficial mutations from the tyranny of the genetic environment in which they arose so that in new combinations they can work to best advantage. Sexual reproduction also gives whole chromosomes this opportunity, too, since each generation unites the genetic heritages of two different individuals. But then it also splits them up again at the next generation; so each combination, the good and the bad, is unique and short-lived. A more recent speculation by M. Kirkpatrick and C. Jenkins holds that the ready achievement of homozygosity is the "purpose" of sexual reproduction. They argue that many genes are most advantageous only when they are homozygous. In asexually reproducing organisms, once the first mutation has occurred, it could be a very long time before another independent event on the homologous chromosome results in homozygosity, whereas homozygosity can occur much faster through sexual reproduction. Maybe so, but then one could argue that diploidy itself is made necessary only because of sex and that in an asexual world each gene would be free to express its full potential on its own. Of course, a gene could find itself the proverbial "lonely little petunia in an onion patch" unable ever to associate with more agreeable neighbors because there would be no recombination. A fully satisfactory answer to the question posed at the beginning of this paragraph has yet to be found.. Equally puzzling, while we are on this subject, is why there are commonly only two sexes. This arrangement leaves only 50% of the population as possible sexual partners, whereas if there were more sexes, a much larger proportion of the population would be of a different mating type. This question, too, awaits a satisfactory explanation.

REFERENCES

1. Sapienza, C., 1990. Parental imprinting of genes. Sci Amer. 263:52-60 (Oct.).

2. Cattanachi, B. and Kirk, M., 1985. Nature 315:496.

LECTURE 18

THE CYTOPLASM AND ITS MEMBRANES

Everything inside the cell membrane (sometimes called the **plasma membrane**) and outside the nucleus constitutes the **cytoplasm** of the cell. It used to be thought of as a disorganized colloidal solution, but in the past 40 years, and especially since the advent of electron microscopy, we have come to appreciate the cytoplasm as a highly structured medium wherein few interactions are left to chance encounters. The cytoplasm houses several membrane-bound organelles, as well as free-floating proteins and ribosomes. The term **cytosol** is often used to refer to the less structured part of the cytoplasm outside any of the membrane-bound organelles. Three very prominent cytoplasmic inclusions—chloroplasts, mitochondria, and ribosomes—will be discussed in detail in later lectures, but here we shall examine what keeps everything both together and apart: biological membranes.

COMPONENTS OF MEMBRANES[1] *exam question exists*

Membranes are still very much an enigma, despite having been the subject of intense investigation since at least the 1930s. There is still, for example, no completely satisfactory model to explain their structure and properties. Not that one expects all biological membranes to be identical—we know, in fact, this is not the case—but even the structure of particular membranes remains very much *sub judice* for the moment. Nevertheless, some general observations can be made, such as the fact that the two principal components of a membrane are clearly **proteins** and **lipids**, particularly **phospholipids**; the steroid **cholesterol** is found in many membranes, as well. The general structure of proteins is dealt with in a later section; phospholipids are described below.

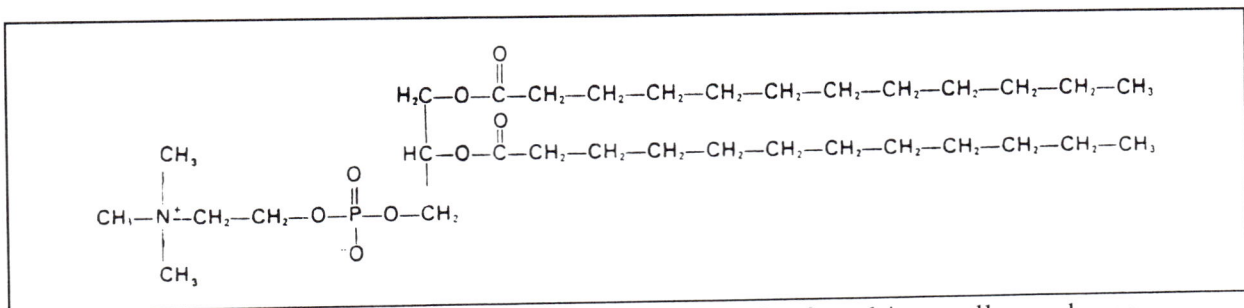

Figure 18.1 A typical diglyceride of the type that might be found in a cell membrane.

Lipids in general are difficult to categorize by a neat definition, but for our purposes will be treated as esters of alcohols and fatty acids. **Fatty acids** are long-chain carbon structures terminating in a carboxyl group. The number of carbons in these acids varies from four (butyric acid) to two dozen or more, but the number is always even. If the carbon-carbon linkages are

C—C →Saturated
C=C → unsaturated.

single, the fatty acid is said to be **saturated**; one or more double carbon-carbon bonds renders it **unsaturated**. These fatty acids can react with the alcoholic groups (-OH) of alcohols to form an **ester**. The alcohol commonly selected for this reaction is **glycerol**, which has three alcoholic groups available for reaction. The reaction between glycerol and three fatty acids (which may or may not be identical) produces the commonest type of lipid: a **triglyceride**.

Although triglycerides are the commonest of the lipids, they are not the type usually found in membranes. Membrane lipids are **diglycerides** in which the third alcoholic group, instead of esterifying with a fatty acid, is joined to a phosphate group, which in turn is linked to the nitrogen-containing compound, **choline**. Chemically, this over-all combination produces a phosphatidyl choline, of which the most common is **lecithin**, a very important component of most membranes. The double stem or "tail" of this molecule is composed of the electrically neutral fatty acids, but the "head", the choline portion, bears an electrical charge. In conse-

Figure 18.2 Phosphatic acid + choline yields lecithin + water.

quence, the "head" is hydrophilic and the tail is hydrophobic; when layered on water, a phospholipid will spontaneously orient itself so that the "head" protrudes into the water and the tail extends away from the surface.

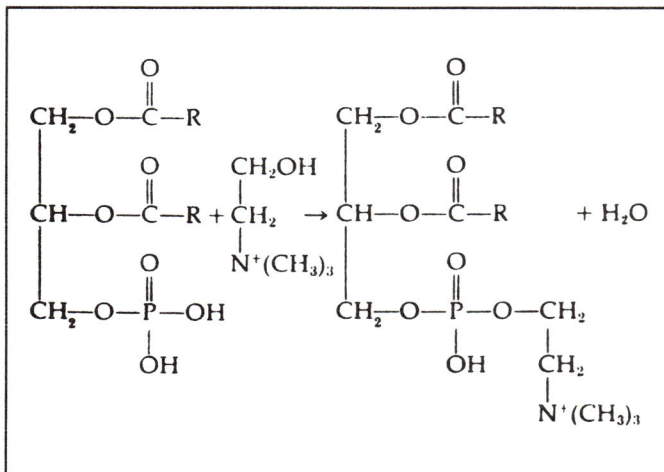

Early studies showed that there was about twice the amount of phospholipid in a membrane than would be needed for a single layer over a given surface, and later non-biological studies demonstrated that phospholipids often spontaneously orient themselves at an aqueous surface in **bilayers**. In a bilayer the fatty acid tails interdigitate so that the heads of the lipids point out from both the upper and lower surfaces of the bilayer. Such bilayers form the basis of all current theories of membrane structure.

Under high resolution electron microscopy, a membrane can be seen as two thin dark-staining lines each about 25Å wide and separated by a clear space also about 25Å wide. The clear space is generally taken to be the interleaved fatty acid chains, and the dense layers to be protein layers associated with the hydrophilic heads of the lipid layer. Over-all, therefore, a biological membrane is about 75Å thick. It is also obvious from electron microscopy and other studies that the membrane is sometimes perforated by tiny pores that permit larger molecules passage through. This model of the membrane is sometimes called J. Robertson's **unit membrane hypothesis**.[2]   selectively semi permeable

Although the unit membrane hypothesis is a useful abstraction, few people seriously believe it reflects the true structure of many (or perhaps any) biological membranes, and a number of

102

alternative views have been advanced, especially in reference to particular membranes. One that is much talked about nowadays sees the membrane not just as a protein-coated bilayer, but as a phospholipid bilayer whose regular structure is frequently interrupted by chunks of protein at regular or irregular intervals in the interior parts of its structure. Some of these proteins are on the surface only, whereas others penetrate the lipid bilayer to varying degrees. Moreover, many of the proteins are mobile within the membrane, "floating" through the lipid like icebergs at sea. This view of membrane structure, which has widespread acceptance, goes by the name of the **fluid mosaic model**. The "fluidity" of membranes has been demonstrated by the fusion of mouse and human cells to form a single hybrid cell. Each of these cells has unique surface proteins that can be detected and distinguished visually by means of fluorescent dyes. When a hybrid cell is created, the mouse and human surface proteins are at first grouped separately around the half of the hybrid contributed by their parental cell, but soon both mouse and human proteins become thoroughly intermixed all around the membrane.[3]

MEMBRANOUS CELL ORGANELLES

The two membrane systems found in the cytoplasm of eukaryotic cells, but not of prokaryotes, are the **endoplasmic reticulum** and the **golgi apparatus**. The endoplasmic reticulum, abbreviated ER, is an extensive membrane system that seems, in parts, to be continuous with both the nuclear and plasma membranes. Functionally, the ER is divided into two types, **rough** and **smooth**, depending upon whether or not small granules of ribonucleoproteins, the **ribosomes**, are attached to the membrane. The membranes most frequently resemble elongated, much flattened balloons, the enclosed spaces being called **cisternae**. This enclosed space is also called the **ER lumen**. Although in microscopic cross section, the ER would give the appearance of being composed of many individual flattened structures, it is likely that the ER is a single highly convoluted membranous structure and that the lumen, in consequence, is really just one highly convoluted but continuous space. The ribosomes adhere to the outer surfaces of the rough ER membrane and do not appear in the lumen. The smooth ER has somewhat thicker

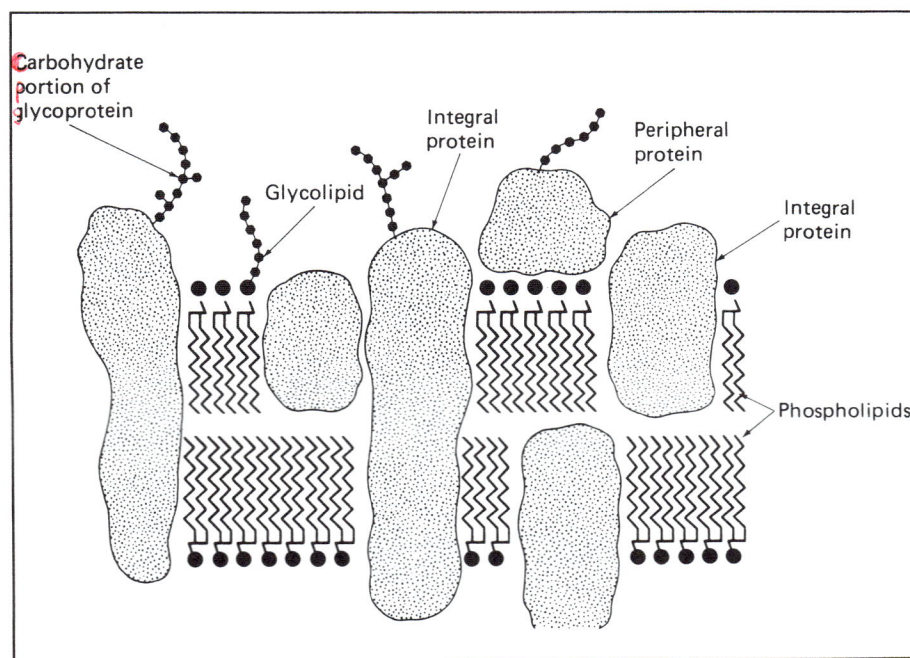

Figure 18.3 The fluid mosaic model of membrane structure. From BASIC CELL BIOLOGY, 2ND EDITION by C.J. Avers. Copyright © 1982 by Wadsworth Publishing Company. Reprinted by permission of the publisher.

Table 18.1 Not all life forms contain all the organelles discussed in this account. This table will help sort out which go where.

STRUCTURE	PLANT	ANIMAL	PROKARYOTE
CELL MEM- BRANE	present	present	present
CELL WALL	cellulose	absent	non-cellulose
NUCLEUS	present	present	absent
CHROMOSOMES	multiple	multiple	genophore
ENDOPLASMIC RETICULUM	usually present	usually present	absent
MITOCHONDRIA	present	present	absent
RIBOSOMES	present	present	present
GOLGI APPAR- ATUS	present	present	absent
LYSOSOMES	usually absent	usually present	absent
CENTRIOLES	absent in higher plants	present	absent

membranes than the rough and is more tubular than vesical-shaped.

The rough ER certainly plays a role in the synthesis of protein, for as a later section of these notes will make clear, the ribosomes are the sole sites for all protein synthesis in the cell. The full role of the smooth ER is still to be determined. It seems to play a role in the transport of proteins made on the rough ER but may also be involved in the synthesis of lipids and in glycogen metabolism. A number of enzymes used specifically in fatty acid, lipid, and steroid metabolism have been isolated from the smooth ER.

About half of the ribosomes in a typical eukaryotic cell would be found attached to the ER. Proteins synthesized on the bound ribosomes are threaded through the ER membrane into the lumen. In contrast to proteins that are found in the cytosol, those in the ER lumen have had sugar molecules attached to them. It would appear, therefore, that **glycosylation** of proteins is one function of the ER.

The other membranous organelle, the golgi apparatus, consists of a stack of four to eight very much flattened **saccules** whose edges are sometimes a bit more ballooned out. The end of the stack facing the plasma membrane is called the **mature face** and the opposite end, the **immature face**. Small spherical vesicles can sometimes be seen around the immature face and larger ones, called **condensing granules**, occur beyond the mature face. The golgi apparatus is involved in "packaging" proteins and making protein derivatives for export to the outside of the cell. These proteins apparently arrive at the immature face in the vesicles, which seem in turn to be blebbed off from the endoplasmic reticulum. These vesicles fuse to form the saccules at the immature face of the golgi, and as more saccules are added in this manner, the material, perhaps chemically modified in transit, finally reaches the mature face. Here the saccules break into the condensing vacuoles and travel to the plasma membrane. The fusion of the condensing vacuoles with the plasma membrane liberates the enclosed protein (or mucoprotein or glycoprotein) into the extracellular space. In plants, some of this material is certainly used in the synthesis of the cell wall.[4]

The golgi, therefore, transport and perhaps modify in the process proteins made originally on the rough ER. They also contribute to the plasma membrane and seem additionally involved in the synthesis of polysaccharides. They also bud off another membrane-bound cell organelle, the **lysosome**, which contains a collection of potentially very destructive digestive enzymes.[5] These enzymes, were they free in the cytoplasm, would quickly destroy the cell, but they are kept at bay within the lysosome until needed. When the cell ingests a food particle, it is kept surrounded by a membrane derived from an invagination of the plasma membrane. The lysosome fuses with the membrane of this **phagosome**, and the enzymic Furies it carries are then released to degrade the food particle to small molecules useful in the synthesis of cell components. Lysosomes are not found in plant or prokaryotic cells. *phagosome = ◯ → cell membrane reforms around food particle...*

One should not leave the subject of membranes without at least briefly standing in awe of their remarkable properties, for it is the need to explain these astonishing activities that makes speculating about their structure so difficult. The cell or plasma membrane is selectively permeable. Most substances that are lipid-soluble enter the cell quite readily, but so do many other substances (e.g. certain sugars, and water) that ought not to pass readily through a lipid barrier. The water molecules presumably enter by diffusion through minute pores in the membrane. These same channels ought also to facilitate the passage of many ions, but the fact is that membranes often remain impermeable to them. In part, this is likely because ions are often surrounded by "shells" of water are, in consequence, much bulkier than the unhydrated ion. Some other molecules, which might be capable of diffusing through the lipid bilayer very slowly, actually do so at a rate that cannot be accounted for by passive diffusion. It is then postulated that there exist specialized "carrier" molecules within the membrane that play a role in bringing specific substances inside the cell by **facilitated diffusion**. There are, however, many instances where molecules or ions are maintained inside or outside the cell against enormous concentration gradients that would normally favor very rapid equalization of these concentrations across the plasma membrane. Certain algae, for instance, concentrate iodine to a level hundreds of times that of the surrounding seawater, and animal cells expel sodium ions to reduce their concentration to a level ten times less than that of the extracellular space. To perform these feats of **active transport** across cell membranes, energy must be expended, and special enzymes known as

permeases become involved. Because of the involvement of enzymes, these active transport systems are very specific. One type of sugar molecule, for example, may be selected and concentrated within the cell whereas another of very similar structure may never be able to cross the cell membrane at all.

REFERENCES

1. Bretscher, M., 1985. The molecules of the cell membrane. Sci. Amer. <u>253</u>: 100-109.

2. Lodish, H. and J. Rothman, 1979. The assembly of cell membranes. Sci. Amer. <u>240</u>: 48-63.

3. Unwin, A. and R. Henderson, 1984. The structure of proteins in biological membranes. Sci. Amer. <u>250</u>: 78-94.

4. Satir, B., 1975. The final steps in secretion. Sci. Amer. <u>233</u> (Oct).

5. de Duve, C., 1963. The lysosome. Sci. Amer. <u>208</u>: 64 (May).

LECTURE 19

CHEMICAL THERMODYNAMICS AND CELL ENERGETICS

> It is a pity that the whole idea of energy has been confused in many people's minds by the way in which the word is often used colloquially. Like "stress" and "strain", "energy" is used to refer to a condition in human beings: in this case one which might be described as an officious tendency to rush about doing things and pestering other people.
>
> J. Gordon[1]

Actually I chose this lecture title maliciously, just to startle the somnolent reader; what we shall be examining in this section is not nearly so fearsome as the title adumbrates. Nevertheless, it does serve to draw attention to the important fact that chemical interactions occurring within the cell obey the same principles that govern chemical reactions occurring anywhere. There are, in other words, no peculiarly "biological" chemical or physical laws.

*THE FIRST AND SECOND LAWS**

The starting point for any discussion of cell energetics is invariably—implicitly or explicitly—the two laws of thermodynamics. ("Laws" in this case is a descriptive rather than a prescriptive term.) The **first law** states simply that the total amount of energy in the universe remains constant. Energy can, of course, be changed from one form to another, and it may leave one defined system and enter its surroundings or *vice versa*, but it cannot be got rid of. For this reason, the first law is often termed "the law of conservation of energy". The **second law** can have a number of different statements, but for our purposes will be phrased that "chemical and physical interactions occur in such a way that the amount of **entropy** in the defined system plus its surroundings always increases". Entropy is an energy measurement and refers specifically to energy no longer available to do work. It is unavailable to do work because it is dissipated in generating random molecular motion. Since all chemical reactions occurring above absolute zero (-273°C) *must* contribute to the increase in entropy, it stands to reason that this obligation

---

*A good many times I have been present at gatherings of people who, by the standards of traditional culture, are thought highly educated and who have with considerable gusto been expressing their incredulity at the illiteracy of scientists. Once or twice I have been provoked and have asked the company how many of them could describe the Second Law of Thermodynamics. The response was cold; it was also negative.

C. Snow
The Two Cultures (Cambridge Univ. Press, London. 1959.)

107

becomes the driving principle of all chemical reactions, determining both their direction and extent. And not just chemical reactions: *every* change in the physical realm occurring anywhere occasions an increase in entropy[*]. The inexorable increase in entropy foretells that the outlook for the universe over the long haul is rather bleak and uninspiring.[2]

FREE ENERGY CHANGES

When a chemical reaction occurs, not all energy is lost as entropy; indeed a good fraction of it is usually salvaged in what is called **free energy**. If entropy is a measure of the disorder of a system, free energy conversely is a measure of its remaining order. This energy survives, generally, by dint of the structure or concentration of the molecules undergoing the change, since highly structured or concentrated molecules have a greater inherent energy content than those less structured or concentrated. In changing from a more to a less structured condition, entropy is always increased but some of the energy, the free energy, can be harnessed to do work.

From this description one must not get the idea that all systems must, at a given time, be in the process of disintegration. The universe as a whole is, indeed, a closed system (by definition) that is continuously winding down, but within that large system one can define countless subsystems enclosed by their particular surroundings[**]. These subsystems—of which living matter may be considered one—can actually decrease their entropy (or, conversely, increase their free energy), but only at the expense of an accelerated increase in the entropy of their surroundings. There is nothing magical about this principle, for it is simply a manifestation of the commonplace observation that, given a finite number of things, if you concentrate them in one place, you have to take them away from some place else. Since the total energy is the sum of entropy and free energy, these quantities must bear, therefore, a reciprocal relationship to each other; and hence the importation of free energy from the surroundings into a system must be at the cost of increasing the entropy of the surroundings. Living things, as we shall presently learn, ultimately

---

[*]Something there is that doesn't love a wall,
  That wants it down.
              Robt. Frost
              "Mending Wall"

That "something" is a reflection of the Second Law that seeks to randomize the ordered stones of the wall and, later, to randomize the constituents of the stones themselves.

[**]"Surroundings" as used here is a technical term meaning everything not included in the definition of a particular system or subsystem.

steal their free energy from the sun. This makes many nice things happen here on earth, but in so doing it makes quite a mess of the sun.*

To quantitate these concepts, it is best to examine in detail a hypothetical simple chemical reaction:

$$A \quad \text{--------} \rightarrow \quad B$$

This reaction will proceed in the direction indicated only if the free energy of the product, B, is less than that of the reactant, A. The change of A into B, therefore, results in a change of free energy, a change denoted by the symbol delta-G (ΔG). Since the change has resulted in less free energy at the end of the reaction than there was in the beginning, the sign of this quantity is negative (i.e., $-\Delta G$). The actual magnitude of this $-\Delta G$ (read: "negative change in free energy") will depend upon the initial concentrations of A and B as well as upon the temperature, volume and pressure under which the reaction was carried out; nevertheless, the direction of the reaction is implicit in the sign of the change. As an isolated reaction, any individual reaction for which the sign is negative can proceed only to the right; any for which it is positive will proceed only to the left, i.e., B becomes the reactant and A the product. When $\Delta G = 0$, the reaction is at equilibrium, and no work at all can be derived from the system.

Regardless of any other considerations, every chemical reaction has an equilibrium value which is calculated from the final ratio of products to reactants. In our example, the final concentration of B in moles divided by the final concentration of A in moles yields a value, K_{eq}, which is a constant for a given temperature, volume and pressure. Generally, the temperature is set by definition at 25°C and the pressure at one atmosphere.

The fact that the magnitude of ΔG (in dimensions of calories per mole) is dependent upon the initial concentrations of A and B, together with the fact that when the reaction is at equilibrium $\Delta G = 0$, suggest that the magnitude of ΔG will depend upon how far the initial concentrations of reactants differ from the concentrations that will exist once equilibrium is attained.

It would seem, therefore, that ΔG and K_{eq} ought somehow to be related, and they indeed are by the expression:

$$\Delta G = \Delta G^\circ + RT \ln [B]/[A]$$

---

*"...a living organism...feeds upon negative entropy.... Thus the device by which an organism maintains itself at a fairly high level of orderliness (=fairly low level of entropy) really consists in continually sucking orderliness from its environment."
E. Schrodinger, in *What Is Life?*
(Cambridge Univ. Press, London, 1964)

where R is the universal gas constant (= 1.987 cal/mole/ degree), T the absolute temperature, and ΔG° a new and very useful chemical constant, the standard free energy change of the reaction. At equilibrium, [B]/[A] = K_{eq}, ΔG = 0, and ΔG° then becomes equal to -RT ln K_{eq}.

The standard free energy change (ΔG°) for a reaction is the energy change that results when one mole of reactant is converted to one mole of product, when all products and reactants are continuously at one mole concentration and when the temperature volume and pressure are kept constant at a standard value. If the reaction proceeds to the right, the K_{eq} will be greater than one and the ΔG° will be negative; conversely, if the reaction were to proceed to the left, the K_{eq} would be less than one, and the ΔG° would be positive. The sign of the ΔG° again indicates the direction in which the reaction will proceed—if the reactants and products are kept in the standard (1 M) quantities and the temperature, volume and pressure are standard and constant.

As it is fairly easy to measure the K_{eq} of many reactions—all one needs to do is determine the products/reactants ratio at equilibrium and plug this value into the equation—the value of ΔG° rather than ΔG is most often cited in biochemical discussions. But actually ΔG° is equal to ΔG only when dealing with molar quantities under standard conditions--an impossible eventuality within a cell. Hence, one cannot always tell the direction of the reaction from the ΔG° alone, since it is possible to have, for example, a positive ΔG° but a negative ΔG when the actual intracellular concentrations of the reactants and products are taken into account. The direction of the reaction is really dependent upon ΔG and not ΔG°. Nevertheless, under most reaction conditions the sign and magnitude of ΔG° are sufficient in themselves to predict the direction of the reaction, and hereafter all energy values will be given in terms of the ΔG°.

Another important point concerning the value of ΔG° has to do with its dependence only upon the difference in standard free energies of the reactants and the final products and not in any way on the actual biochemical course this reaction takes. For example, the standard free energy change for the decomposition (oxidation) of glucose to carbon dioxide and water is -686 Kcal. This value is the same whether the glucose is burnt all at once in a chemist's crucible or, in Frostian terms, metabolized by "the slow, smokeless burning of decay". On the other hand, the course of the reaction makes an enormous difference in our ability to harvest the free energy released in the process. Most of the free energy will be lost if it is released all at once, since the common biochemical "receptacle" for this energy "carries" no more or less than about seven Kcal. The strategy employed by the cell, therefore, is to degrade glucose by a large number of intermediate steps so that whenever the ΔG° of any particular step is around -7 Kcal, another "receptacle" can be filled. The "common biochemical receptacle", as we all know, is a molecule of **ADP (adenosine diphosphate)** or **AMP (adenosine monophosphate)**. But more about this later.

The fact that the ΔG° values for a series of sequential biochemical interactions are strictly additive is routinely exploited by cells in order to power individual steps within the pathway that by themselves would be thermodynamically untenable. The conversion of A to B, for example, could have an appreciable positive ΔG° and consequently a very small K_{eq}. Therefore, it is really the reverse reaction, the conversion of B into A, that is strongly favored. The conversion of A

110

into B could still proceed, however, if B were a reactant in a subsequent step whose $\Delta G°$ was negative and larger in absolute value than that for the first reaction. This second reaction could be written B + C ---→ D. The sum of the two reactions becomes A + C ---→ D, a reaction whose over-all $\Delta G°$ is the arithmetical sum of the two separate reactions given above. As long as this value is negative, the reaction will go forward under the standard conditions inherent in the definition of $\Delta G°$. Such sequential reactions are said to be **coupled** through a **common intermediate**; the common intermediate cancels out in the final equation given for the over-all reaction.

It is important to recognize that changes in free energy in chemical reactions will be governed above all by two considerations: the inherent chemical nature of the putative reactants, and the concentrations of the putative reactants and products. The distribution of electrons around atoms and molecules imposes restrictions on which electrons can be donated, transferred or shared with which other atoms or molecules. This much is obvious from the discussion of atomic and molecular structure in earlier chapters. Less obvious is the fact that the direction of a chemical reaction can be manipulated according to the relative concentrations of reactants and products. As pointed out in the previous paragraph, removing a product in a subsequent reaction with a strongly negative free energy change will drive the first reaction despite its positive $\Delta G°$.

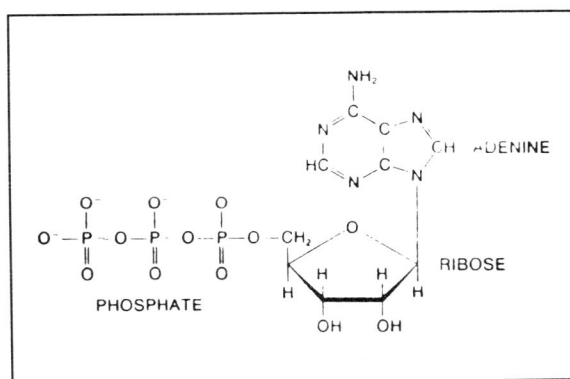

Figure 19.1 Adenosine triphosphate (ATP)

ADENOSINE TRIPHOSPHATE (ATP)[3]

The molecule most commonly involved in energy transfers in the cell is **ATP**, and since much of metabolism centers around either the production or utilization of ATP, we ought to become familiar with its basic properties.

ATP is ribonucleoside-5'-triphosphate, the same one which also serves as a substrate in the synthesis of RNA mediated by RNA polymerase. It can be hydrolysed (combined with water) to yield adenosine diphosphate (ADP) and inorganic phosphate, $(PO_4)^{-3}$,[*] with a standard free energy change of about -7.3 Kcal/M. Conversely, ADP and inorganic phosphate will react to form ATP if at least 7.3 Kcal/mole is supplied in a coupled reaction. In most cellular reactions involving ADP and ATP, the phosphate group is acquired from or donated to another organic molecule, but this is not relevant with respect to the free energy difference between the ATP and ADP molecules being concomitantly generated in the reaction. This standard free energy change (7.3 Kcal/mole) is large but nonetheless about in line with those energy changes that occur when the phosphate group is transferred in reactions involving other organic phosphate compounds in the cell. In fact, it is precisely because the $\Delta G°$ of the ATP/ADP interconversion is of an

---

[*]Inorganic phosphate is often written P_i and pyro-phosphate as PP_i.

111

intermediate value that makes ATP so useful as an energy transducer. If the $\Delta G°$ for the hydrolysis of ATP were much more negative, there would be few reactions occurring in the cell that could supply sufficient energy to generate ATP from ADP. If, on the other hand, the $\Delta G°$ were more positive (less negative), the hydrolysis of ATP to ADP would entail a free energy change insufficient to power the energy-requiring reactions with which it becomes coupled in metabolism. The role of ATP in cellular metabolism now becomes clear. Energy-producing (**exergonic**) reactions for which the $\Delta G°$ is greater than -7.3 can be coupled with ATP synthesis from ADP; at a later stage the hydrolysis of ATP can be coupled with energy-requiring (**endergonic**) reactions, and as long as the additive $\Delta G°$ values are negative, the reaction will proceed. In order to preserve the maximum free energy derived from the metabolism of foodstuffs, the optimum metabolic strategy should now too be obvious: liberate this free energy in steps so that as many of the intermediate free energy changes as possible occur in reactions where the $\Delta G°$ is at least, but not much greater than, -7.3 Kcal. These conversions can then be coupled with ATP production and this ATP, in turn, can be used elsewhere to power endergonic reactions.[4]

The central role of ATP in cellular energy transformations was first recognized by F. Lipmann and H. Kalckar in 1941. Since that time it has become common to refer to ATP as a "high energy compound", and to denote the terminal phosphate-phosphate diester bond with a squiggle ($\sim$). This has given rise to a number of misconceptions about the molecule. We have already noted, for instance, that the "high energy" refers to the $\Delta G°$ involved in the conversion of ATP into ADP + P_i and that it is not, and must not be, all that great. Anent the squiggle, the implication is that this bond is of a qualitatively different type than others that occur in this and other molecules. This is not so. It is a perfectly ordinary covalent bond. The terminal phosphate group is cleaved preferentially because, first, the close proximity of the repulsive negative charges on the adjacent phosphate groups tends to destabilize this terminal phosphate, and, second, because the hydrolysis of ATP to ADP and P_i results in a more stable electron resonance configuration in the products than existed in ATP itself.

It might be more correct to say that ATP has a "high group donor potential", meaning that it readily transfers phosphate groups (or sometimes the AMP "group") to other molecules. The potential to transfer groups is a property of the entire molecule and doesn't reside in a single bond. In the hydrolysis of ATP, the terminal phosphate is transferred to water, forming ADP and H_3PO_4 (P_i, inorganic phosphate). When the terminal phosphate (or the two terminal phosphates) is transferred to another molecule, this second molecule often itself acquires a high group transfer potential; that is, it becomes activated. This is not an invariable result, however. For example, when in glycolysis, a phosphate is transferred to glucose to form glucose-6-phosphate, this latter molecule displays little tendency to pass on the phosphate to other molecules.

Cells utilize an enormous amount of ATP in the course of their metabolic activities. A bacterium like *E. coli*, for example, which appears pretty much like a paradigm "couch potato" as far as overt physical activity is concerned, still uses about 2.5 million molecules of ATP—per second. And since it has only about 5 million ATP molecules available—i.e, two seconds of reserve—it must continuously regenerate them at this same rate. A human being is estimated to go through about 40 kilos (88 pounds) of ATP per day.

112

REFERENCES

1. Gordon, J. *Structures or Why Things Don't Fall Down.* (Penguin Books, 1978).

2. B. Commoner has an inspired layperson's explanation of the First and Second Laws in *The New Yorker* (2 Feb., 1972, pp 39-44).

3. Lehninger, A., 1961. How cells transform energy. Sci. Amer. 205: 62 (Sept).

4. Hinkle, P. and R. McCarthy, 1978. How cells make ATP. Sci. Amer. 238: 104-123 (Mar.).

LECTURE 20

ENZYMES

REDOX REACTIONS

> What chemical feature most clearly enables the living cell and organism to function, grow, and reproduce? Not the carbohydrate stored as starch in plants or glycogen in animals, nor the deposits of fat. It is not the structural proteins that form muscle, elastic tissue, and the skeletal fabric. Nor is it DNA, the genetic material. Despite its glamor, DNA is simply the construction manual that directs the assembly of the cell's proteins. The DNA itself is lifeless, its language cold and austere. What gives the cell its life and personality are enzymes. They govern all body processes; malfunction of even one enzyme can be fatal. Nothing in nature is so tangible and vital to our lives as enzymes, and yet so poorly understood and appreciated by all but a few scientists.
>
> Arthur Kornberg[1]

ENZYMES

No biochemical reaction taking place in the cell proceeds without the participation of a specific protein catalyst called an **enzyme***. This is true even when thermodynamic and other considerations would suggest that the reaction could proceed spontaneously and quickly without a catalyst. It is, therefore, the participation of enzymes in cellular chemical reactions that, more than anything else, distinguishes them from non-biological reactions taking place outside the cell. It is also the presence of these catalysts that determines which one of a number of thermodynamically equally probable reactions will actually happen, and which others won't.

A catalyst for any reaction participates in the chemical transformations it mediates but is not permanently altered as a result. For this reason only a small amount of the catalyst need be present, for these same molecules can be reused many times without loss of their initial activity. The magnitude of this activity can stagger the imagination: some enzymes can perform in a few

---

*Recently it has been discovered that very rarely RNA can serve as an enzyme. Such an enzyme is called a **ribozyme** to distinguish it from the vast majority of enzymes, which are proteins.

seconds what would take centuries to achieve in their absence; the conversion of 10^4 or even 10^5 molecules of reactant may be mediated per second. When discussing enzyme-catalyzed reactions, the reactant is usually called the enzyme **substrate**; the number of substrate molecules altered per minute or second is the **turnover number**.

Enzymes are also remarkable in the range of their specificities. Some enzymes, for example, are so selective that a seemingly very trivial difference in the structures of two alternative substrate molecules will completely exclude one from participating in the enzyme-catalyzed reaction. We have already seen an example of this in the various enzymes involved in nucleic acid synthesis. The only difference between, say, GTP and dGTP lies with the presence or absence of one hydroxyl group at the 2' position of the sugar moiety; nevertheless, the difference is sufficient to exclude GTP as a substrate for DNA polymerase I and to exclude dGTP as a substrate for RNA polymerase. By contrast, other enzymes are specific not for particular molecules but for particular linkages or groups of atoms within molecules. The molecules can be otherwise quite different structurally, but the one enzyme will nonetheless use them equally well as substrates.

In catalyzing a biochemical reaction an enzyme can in no way set aside the underlying thermo-dynamic principles that govern the ΔG° and K_{eq} of the reaction. In other words, an enzyme cannot coerce a reaction into proceeding that is thermodynamically impossible, nor can it alter the final equilibrium ratio of reactants and products. What it can do, however, is lower the **activation energy** needed to get the reaction under way. In order to react (i.e., to break or rearrange chemical bonds) molecules must collide with the requisite energy and in the requisite orientation. For most biological reactions, the likelihood of this happening at body temperature is slight. (Indeed, for most compounds familiar to us in our environment, this is also true—or else they would have already reacted with many of the other things in our environment and disappeared long ago.)

An enzyme facilitates a reaction by providing a precisely defined surface on which the reactants are brought into effective contact. In the absence of an enzyme, the reacting molecules may only infrequently collide with sufficient energy and in an appropriate orientation in order to react. On the enzyme surface, by contrast, the reacting molecules are held in the orientation which most strongly favors their interaction. The input of energy (usually in the form of heat) that would normally be necessary to initiate the reaction is thus obviated so that reactions that might otherwise proceed only at very high temperatures can then take place in the relatively cool cellular environment.

Enzymes, therefore, serve much the same function as wax on a toboggan. A well-waxed toboggan on a gentle grade can get started without the input of activation energy (a push) that might otherwise be needed. The wax, during the downhill slide, then speeds up the journey, but does not alter the final destination, the foot of the hill (K_{eq}), where the toboggan has yielded the greatest amount of kinetic energy (ΔG°). (The friction, I suppose, is analogous to the entropy of a chemical reaction, and the kinetic energy, which is analogous to the free energy, has been spent in getting the otherwise dour tobogganer squealing with delight.)

Some enzymes act totally autonomously in catalyzing a reaction, whereas others require a **co-enzyme** for activity. The co-enzyme is generally not a protein and by itself has no catalytic properties. Most vitamins, for example, behave as co-enzymes for specific biochemical reactions. Enzymes are also very sensitive to changes in temperature, pH and certain enzyme inhibitors and poisons. Usually for every enzyme there is an optimal narrow temperature and pH range within which it functions and outside of which it functions poorly or not at all.

The rate of an enzyme-catalyzed reaction will depend, therefore, upon temperature and pH, but if these are kept constant, it will then become dependent, up to a point, on the concentration of the substrate.

The maximum rate of substrate conversion is called the Vmax and the concentration of substrate necessary to achieve one-half this rate (Vmax/2) is called the K_m (**Michaelis Constant**) of the enzyme. Every enzyme has, for a given pH and temperature, a specific K_m for each of its substrates. The determination of the K_m value is predicated on the belief that the first step in an enzymatic reaction is the reversible formation of an **enzyme-substrate complex** within which takes place the actual biochemical conversion.[2] This is followed by a second reversible reaction which dissociates the product and the enzyme. The site where the substrate binds is called the **active site**, and it comprises only a small portion of the total enzyme volume. Sometimes it is possible to alter some parts of structure of an enzyme without affecting this active site. Even small changes at still other locations, however, can drastically alter the enzyme activity. This may be because the part tinkered with is physically part of the active site itself or because a change in one location can influence how the protein is arranged somewhere else, in this case at the active site. Changes in one part of an infactuously folded protein which influence the structure of other parts of the same molecule are called **allosteric effects**.

Allosteric effects are fundamental to the way in which the activity of many enzymes is regulated. It is frequently observed that the concentration of the final product in a series of enzyme-mediated biochemical conversions determines the activity of the first enzyme in that particular sequence. The accumulation of the final product thus indirectly determines the concentration of all intermediates in the pathway. This control is exercised by the binding of the end-product to some site on the first enzyme in this particular biochemical sequence. The binding site for the end-product is different from the active site occupied by the normal substrate. The resulting conformational change in the enzyme allosterically alters the active site and thus inhibits the enzyme activity. As less of the end-product is synthesized as a result, its inhibitory influence on the first enzyme is proportionally lessened, and its synthesis is then accelerated. This common control mechanism is termed **feed-back inhibition**. Cases are also known of allosteric enzyme activation.[3]

REDOX REACTIONS

Oxidation-reduction reactions (**redox reactions**) are very common in metabolism and warrant a few paragraphs of explanation. The basis for saying that a compound has been **reduced** in a biological reaction is that it has acquired a pair of electrons; the compound that gives up these electrons is said to be **oxidized**. Since electrons gained by one molecule are, from a different perspective, electrons lost by another it is obvious that every reduction must be accompanied by a simultaneous oxidation. Also, since the acquisition or loss of electrons will change the electrical charge of the molecules involved, movement of electrons will be accompanied by movement of protons (hydrogen ions). From the point of view of the structure of the molecules, it is the movement of protons which is the more tell-tale. Dehydrogenation reactions are oxidations, hydrogenations are reductions.

Some compounds have a strong tendency to pass off electrons and are hence known as good **reductants** or **reducing agents**; other compounds are always on the look-out for electrons and hence are good **oxidants** or **oxidizing agents**. Given this terminology, it should come as no surprise that oxygen has a strong tendency to attract electrons from other compounds; that is to say, oxygen is a strong oxidizing agent. In a very general way, metabolic synthesis (**anabolism**) depends upon biochemical reductions and is hence dependent upon strong reducing agents, whereas metabolic decomposition (**catabolism**) depends upon oxidations. It is, of course, part of our every-day experience that exposing many biological products to the air, the active ingredient of which is oxygen, causes them to "spoil" or decompose.

In the cell the molecules most often involved in the removal and donation of electrons are the **pyridine nucleotides**, NAD, NADP and FAD (oxidized notation) or NADH, NADPH and FADH (reduced notation). The passage of a pair of electrons from NADPH to 0_2 results in a huge standard free energy change of -52.7 Kcal[*]. As we shall learn presently, the extraction of electrons from foodstuffs during their catabolism and the subsequent passage of these electrons to oxygen provides the rationale for aerobic respiration.

REFERENCES

1. Kornberg, A. *For The Love of Enzymes.* Harvard Univ. Press (Cambridge, 1989).

2. Frieder, E., 1959. The enzyme-substrate complex. Sci. Amer. 201: 119 (Aug.)

3. Changeux, J-P., 1965. The control of biochemical reactions. Sci. Amer. 212: 36 (Apr.).

---

[*]The strength of a reducing agent—its tendency to donate electrons—is actually measured in **volts**. There is a convenient formula to concert the potential difference between reactants and products, measured in volts, to the standard free energy change, measured in Kcal.

LECTURE 21

PHOTOSYNTHESIS

It is common knowledge that the ultimate source of all our energy and negative entropy is the radiation of the sun. When a photon interacts with a material particle on our globe it lifts one electron from an electron pair to a higher level. This excited state as a rule has but a short lifetime and the electron drops back within 10^{-7} to 10^{-8} seconds to the ground state giving off its excess energy in one way or another. Life has learned to catch the electron in the excited state, uncouple it from its partner, and let it drop back to the ground state through its biological machinery utilizing its excess energy for life processes.

A. Szent-Gyorgyi[1]

INTRODUCTION

All of the ATP needed by a green plant is generated, directly or indirectly, from sunlight, and all the ATP needed by animals is obtained by consuming, directly or indirectly, green plants. In generating useful chemical energy from the radiant energy of the sun, green plants give off a gaseous "pollutant", oxygen, which in turn is needed to catabolize the complicated hydrocarbon compounds anabolized during photosynthesis. In oxidizing these compounds, carbon dioxide is given off, and it so happens that this same compound is needed as the key reactant in photosynthesis. Photosynthesis and respiration are thus intimately interconnected in an endless Sisyphean drama, the former putting carbon dioxide and water together to make something useful (hydrocarbons), the latter taking the hydrocarbon molecules apart again. The solar energy which was captured to power the photosynthetic anabolism is later bled out as chemical energy during respiratory catabolism. This energy then becomes available to fuel those processes that are uniquely "living": growth, differentiation, movement and consciousness.

THE GENERAL PLAN

The over-all chemical equation for photosynthesis can be given quite simply as:

$$6CO_2 + 6H_2O \longrightarrow C_6H_{12}O_6 + 6O_2$$

The standard free energy change for the reaction is +687 Kcal/mole; obviously it could not proceed in the direction written without the supply of an enormous amount of energy. It would, of course, be self-defeating if this energy were derived from the normal energetic processes of the cell since these processes ultimately derive their own energy from catabolizing this same

ADP → ATP

molecule of hydrocarbon that is being synthesized. For the same reasons that exclude perpetual motion machines in the physical world, perpetual chemical circuits are excluded in chemistry. Consequently an external energy source, the sun, has to be enlisted.

Chemically, the gain of electrons by CO_2 to make carbohydrate is, in effect, a reduction reaction and their loss from the H_2O an oxidation. To reduce CO_2 requires an extremely powerful reducing agent as well as other energy inputs; in the cell, these are, respectively, NADPH and ATP. To oxidize water requires a very strong oxidant, the precise identification of which has still to be made. The production of the reductant, the oxidant, and the ATP all occur in the presence of light. The oxidant seems short-lived and is used immediately to scrounge electrons out of water, leaving the bereft oxygen to react with another and escape as molecular oxygen. However, the ATP and the reducing agent, which has ultimately acquired these electrons that were pulled out of water, need not be used at once. Having been generated in the presence of light, they are subsequently used in reducing CO_2, a synthetic process that may proceed either in the light or the dark. In other words, photosynthesis is really composed of two distinct events: the **light reactions** that split water and generate NADPH and ATP, and the **dark reactions** that reduce CO_2 into hydrocarbon. The obvious syntheses (of hydrocarbons) to which the term **photosynthesis** refers actually occur, therefore, in the dark.

The complete independence of the light and dark reactions was demonstrated by D. Arnon in 1958. First he isolated chloroplasts from spinach and illuminated them in the presence of NADP and ADP. Molecular oxygen was evolved and (reduced) NADPH and ATP synthesized. He then discarded the green portion of the chloroplasts and demonstrated the incorporation of $C^{14}O_2$ into hydrocarbons in complete darkness. During this synthesis the NADPH and ATP manufactured in the earlier light reaction was reconverted into NADP and ADP.

THE CHLOROPLAST

The cell organelle in which both the light and dark reactions occur is the **chloroplast**. The pigment that captures light and lends plants their green color is **chlorophyll**.[2] In a seeming fault of engineering, the chlorophyll molecule absorbs light in the red and violet parts of the solar spectrum, whereas the sun gives off more light in the yellow and green portions of the spectrum. Many plants have evolved accessory pigments to capture energy in these parts of the spectrum, as well. (They can be seen in the fall foliage, when the masking chlorophyll molecules disappear.)

The chloroplast is surrounded by a double membrane. The interior space of the chloroplast is called the **stroma**. Within the stroma are located the chlorophyll-containing **grana**, which resemble stacks of saucers. Each "saucer" is in reality a flattened membranous sack called a **thylakoid**; the chlorophyll molecules are arranged within (or attached to the inner surface of) the thylakoid membrane in tiny (155 x 180 x 100 Å) oblong particles called **quantasomes**. The grana seem to be interconnected in places by structures called **stromal lamellae** or **fret membranes**.

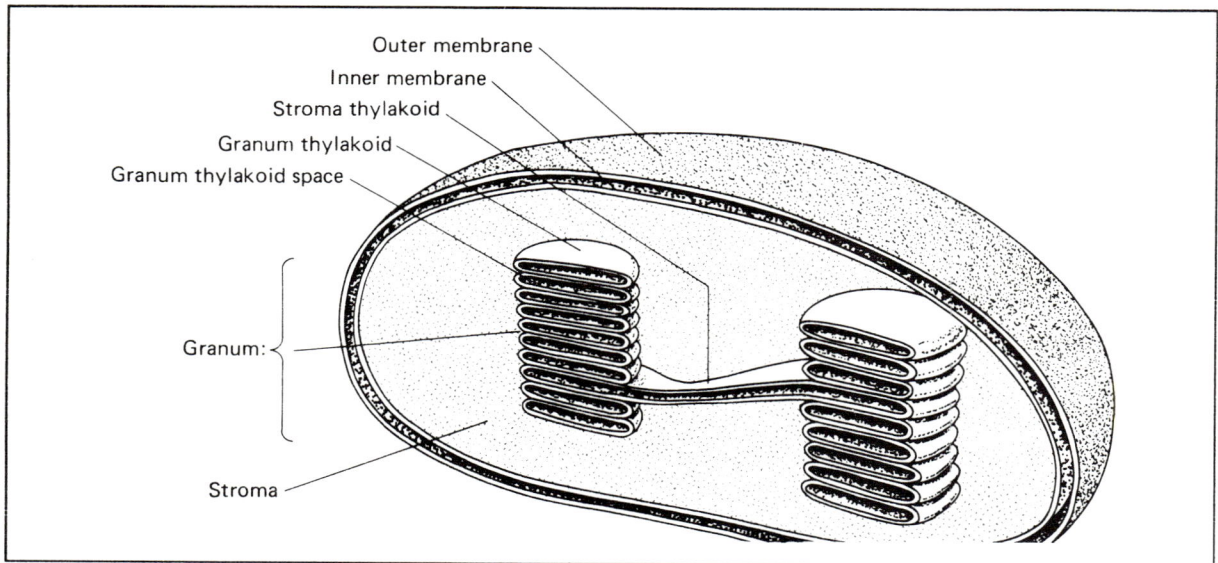

Figure 21.1 The chloroplast. From BASIC CELL BIOLOGY, 2ND EDITION by C.J. Avers. Copyright © 1982 by Wadsworth Publishing Company. Reprinted by permission of the publisher.

One surprising component of a chloroplast is DNA. Chloroplasts, it seems, have a certain degree of independence from the remainder of the cell and can synthesize, on their own autochthonous ribosomes, and about 80 endogenous proteins. In addition, the chloroplast DNA encodes the information for the r-RNA (ribosomal RNA) of chloroplast ribosomes and for the various t-RNA (transfer RNA) molecules needed to translated chloroplast genes. This independence is not permitted to get out of hand, however, for the synthesis of many important chloroplast proteins remains under the exclusive control of the nuclear genetic material. There are between 20 and 80 DNA molecules in each chloroplast, depending upon the plant species. Further details on protein synthesis in chloroplasts are given in the lectures on the mechanism of protein synthesis.

Chloroplasts are now viewed as the remnants of formerly free-living organisms that became endosymbionts within other cells that originally lacked the ability to photosynthesize. Until recently, it was thought that modern photosynthetic bacteria are evolved from the primitive photosynthetic prokaryotes that were engulfed by a non-photosynthetic eukaryotic cell. This may be true in some instances, but more recent research indicates that photosynthetic eukaryotes may have arisen several times in the course of evolution and that the engulfed cells may have included other photosynthetic eukaryotic cells, not just prokaryotes. The chloroplasts of some cells appear to harbor a nucleus-like organelle that contains both DNA and RNA and that may well be the remnant of a fully functional nucleus which had been present in the original engulfed cell.

THE LIGHT REACTIONS

There are two primary light reactions in photosynthesis, both of which involve the displacing of an electron from chlorophyll. In the first set of reactions (called **photosystem I**), light strikes "antennae" molecules of chlorophyll (or accessory pigments), which in turn pass the absorbed energy to the **reaction center**. The reaction center consists of two chlorophyll molecules and associated proteins, known as the **P700 site**. This causes an electron to be kicked out of the chlorophyll molecule at the reaction center—or, in more correct and prosaic terms, moved to a higher energy level. From this point, the "energized" electron is captured by an as-yet unknown molecule called FRS (ferredoxin reducing substance), which immediately transfers it to ferredoxin, an iron-containing protein. The chemical effect, of course, is the reduction of ferredoxin. The reduced ferredoxin can, at this point, do one of two things with its electron. Together with the electron from another ferredoxin molecule, the two electrons can be transferred enzymatically to NADP to produce the strong reducing agent used in the dark reactions, NADPH. Alternatively, the two electrons can be returned to the P700 molecule by a route (to be discussed in greater detail later) which results in the generation of one ATP molecule from ADP and P_i. In other words, if there is no oxidized NADP acceptor, the energized electrons may return to fill the "electron hole" in P700, but in travelling down this energy gradient some of their energy is harnessed (indirectly) to produce ATP. Because electrons may return to the P700 molecule, this reaction system is often called **cyclic photophosphorylation**. Cyclic photophosphorylation is common in photosynthetic bacteria, but in higher plants the electron is almost always used to generate NADPH.

When NADPH is produced in PSI, the P700 molecule is left with an "electron hole" (deficit) that must be filled, and to do so there is a separate **photosystem II**. In PSII light again drives an excited electron out of a (separate) reaction center (called the **P680 site)** into the embrace of an unknown acceptor called "Q". Reduced Q then passes it to a series of protein acceptors in which each becomes first reduced and then oxidized as it passes the electron to the next acceptor. At each transfer the electron loses energy, and one molecule of ATP is generated in an indirect reaction whose mechanism is still the subject of lively controversy (see below). By the time it has reached the end of this transport chain, the "tired" electron is delivered to the P700 site in exchange for the one that had been captured in NADPH.

If PSI regains electrons from PSII, how does PSII regain its electrons? By an unknown mechanism, the oxidized (i.e. electron deficient) PSII pigment steals electrons from a molecule of water, which then goes to pieces: protons and atomic oxygen. Two atoms of oxygen combine to form a molecule of oxygen, and the protons enter the environment in exchange for those picked up by NADP and the other intermediate carriers when they become reduced. The real "miracle" of photosynthesis, if there is one, is this generation of an oxidant powerful enough to oxidize water, since the avidity of oxygen for attracting and keeping electrons is what led to the term "oxidation" for this process in the first place.

The mechanism by which ATP is generated is known by the name of **chemosmosis**. Although the exact details are still unknown—or at least highly controversial—the broad principles are well understood. Basically, the movement of electrons and their associated protons (H+ ions) leads to an accumulation of protons inside the thylakoid and, in consequence, an electrochemical gradient across the thylakoid membrane between the interior of the thylakoid and the stroma. In essence, the thylakoid becomes a sort of battery, except that protons instead of electrons furnish the potential. The ATP is generated by a complex called the **CF_1 complex** which, in

121

permitting the protons to travel through the thylakoid membrane into the stroma, harnesses the electrochemical potential to make ATP from ADP. It is the details of this harnessing that are still obscure.

THE DARK REACTIONS

The identity of precursors used by plants to increase their mass has long intrigued biologists. In 1648 J-B. Van Helmont, a Dutch botanist, devised a simple experiment which he hoped would provide an answer. A small, carefully weighed willow seedling was planted in a pot of earth of known weight. The tree was watered for several seasons until it had matured into a small tree. It was then cut down and reweighed. The roots, too, were dug up and weighed (although no attempt was made to weigh the leaves that had dropped each year). At the end of five years, the tree had gained 164 pounds, but the soil had lost only two ounces. Since the only obvious source of this increase in mass was the water that had been poured into the pot over the years, Van Helmont concluded, reasonably enough, that the mass of the tree was derived from this water. He was wrong. The mass of the tree was actually derived from something Van Helmont could have had no knowledge of at the time—the gas, carbon dioxide.[*]

The series of biochemical reactions that results in the formation of carbohydrate from carbon dioxide and protons (hydrogen ions) is eponymously named the Calvin-Benson cycle after the investigators who first worked it out. In 1945, these investigators began what turned into several years of work with the unicellular alga, *Chlorella*. Their plan was to administer a brief "pulse" of $C^{14}O_2$ to actively photosynthesizing algae and then to extract and identify any radioactive carbon derivatives from the cells. The cell extracts were chromatogrammed and the position of the radioactive substances revealed by pressing the chromatogram against photographic paper. The chromatogram derived from a one-minute exposure to radioactive CO_2 was a mass of different radioactive spots and provided no clue, therefore, to the identity of the first-formed intermediate in photosynthesis. When the time interval in $C^{14}O_2$ was reduced to five seconds, only one prominent radioactive spot appeared on the chromatograms. Chemical analysis revealed this substance to be 3-phosphoglyceric acid, a three-carbon compound.[3]

This might suggest that the radioactive carbon was first attached to a two-carbon acceptor molecule, but further analysis showed that, whereas 3-phosphoglyceric acid was the first detectable intermediate in CO_2 fixation, the first actual intermediate was a highly unstable six-carbon compound (usually left unnamed) which almost immediately on formation was converted into two molecules of 3-phosphoglyceric acid. The acceptor molecule, therefore, is a five-carbon compound now known to be the sugar, ribulose- 1,5-diphosphate. The enzyme that

---

[*]Not all of Van Helmont's scientific work was carried out with such attention to detail. He reported that mice arose spontaneously in a mixture of sweaty shirts and grains of wheat, if these reagents are kept in a dark corner for awhile. It has not proved possible to reproduce these observations, as such essential details as the color of the shirts, the species of wheat, the number of grains, the angle of the corner, etc. were not adequately specified.

attaches the CO_2 to the acceptor is known as **ribulose diphosphate carboxylase**[*]. This protein makes up more than 25% of all protein in the chloroplast and in consequence is probably the most common protein on earth.

By giving brief pulses of $C^{14}O_2$ followed by increasing synthetic periods in non-radioactive CO_2—technically known as a **pulse-chase experiment**, a procedure very widely employed in radiotracer work—it was possible for M. Calvin and his associates to identify the metabolic pathway followed by the radioactive intermediates derived from the first pulse of carbon dioxide. It is not useful for us here to trace this network of interconversions in detail, but some general observations about the scheme can be made.

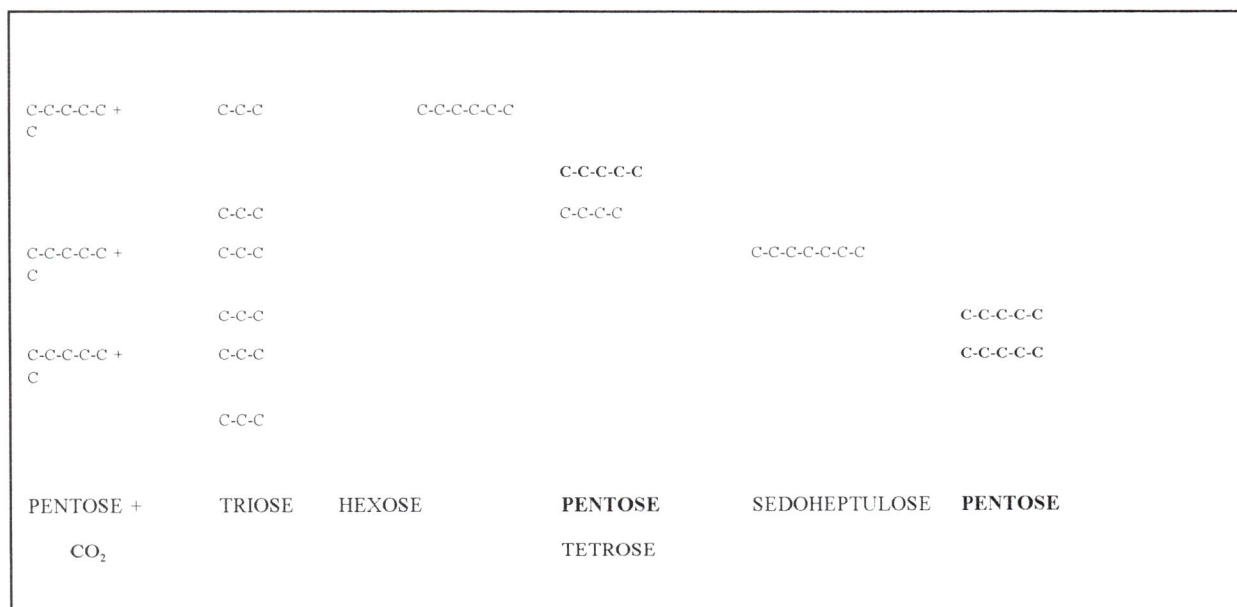

C-C-C-C-C + C	C-C-C	C-C-C-C-C-C				
			C-C-C-C-C			
	C-C-C	C-C-C-C				
C-C-C-C-C + C	C-C-C			C-C-C-C-C-C-C		
	C-C-C				C-C-C-C-C	
C-C-C-C-C + C	C-C-C				C-C-C-C-C	
	C-C-C					
PENTOSE +	TRIOSE	HEXOSE	**PENTOSE**	SEDOHEPTULOSE	**PENTOSE**	
CO_2			TETROSE			

Figure 21.2 The path of carbon in photosynthesis, viewed only as changes in the number of carbons in the "skeletons" of the compounds involved. See text for details and explanation.

The accompanying diagram traces the metabolic events from the vantage point of the carbon skeletons of the intermediates. To make the events easier to follow, the diagram shows three molecules of CO_2 being transferred to three pentose sugars, which then break down into six molecules of triose, the 3-phosphoglyceric acid mentioned above. At this point, the three 3-PGA molecules are converted to a number of other three-carbon molecules (not shown because the number of carbons in the skeleton does not change), and it is in these interconversions that all

---

[*]This enzyme is composed of two subunits, one of which is encoded by a nuclear gene and one by a chloroplast gene. Ribulose diphosphate is commonly also called ribulose biphosphate, and the enzyme correspondingly called ribulose biphosphate carboxylase.

of the NADPH and some of the ATP produced in the light reactions are used up. Two of these trioses combine to form a six-carbon hexose, which then combines with another of the trioses to form a pentose and a tetrose. The tetrose combines with a fourth triose to form the seven-carbon sugar, sedoheptulose. Finally, the sedoheptulose combines with a fifth triose to form two pentoses. The effect of all this shuffling of Carbon is to regenerate the three molecules of the accepting pentose, ribulose 1,5-diphosphate. In re-synthesizing these three pentoses, however, there is one molecule of triose left over; that is, for every three molecules of CO_2 fixed, one spare triose is generated. Two of these "surplus" trioses can then combine to form a six-carbon sugar like glucose, a molecule which is the usual starting point in respiration. In summary, therefore, these chemical events constitute a cycle whereby the three acceptor molecules of pentose are regenerated and in effect (although not in chemical reality) the three single carbons of carbon dioxide are joined to make one molecule of triose. And, as mentioned, two "turns" of this cycle generate two molecules of triose, which can join to make a six-carbon glucose molecule.

For every molecule of CO_2 fixed, two molecules of NADPH must be oxidized to NADP and three molecules of ATP hydrolysed to ADP. The true photosynthetic equation can thus be written:

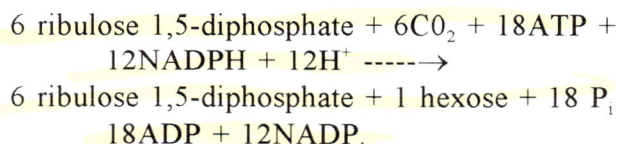

$$6 \text{ ribulose 1,5-diphosphate} + 6CO_2 + 18ATP +$$
$$12NADPH + 12H^+ -----\rightarrow$$
$$6 \text{ ribulose 1,5-diphosphate} + 1 \text{ hexose} + 18 \text{ P}_i$$
$$18ADP + 12NADP.$$

As a rough approximation, we can set the ΔG° value of each reduction as -53 Kcal/mole and of each ATP hydrolysis as -7.0 Kcal/mole. The formation of a six-carbon sugar then requires 6 X -127 or -762 Kcal/mole. Since the complete catabolism of glucose yields -687 Kcal of free energy per mole, this represents an efficiency of 90% in the utilization of chemical energy.

"C-4" PLANTS

More recently a variation of the Calvin cycle was found to occur in corn and some tropical plants like sugarcane. This variation, called the **Hatch-Slack cycle**, employs a 3-carbon acid as the CO_2 acceptor molecule. The added carbon is then passed to ribulose 1,5-diphosphate and enters the normal Calvin cycle thereafter; the still ill-defined acceptor molecule is then converted into pyruvic acid as a result of the transfer. Plants employing this alternative cycle are called **C4 plants** to distinguish them from **C3 plants**, the 3 and 4 denoting the number of carbons in the first isolable radioactive carbon compound.[4]

The necessity for the C4 pathway stems from the fact that ribulose diphosphate carboxylase requires a carbon dioxide concentration of at least 0.005% in order to efficiently attach the CO_2 to the 5-carbon acceptor. Should the concentration of carbon dioxide fall below that threshold level, carboxylase will attach O_2 instead, and this causes the ribulose diphosphate to break down, releasing CO_2. In essence, this is the reverse of photosynthesis, and is obviously counter-productive. Such a situation could arise during periods of hot, dry conditions when the plant, in order to minimize the loss of water, closes its stomata. The stomata, however, are the openings by which carbon dioxide enters the tissues, as well as those by which water vapor exits. Prolonged closing of the stomata will reduce the internal level of carbon dioxide below the critical value of 0.005%. Plants adapted to lengthy hot, dry periods can transfer carbon dioxide temporarily to a three-carbon acceptor in a reaction that can occur even in very low concentrations of carbon dioxide.*

To carry out photosynthesis enormous quantities of air are required, since only 0.03% of air is carbon dioxide. For example, about 10,000 cubic feet of air are needed to make about 1/4 pound of glucose. About 10^{10} tons of carbon is assimilated annually into carbohydrate and derived organic matter. This represents about 10^{12} Kcal of free energy converted from sunlight.

Table 21.1 The energy requirements and solar efficiency of photosynthesis

---

Total glucose formed by one acre of corn in one growing season: 8,700 kg.

Energy required to synthesize 1 kg. of glucose: 3,800 kCal.

Energy required to synthesize 8700 kg: 33,000,000 kCal.

Total solar energy available for one acre: 2,040,000,000 kCal.

$$\text{Available energy used} = \frac{33 \text{ million kCal.}}{2040 \text{ million kCal.}} \times 100 = \textbf{1.6 \%}$$

REFERENCES

1. in *Light and Life*, W. McElroy and B. Glass, eds. (Johns Hopkins Press, Baltimore, 1961.

2. Rabinowitch, E., 1965. The role of chlorophyll in photosynthesis. Sci. Amer. 213: (July)

3. Bassham, J., 1962. The path of carbon in photosynthesis. Sci Amer. 206: 88 (June).

4. Bjorkman, O. and J. Berry, 1973. High efficiency photosynthesis. Sci. Amer. 229: 80 (Oct).

---

*Suburbanites who do not compulsively water their lawns during summer droughts may have noted the gradual replacement of their Kentucky bluegrass by the much less attractive crabgrass. The bluegrass, a C3 plant, is much more vulnerable to drought than is crabgrass, a C4 species.

LECTURE 22

CATABOLISM OF CARBOHYDRATE (RESPIRATION)

INTRODUCTION

Once a six-carbon sugar (e.g., glucose) has been synthesized in a plant or ingested by an animal its energy can then be extracted at a convenient time and harnessed to drive the cell syntheses and other metabolic activities*. This extraction process is called **respiration**. Respiration, despite the obvious association the term has with the physical phenomenon of breathing, can be either **anaerobic** (occurring without oxygen) or **aerobic** (requiring oxygen), depending upon the type of cell and the circumstances. Even when aerobic respiration is taking place, the first many steps in the catabolism of glucose do not require oxygen and in effect constitute most of the pathway of normal anaerobic respiration. This common pathway goes by the name of **glycolysis** or the **Embden-Meyerhof pathway**, after the two workers who played the major role in working it out in the early 1940s.

If respiration for this particular cell is to remain anaerobic, glycolysis is terminated by reactions that produce either alcohol or lactic acid (again depending on the type of cell), and the sequence of biochemical conversions as a whole is often called **fermentation**. Alternatively, if aerobic respiration is going to succeed glycolysis, the final metabolite of glycolysis is fed through an intermediate into what is called the **citric acid cycle** or **Krebs cycle**, in honor of the person, H. Krebs, who did much of the pioneering work on it in the mid-1930s.

The over-all equation for the complete catabolism (glycolysis plus Krebs cycle) of a molecule of glucose is simply the reverse of that for its synthesis:

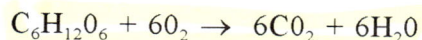

$$C_6H_{12}O_6 + 6O_2 \rightarrow 6CO_2 + 6H_2O$$

The $\Delta G°$ for this reaction is negative but of the same absolute value as that for its reverse (i.e., for photosynthesis): 687 Kcal/mole. In drawing these parallels, however, do not be misled into

---

*The hexose, glucose, is conventionally given as the end-point for photosynthesis and the beginning point for respiration. It should be noted, however, that cells, especially plant cells, have little free glucose lying around. Instead, the glucose generated in photosynthesis is usually polymerized into a **polysaccharide** such as starch for storage in plants. In animal cells, glucose is polymerized into glycogen. When energy is required, the glucose can be regenerated and fed into glycolysis.

believing that the entire pathway for the catabolism of glucose is simply the reverse of that followed in its synthesis from CO_2. They do overlap in part—the assembly of glucose from triose involves the simple reversal of many of the steps used in the breakdown of glucose to triose—but below the level of triose, the respiratory pathway follows a completely different course from the photosynthetic one. It is also important to remember, as will be explained presently, that most of the energy derived from the catabolism of glucose does not come from ATP formed in coupled reactions within the catabolic pathway itself. Some ATP is generated in this way—generated "at the substrate level", as sophisticated biochemists say—but much more is generated through **oxidative phosphorylation**, a process whereby ATP is generated from "high energy" electrons that are pulled out of various intermediates during their interconversions within the Krebs cycle.

The foregoing represents the basic plan for metabolic respiration. It is not necessary for our purposes here to go into great biochemical detail; as with photosynthesis, it is more important that the reader understands the outlines of the process as reflected in changes in the carbon skeletons of the respiratory intermediates and thoroughly comprehends the significance of these events for the energy budget of the cell.

Before the pathways are further discussed it should also be emphasized that glycolysis and the citric acid cycle are very important sources of biochemical precursors for many important synthetic pathways in the cell. This means that many of the respiratory intermediates will be bled away from their catabolic pathway and used as the starting point for the synthesis of various molecules like amino acids, fats, and so on. Whether a molecule of glucose is completely catabolized to carbon dioxide and water or only partially degraded will depend upon the actual metabolic needs at the time.

ANAEROBIC RESPIRATION

Beginning with the six-carbon sugar, glucose, the pathway proceeds through three intermediate steps to the six-carbon phosphorylated sugar, fructose 1,6-diphosphate. Putting the two phosphates on to the sugar required the expenditure of two ATP molecules; so at this point the energy balance of respiration is actually in the red, so to speak. The fructose then splits into two inter-convertible trioses, and the biochemical pathway continues on from one of them, glyceraldehyde 3-phosphate. Five further conversions follow to terminate the pathway in pyruvate (pyruvic acid)[*]. Since both glyceraldehyde 3-phosphate and pyruvate have three carbons, these latter conversions have not altered the carbon skeleton, but they have resulted in the formation of one NADH and two ATP molecules. Since two pyruvates are produced from each molecule of glucose, a total of two NADH and four ATP molecules are generated by the conversions involving the various trioses. The net result of the pathway as a whole is then two NADH and two ATP molecules.

---

[*]At the pH of the intracellular milieu, the acidic respiratory intermediates would usually be fully ionized. It is common, therefore, to speak of the ion (e.g. "pyruvate") rather than the acid.

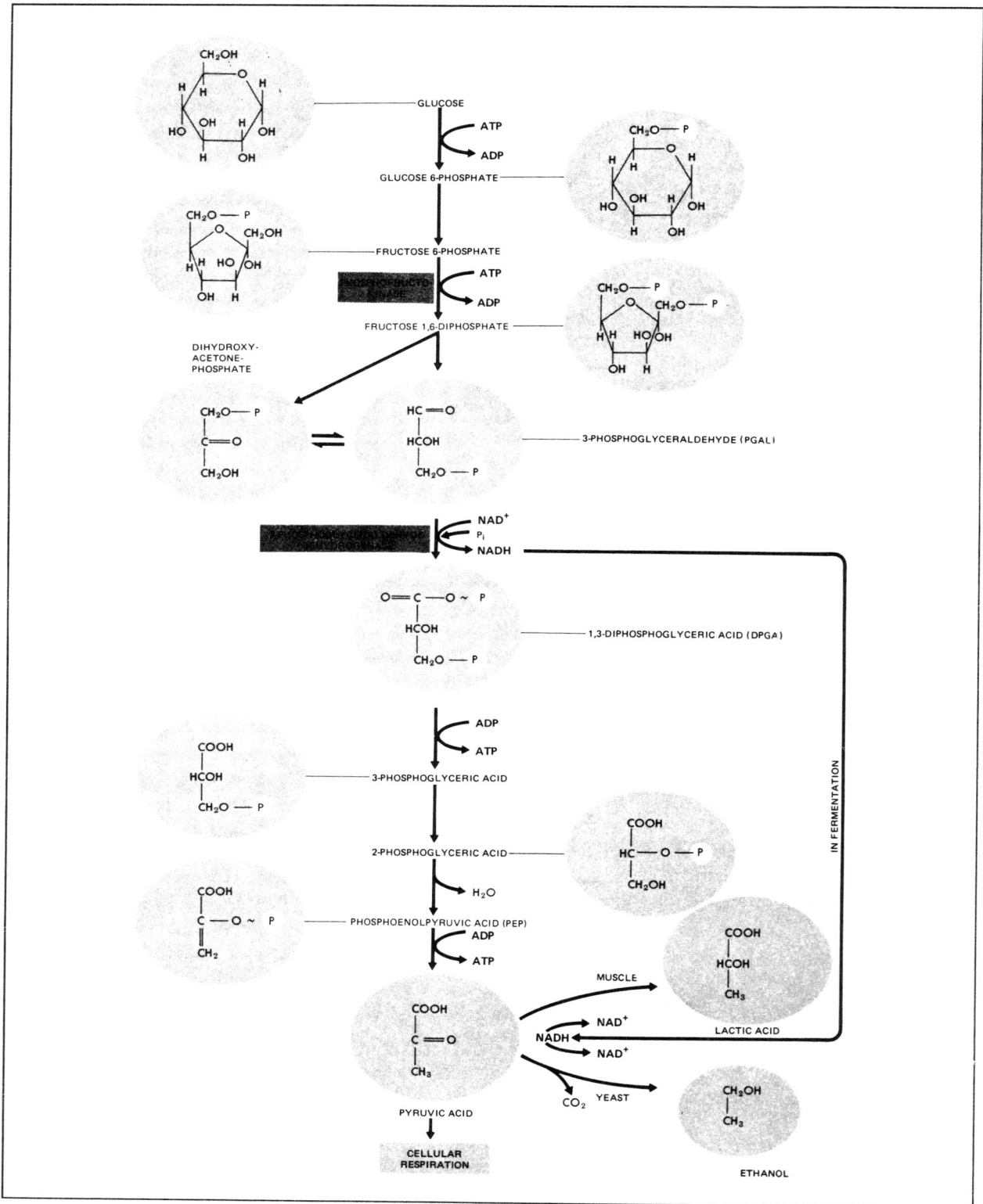

Figure 22.1 The pathway of glycolysis. From CELL BIOLOGY, THIRD EDITION by John W. Kimball. Copyright © 1984 by John W. Kimball. Reprinted by permission of the author.

In the absence of oxygen, the pyruvate may be converted into a two-carbon compound whose synthesis consumes the NADH generated in glycolysis and yields CO_2. In microorganisms like yeast, this two-carbon fermentation product is ethyl alcohol (ethanol); in animal tissues temporarily deprived of oxygen, pyruvate is converted into a 3-carbon compound, lactic acid. Since for every glucose molecule catabolized two molecules of pyruvate must be converted to the fermentation product, all the reduced NADH produced in glycolysis becomes re-oxidized to NAD; the net energy balance from fermentation as a whole is therefore just two molecules of ATP. Anaerobic respiration, in consequence, leaves most of the potential free energy of glucose locked up in molecules that are secreted from the cell as metabolically useless.

AEROBIC RESPIRATION: THE MITOCHONDRION

If oxygen is present, aerobic eukaryotic organisms take up the pyruvate into a special cell organelle, the **mitochondrion**, where it reacts with an adenine-containing co-enzyme known as **co-enzyme A (CoA)**. The products of this reaction are **acetyl-CoA**, carbon dioxide and NADH. The pyruvate has thus been diminished to a two-carbon molecule, the acetyl group, which becomes attached to the coenzyme.

This reaction and all the Krebs cycle reactions that follow from it occur within the mitochondrion, an organelle found in all plant and animal cells (but not bacteria) that must now be described. Cells have a variable number of mitochondria, depending upon the energy needs of the cell type. Nerve and muscle cells can have up to 100 or so, whereas other cells with lower energy needs may have just one or two. The shape of mitochondria is extremely variable, but in general they are peanut-shaped particles with a diameter of 0.5 to 1.0 micron and a length of 3.0 to 10 microns. They are surrounded by two membranes, the inner one being elaborately folded into finger-shaped **cristae** that project into the interior of the mitochondrion. Many **elementary particles** line the cristae and project like little mushrooms into the inner cavity (**matrix**).

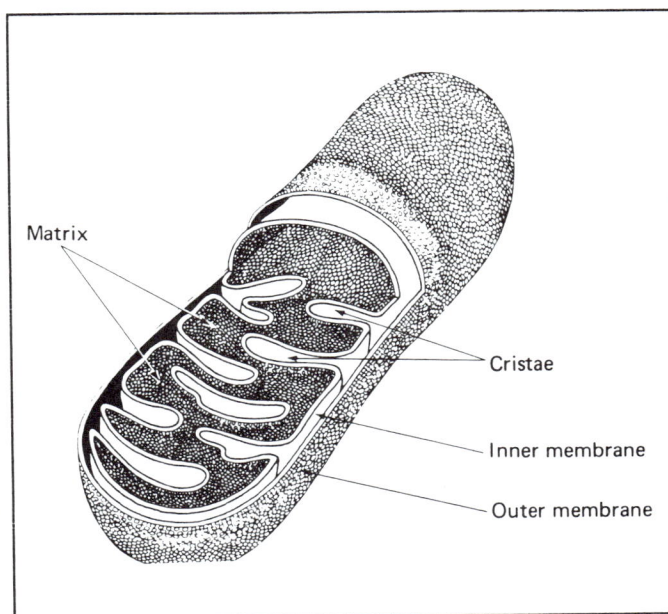

Figure 22.2 The mitochondrion. From BASIC CELL BIOLOGY,2ND EDITION by C.J. Avers. Copyright © 1982 by Wadsworth Publishing Company. Reprinted by permission of the publisher.

Like the chloroplast, the mitochondrion has a small piece of DNA and some ribosomes of its own. The DNA is present in from two to 10 copies in each mitochondrion, each molecule has about 16500 base-pairs. Mutations in mitochondrial DNA are known to be the cause of certain

129

rare human diseases—the mitochondrial myopathies and encephalopathies. In most cases the defect as far as is known lies with some aspect of oxidative phosphorylation. In higher animals, mitochondria are inherited exclusively through the egg, that is, maternally[*]. Mitochondrial diseases, therefore, would show the same mode of maternal inheritance.

The mitochondrion lives in "sovereignty-association" with the cell, making some of its own proteins, perhaps about a dozen or so of the more than 50 known to occur in mitochondria,[**] and appears to replicate largely on its own initiative. On this account, most evolutionists now believe that mitochondria and chloroplasts are vestiges of formerly autonomous organisms that invaded plant and animal cells and established a symbiotic relationship with (or rather, within) the host.[1] (See also the discussion of chloroplasts in Lecture 21.)

THE KREBS (CITRIC ACID) CYCLE

In the Krebs cycle proper, the two-carbon pyruvate derivative attached to CoA reacts with a four-carbon acid to produce the six-carbon molecule, citric acid. In a series of succeeding decarboxylation reactions, two molecules of CO_2 are blown off and the initial four-carbon acceptor (oxaloacetate) is regenerated. Basically, therefore, the Krebs cycle is a way to convert the remaining two carbons of the original glucose into CO_2—although they are not the same two carbons which entered the cycle as acetyl-CoA. In the process, three molecules of NADH, one of FADH and one GTP (interconvertible with ATP) are produced at the substrate level. Again, since there are two acetyl-CoA molecules produced from every glucose molecule being respired, the net result is two "turns" of the Krebs cycle that generate six NADH, two FADH and two ATP molecules. And then there are, in addition, the two NADH produced in making acetyl-CoA itself.

OXIDATIVE PHOSPHORYLATION

The astute reader may have noticed that no oxygen is consumed in any of the reactions occurring within the Krebs cycle proper—this despite the fact that oxygen is absolutely required in order to have any acetyl-CoA flowing into it. The need for oxygen arises when the electrons of NADH and FADH are got rid of, for if these electrons can be passed to molecular oxygen, each pair results in a standard free energy change of -53 K/cal. Some of this energy is trapped in the generation of ATP, a process called **oxidative phosphorylation**.

---

[*] In some organisms, such as mussels (genus *Mytilus*), the mitochondria can without doubt be inherited from both parents. There is some indication that "paternal leakage" of mitochondria (or of mitochondrial DNA) is more prevalent than previously thought and may include even the *Drosophila* fruit fly and the mouse.

[**]The mitochondrial genome also encodes two of the RNA molecules found in their own ribosomes and 22 of the transfer RNAs used in the synthesis of mitochondrial proteins. The significance of ribosomes and transfer RNA is discussed in the sections on protein synthesis.

Oxidative phosphorylation occurs as the electrons are passed along a chain of **respiratory cytochromes**, iron-containing proteins that are believed to constitute the elementary particles lining the cristae. As electrons are passed from NADH to 0_2, there are about seven intervening redox reactions. In a manner very similar to the way in which ATP is indirectly made during the light reactions of photosynthesis, three molecules of ATP are produced for each pair of electrons passed down the chain. In the case of mitochondria, the electrochemical gradient is established between the inner and outer compartments of the mitochondrion. Here, the protons accumulate in the outer compartment and the F_1 complex generates ATP by bringing protons through the membrane to the inner compartment.

The electrons passed down the chain from FADH enter the chain at a point that bypasses the first phosphorylation, and hence only two ATP molecules are generated from FADH. The net NADH that is generated in glycolysis can also be used to generate ATP, but the mitochondrial membrane will not allow NADH itself to pass through. In bringing the electrons through the membrane by means of alternative carriers some energy is lost and the NADH-equivalent molecules produce only two ATP molecules each.

These data now allow an enumeration of the useful energy the cell derives from the complete respiration of glucose.

Glycolysis:	Substrate-level phos-phorylations	= 2 ATP
	2 NADH outside the mitochondrion	= 4 ATP
Synthesis of acetyl-CoA from pyruvate:		= 6 ATP
Substrate-level phosphorylations in Krebs cycle:		= 2 ATP
2 FADH		= 4 ATP
6 NADH		= 18 ATP

At -7.3 Kcal/mole of ATP, the complete respiration of a mole of glucose salvages -263 Kcal of useful energy. Since the ΔG° of glucose is -636 Kcal/mole when its conversion to CO_2 and H_2O is looked at as a purely chemical event, the figure of -263 Kcal/mole represents a metabolic efficiency in harnessing this potential energy of 38%.

RESPIRATION AND ITS TIE-IN WITH GENERAL CELL METABOLISM

It was mentioned earlier that many other metabolic pathways are closely tied to those of glycolysis and the Krebs cycle. It is not consonant with the purposes of this narrative to investigate the metabolism of all biologically important compounds; nevertheless, one should become familiar in a general way with their relationships to respiratory catabolism.

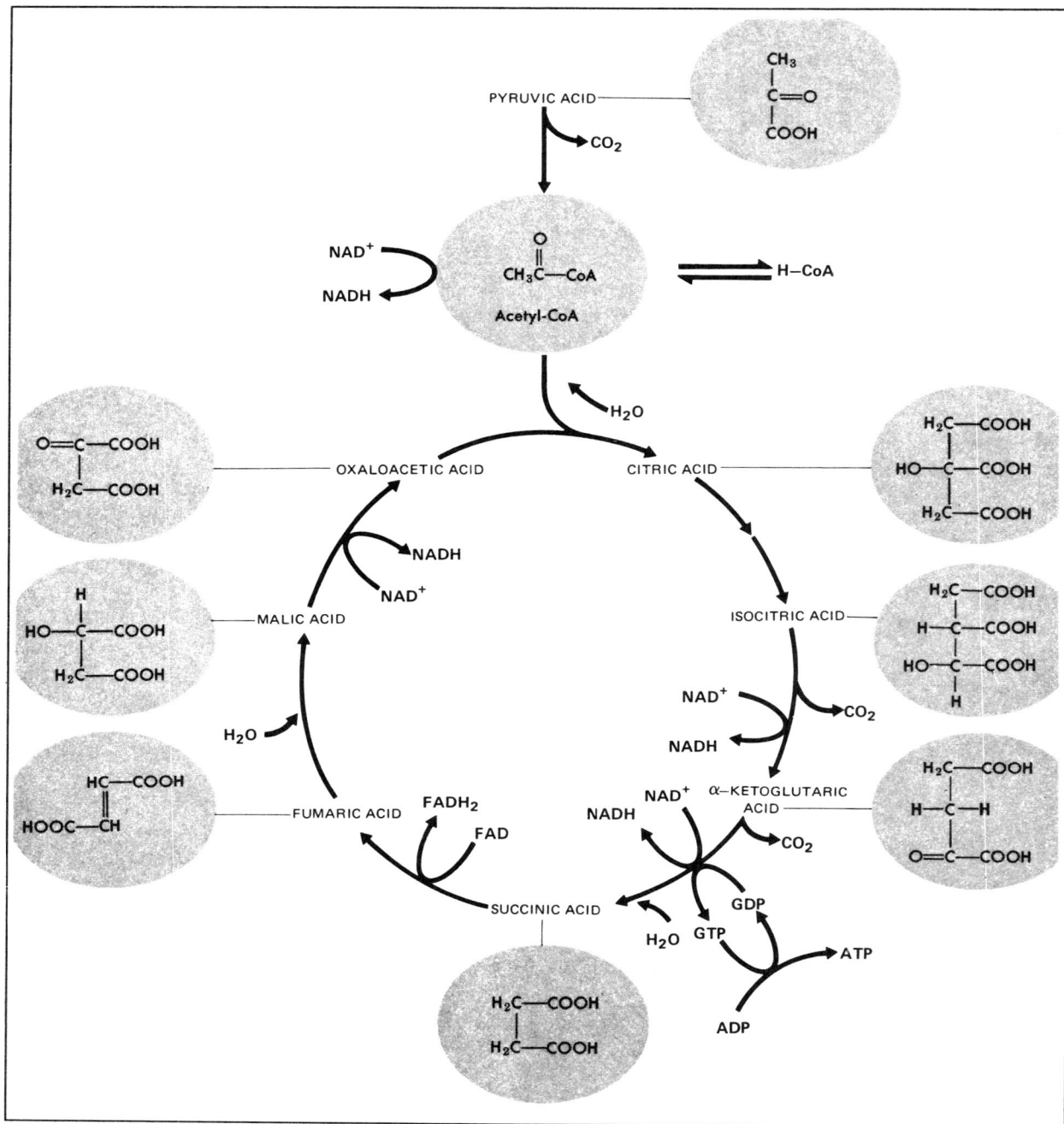

Figure 22.3 The intermediates of the Krebs cycle. From CELL BIOLOGY, THIRD EDITION by John W. Kimball. Copyright © 1984 by John W. Kimball. Reprinted by permission of the author.

The synthesis of polysaccharides (polymers of simple sugars like glucose) would obviously begin with glucose. Fatty acid and hence lipid synthesis begins with acetyl-CoA*, which reacts with CO_2 to form the key molecule in fatty acid synthesis, malonyl-CoA. Acetyl-CoA is also the precursor of cholesterol and other **steroids**. Many of the amino acids are derived from acetyl-CoA and most of the remainder proceed from various other intermediates of the Krebs cycle. From the amino acids, all proteins are derived as well as important parts of the purine and pyrimidine rings found in nucleic acids.

The catabolism of all proteins and fats leads directly back to the Krebs cycle into which the end-products are fed to derive energy for other syntheses. Often, therefore, there is a dynamic balance existing between the egress and ingress of intermediate metabolites within the Krebs cycle. When, however, the drain of these intermediates exceeds their recovery, it is still necessary to ensure that sufficient amounts of oxaloacetate are present to maintain the cycle. To accomplish this, the cell can, if needs be, activate a so-called **anaplerotic reaction** (the **Wood-Werkman reaction**), which generates oxaloacetate by the direct reaction of CO_2 with pyruvate. This mitochondrial reaction, which requires ATP, is a form of non-photosynthetic carbon fixation but is hardly competitive with the chloroplast.

REFERENCES

1. Margulis, L., 1971. Symbiosis and evolution. Sci. Amer. 225: 48 (Aug.).

---

*Acetyl-CoA cannot pass through the mitochondrial membranes to the cytoplasm. When the mitochondrial acetyl-CoA reacts with oxaloacetate to form citrate, the citrate may leave the mitochondrion and become reconverted to oxaloacetate and acetyl-CoA in the cytoplasm. This cytoplasmic acetyl-CoA can then be used in fatty acid synthesis, as outlined above.

LECTURE 23

PROTEINS AND HEREDITY

INTRODUCTION

Up to now we have tacitly accepted as fact that genes control the phenotype of an organism. Genes, however, are just stretches of nucleic acid within the chromosomes, which in turn are found within the cell nucleus. There is certainly nothing obvious about how a certain sequence of nucleotide pairs can specify what an organism will look like and how it will function. Besides replicating, recombining, and sometimes mutating, what do genes actually do with their time?

ALCAPTONURIA: AN IN-BORN ERROR OF METABOLISM

The early attempt to answer this question appears in the work of A. Garrod, a turn-of-the-century physician who was studying the human hereditary disease, **alcaptonuria**. People suffering from alcaptonuria are homozygous recessive for the causative gene. The most striking symptom they evince is the blackening of their urine on exposure to air. This color change is brought about by the presence in the urine of homogentisic acid, a compound not found in the urine of normal people. Most people break down homogentisic acid by means of the enzyme homogentisic acid oxidase; Garrod postulated that this enzyme is either missing or functionally defective in alcaptonurics. Few paid any attention to Garrod's work despite his serious and frequent efforts at publicizing it, but it remains one of the earliest pieces of evidence that ties defective (mutant) genes with missing or defective enzymes. (Strangely, it was not until 1958 that it was clinically proved alcaptonuria patients lack functional homogentisic acid oxidase.)

THE BEADLE & EPHRUSSI EXPERIMENTS[1]

It was not until the well-known studies of G. Beadle and B. Ephrussi of the mid-1930s that the problem was seriously tackled again. These investigators were working with two non-allelic genes of *Drosophila*, cinnabar (cn) and vermilion (v), both of which change the insect's normal dull red eyes to a bright scarlet. Since a normal red eye is really a mixture of two pigments, one a dull brown and the other a bright scarlet, both cn and v must affect the production of the brown (*sic*) pigment, since in mutants homozygous for either, only the scarlet pigment remains. We know now that the brown pigment is a large, non-diffusible polymer whose ultimate precursor is the diffusible amino acid, tryptophan. The first and last molecules in this biochemical sequence are connected by the intermediates, in order: N-formylkynurenine, kynurenine and 3-hydroxykynurenine—all of which are diffusible from cell to cell. At the time Beadle and Ephrussi carried out their work, however, it was known only that the brown pigment was a large non-diffusible molecule which was probably the end-product of a sequential synthesis involving simpler substances as intermediates.

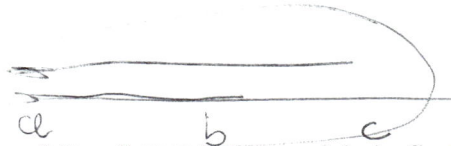

To follow the experiments of Beadle and Ephrussi it is first necessary to pick up a few facts about *Drosophila* embryology. *Drosophila*, like many insects, has a life cycle that begins with an egg, which hatches into a larva, which molts several times as it eats and grows larger, which then enters a stationary pupal stage, where it metamorphoses into an adult fly, which finally breaks its way out of the pupa case to begin the cycle all over again. The two molts that take place during the larval stages divide this period into what are known technically as the first, second and third instars. It is at the end of the third instar larval state that pupation occurs.

The structures of the adult fly obviously bear little resemblance to those of the larvae, and the startling metamorphosis of an ugly grub into a beautiful something-or-other has attracted a great deal of scientific attention along with its share of smarmy prose from lay writers. The development of the adult structures, on closer examination, turns out not to be so unprecedented after all, since each adult structure develops from a separate knot of tissue that existed in the earlier larvae all along. These clumps of tissue in the larvae are called **imaginal discs** (an **imago** is another term for an adult insect), and it is possible to specify exactly which disc will develop into which adult structure during metamorphosis. Sometimes it is even possible to dissect out a particular imaginal disc from one larva and to implant it into a second larva before its metamorphosis. The adult that emerges will then have a double set of structures for those determined by the particular disc that had been transplanted. It was this transplantation technique that Beadle and Ephrussi perfected and applied to their studies of the cinnabar and vermilion mutations.

The particular imaginal disc that interested Beadle and Ephrussi was the one determining the head and eye region of the adult. By careful excision of this disc from a third instar larva and its implantation into the caudal region of another, they first confirmed that the disc would metamorphose during pupation into a seemingly normal head, although one located somewhat inconveniently within the abdomen of the adult fly. (On the other hand, it is difficult to think of a "convenient" location for a second head!) What would happen, they then asked, if the discs were taken from donor larvae homozygous for cn or v, and these were transplanted into wild-type larvae? Would the implanted eyes remain scarlet in color or would influences from the host larvae enable the mutant implants to manufacture the brown pigment? For both cn and v implants, the answer was clear-cut: when transplanted to wild-type larvae, each produced an adult head whose eyes were perfectly normal in color.

The next question should now be obvious: if cn and v are normalized when transplanted to wild type hosts, what happens when a cn/cn disc is transplanted into a v/v host, and a v/_v disc into a cn/cn host? Their results were somewhat unexpected. The homozygous vermilion disc acquired the ability to synthesize the brown pigment and the eyes developing from it became normal in color; but the homozygous cinnabar disc remained (cinnabar) in phenotype. From just these data, Beadle and Ephrussi concluded that the biochemical pathway leading to the brown pigment must include several diffusible intermediates; that the basic effect of the cn and v mutations is to block this pathway at some point; that the location of the blocks is different for each of the two mutations; and finally that the blockage occasioned by homozygous v occurs at an earlier step in the pathway than that occasioned by homozygous cn.

The logic here is close, however, and may require some further explanation. Their conclusions rest primarily on the assumption that normalization (i.e., synthesis of the brown pigment) can occur only if the implanted disc can acquire by diffusion from the host the particular intermediates it needs in order to complete the blocked biochemically pathway all the way to the brown pigment. If the host is wild-type, it can supply any diffusible intermediate to the implant, since, being wild-type, it must be able to synthesize all of them. If, on the other hand, the host is itself impaired in the same biochemical pathway, it could supply a useful intermediate only if its own blockage occurred *beyond* that of the implant. The implant could then take this supplied intermediate from the host and carry it along its own biochemical pathway to the final pigment. Since, therefore, v/v discs are normalized in cn/cn hosts, homozygous cn must cause a blockage beyond the point where the homozygous v discs are themselves blocked.

Although Beadle and Ephrussi claim to have had in mind the hypothesis that missing or defective enzymes would account for these biochemical blockages, they did not explicitly make this connection in their publications at the time. In hindsight, which is always 20/20, we can clearly see the bold outlines of an explanation which would have homozygous v flies missing an enzyme needed for one of the conversions in the pathway, and homozygous cn flies missing an essential enzyme for a different and later step in the same pathway. It is now known definitely that (v) flies cannot convert tryptophan to N-formylkynurenine and (cn) flies cannot convert kynurenine to hydroxykynurenine.

THE BEADLE & TATUM EXPERIMENTS: THE ONE GENE-ONE ENZYME HYPOTHESIS[2]

There is certainly no reason to doubt that Beadle and Ephrussi did indeed have in mind, albeit cautiously, the idea that mutant genes result in defective enzymes, since a few years later Beadle, this time in collaboration with E. Tatum, made this idea explicit in their famous "one gene, one enzyme" hypothesis. To understand the experiments that led to this generalization, it is first necessary to know something about the common bread mould, *Neurospora crassa*.

Neurospora is a pink mould that can readily be cultured in the laboratory on very simple nutrients: a carbon source (e.g. glucose), inorganic salts and minerals, and one complex chemical, the vitamin biotin. Given just these substances, a wild-type *Neurospora* can manufacture all the complex proteins, carbohydrates, nucleotides, lipids, etc. that it needs for normal growth. Because it is pared down to the absolute essentials, such a no-frills growth medium is called a **minimal medium**.

The fluffy mass of the mould is called a **mycelium**; on close examination one sees that it is made up of a dense mat of individual strands called **hyphae**. After a period of growth, the mould will begin producing large numbers of asexual spores on special stalks called **conidiophores**. When these spores germinate, they produce a new colony of *Neurospora* genetically identical to the one that produced the spore. (The spores, in other words, are produced by mitosis, not meiosis; therefore, there can be no recombination or independent assortment to shuffle the genes or chromosomes.)

At the time Beadle and Tatum took up the study of *Neurospora*, nothing was known about its genetics. In fact, it had just become possible to grow large quantities of *Neurospora* in the laboratory, since the vitamin biotin had just recently been synthesized and made available in sufficiently reliable supply for these experiments to be undertaken. Moreover, not only did Beadle and Tatum choose an organism of unknown genetics, but they further decided to limit their attention to a class of mutations not then known to exist in *any* organism: the so-called **nutritional deficiency mutations**. In postulating the existence of such mutations, their reasoning went somewhat as follows:

1. Since normal *Neurospora* can make all their essential complex metabolites from sugar, salts and biotin, there must be a myriad of different enzymes at work in synthesizing them by means of a large number of biochemical pathways.

2. If any of these enzymes were defective or missing in a particular *Neurospora* colony, it would lose the ability to make for itself the end-product of the pathway wherein the enzyme was supposed to function.

3. In order for such a mutant colony to grow, it would have to be provided with that end-product, in addition to the regular nutrients given in a minimal medium.

4. If genes controlled the synthesis of enzymes, then at least some *Neurospora* colonies derived from mutant-containing spores should have lost the ability to grow on minimal medium, but should be able to grow if some particular nutrient was added to it.

Finally, by the way of definition, a minimal medium that has one or a very few particular additional nutrients added to it is called a **supplemented medium**. A medium that has a large array of vitamins, nucleotides, amino acids, cofactors, etc. added—any conceivable nutrient that the most genetically screwed-up mutant *Neurospora* might need—is called a **complete medium**. With these points in mind, we can now examine the actual experiments of Beadle and Tatum, which are very straightforward.

They first collected a large number of spores from wild-type *Neurospora* (i.e., *Neurospora* capable of growing on minimal medium) and irradiated them with ultra-violet light of a wavelength known to be mutagenic. Each irradiated spore was then individually transferred to a complete medium and allowed to germinate and grow. Even if this spore had suffered a nutritional deficiency mutation, it should still be capable of growing on complete medium, which, by definition, was larded with all sorts of goodies to circumvent the lethal effects of whatever deficiency the mould might have had. Of course, at this stage Beadle and Tatum didn't know whether it had such a deficiency or not—wild-type *Neurospora*, remember, will also happily grow in this super-nutrient medium—and so it then became necessary to take some of the spores produced by this pampered colony and attempt to grow them on minimal medium. If they grew on minimal medium, then no nutritional deficiency mutation had been induced in the original irradiated spore; if they didn't grow, they were on to something interesting.

The first irradiated spore which produced a colony capable of growing on complete medium but not on minimal was number 299. To determine precisely the nature of its nutritional deficiency, Beadle and Tatum then plated some of its own spores on various supplemented media and found that they would grow only if vitamin B_6 were present in addition to the usual minimal medium requirements. Spore 299, in other words, had acquired an induced nutritional deficiency mutation that rendered the resultant colony incapable of synthesizing its own vitamin B_6, which must then be supplied exogenously as a condition for growth. The reason for the missing B_6, it turned out, was the absence (or functional impairment) of one of the enzymes needed in B_6 synthesis.

The finding of the mutant spore 299 and the demonstration of its specific defect fully corroborated Beadle and Tatum's *a priori* belief in the existence of nutritional deficiency mutations. Soon they had many more with which to fortify their belief. Spore 1089, as a final example, grew into a colony incapable or manufacturing vitamin B_1, and it too was then shown to lack an enzyme activity essential in B_1 synthesis. In almost all cases, in fact, the cause of the nutritional deficiency was traced to a faulty or missing enzyme. (It was just a coincidence that the first two such mutations were vitamin deficiencies.)[3]

The importance of this work, therefore, is two-fold. First it introduced the concept of nutritional deficiency mutations, which were to become the most important genetic characteristics used in both fungal and bacterial genetic studies thereafter. Indeed, without the potential opened up by the discovery of this type of mutation, it is probable that neither fungal nor bacterial genetics could exist as significant research disciplines. The immediate impact of these studies, however, was on our views about what genes actually do in order to determine a phenotype. Since every nutritional deficiency seemed to be traceable to a defective enzyme function, Beadle and Tatum coined the phrase, "one gene, one enzyme", meaning that each gene in some manner determined the presence or activity of one enzyme. This became one of the most productive generalizations in modern biology and biochemistry—even though, as we shall see anon, it is not altogether correct.

Now that our attention has been directed to enzymes it is time to learn something about the class of important biological compounds to which almost all enzymes belong: proteins.

REFERENCES

1. Beadle, G. and B. Ephrussi, 1936. Genetics 21: 225-47.
   Beadle, G., 1974. Ann. Rev. Biochem. 43: 1-13.

2. Beadle, G. and E. Tatum, 1941. Proc. Nat. Acad. Sci. 27: 499-506.

3. Beadle, G., 1948. The genes of men and molds. Sci. Amer.. (Sept.).

LECTURE 24

STRUCTURE OF PROTEINS

"What is the secret of life?" I asked
"Protein," the bartender declared.
"They found out something about protein."

K. Vonnegut
Cat's Cradle

AMINO ACIDS

Protein is the most abundant substance found in the cell. Just as the nucleotide is the basic building block of nucleic acids, so amino acids are the fundamental units that make up proteins. Although only 20 different amino acids are actually used in the initial synthesis of any protein, some of these may, after incorporation, be modified to other forms, and consequently it is possible to isolate many more than 20 amino acids in nature.[*] The 20 amino acids used in protein synthesis have the common structure given below but differ in the "R" group attached as a side chain. The simplest "R" group is Hydrogen—which yields the amino acid **glycine**. Examples of other "R" groups among the 20 fundamental amino acids are also shown.

The **amino-terminal (N-terminal) end** of one amino acid can react chemically with the **carboxy-terminal (C-terminal) end** of another to form a **peptide bond**. The resultant dipeptide still has an N-terminal and a C-terminal end which can react with other amino acids to form a longer and longer chain called

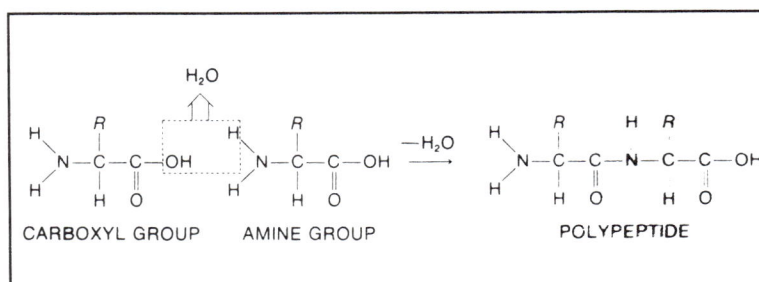

Figure 24.1 The carboxyl group of one amino acid reacts with the amine group of another to form a peptide bond.

---

[*] Before 1954 there was no agreement on which amino acids were primary and which derived from these by later modification. It was Watson and Crick who initially established the now-accepted standard list of 20. They did so over lunch at a Cambridge pub, The Eagle, in 1954, with little more to guide them than inspiration and self-assurance. Their audacious inspirations proved correct.

a **polypeptide**. The formation of each peptide bond liberates a molecule of water. The structures joined through these peptides bonds are then no longer amino acids, strictly speaking, but are instead oftentimes called **amino acid residues**.[*]

THE PRIMARY STRUCTURE

The particular sequence of amino acid residues of a protein is called its **primary structure**[1]. Somewhat arbitrarily, a sequence of fewer than 100 residues is called a polypeptide and one larger than that, a protein. There is no theoretical limit to the number of amino acid residues that can join together to form a chain, but ribonuclease, whose primary structure contains 124 residues, is considered a small protein; and DNA polymerase I, whose primary structure has about 1000 residues, is considered large for a single chain. Taking even the "small" protein as a standard size, we can see that the number of different primary structures possible with 20

---

[*] # 10 Year Old Medical Discovery Finally Available...

Amazing Amino Acids

Free Form Amino Acids are the building blocks of life itself. They produce over 1600 chemical related substances on the body—muscle, blood, hormones, enzymes, skin, hair, even neuro-transmitters....

Taken on an empty stomach with a glass of water, Amino Acids help:
☐ Minimize stress, anxiety, insomia, depression
☐ Increase mental and physical endurance
☐ Build and repair body rapidly
☐ Improve immune system
☐ Stabilize blood sugar
☐ Improve skin and hair texture
☐ Protect body from smoking and drinking
☐ Make dieting easier
☐ Stimulate growth hormone release
☐ Clean toxins from body
☐ Enhance sexual activity ... and much more!!

From an advertizement in *The Advocate* (June 26, 1984).

Humans normally synthesize in their own bodies 11 of the 20 amino acids. The remaining 9—known as **essential amino acids**—(hisdidine, isoleucine, leucine, lysine, methionine, phenylalanine, threonine, tryptophan, and valine) must be supplied through food intake, which is the purpose of protein in the diet. A normal diet would supply all of them in the necessary quantities. (Infants and young children make insufficient quantities of arginine, and for them it is also an essential amino acid.)

different building blocks is astronomically large; in this case 20^{124} different proteins all the same length would be possible. And of course proteins are *not* all the same length, nor, as we shall see, are they necessarily composed of just a single chain of amino acid residues. Since there are "only" 10^{10} or 10^{12} different polypeptides or proteins found in the whole of nature, just a minute fraction of the diversity that is possible with proteins has actually been exploited. On the other hand, if just a single copy of all possible proteins were actually made, the entire known universe would be packed with proteins many times over.[2]

ADDITIONAL LEVELS OF STRUCTURE

A linear sequence of amino acid residues need not imply a linear structure. Spontaneously the primary sequence coils up into an alpha-helix that is stabilized by extensive hydrogen bonding between the carbonyl group of one peptide linkage and the amino group of another four removed from it. This alpha-helix constitutes the **secondary structure** of a protein or polypeptide. It is sometimes irregular as a result of the conflicting requirements of certain "R" groups and as a result of the presence of the atypically constructed amino acid residue, proline, which kinks the helix. The precise secondary structure, therefore, will depend upon the particular primary structure. Since the alpha-helix forms spontaneously, it must be assumed to be the most stable configuration for the polypeptide.

Some proteins, particularly those that compose insoluble fibrous material such as silk and keratin, have a different kind of secondary structure called the **beta configuration** or the **pleated sheet**. In this case, two or more separate polypeptide chains align themselves beside each other and form hydrogen bonds. Alternatively, the same polypeptide can fold back on itself so that hydrogen bonds form between different parts of the same molecule. The "R" groups project above or below the plane of the "sheet" fashioned by the repeated amino-C-carboxy portions of the amino acid residues. The "pleated" refers to the fact that the plane of the sheet is corrugated or zig-zag as, for example, in a pleated skirt.

Few proteins are composed of a simple alpha-helix. Instead we find the alpha-helix folded over on itself in various complicated ways to form a protein with **tertiary structure**. This folding may obliterate evidence of an alpha-helix except for the longer straight runs of amino acid residues within the globular protein structure. This tertiary structure is also stabilized by hydrogen bonds and other weak or strong interactions among the various "R" groups. One very significant "strong" interaction is an actual covalent bond (a **disulphide bridge**) that may form between two **cysteines** in different parts of the molecule. Again, like the secondary structure, the tertiary structure of a polypeptide is precisely determined, and solely determined, by the primary structure (excluding effects of the environment like temperature, pH, etc.); it is assumed spontaneously as the configuration that maximizes the entropy of the molecule and its aqueous surroundings.

Some proteins, especially those with a molecular weight in excess of 50,000D[*], are composed of more than one (but usually an even number of) polypeptide subunits. The constituent subunits of an oligomeric protein are conjoined in a very precise way and stabilized by a concatenation of weak forces or disulphide bridges among the subunits. Such proteins are said to have a **quaternary structure**. Here as in all levels of protein structure, the exact way in which subunits interact depends ultimately upon the primary structures of the constituent polypeptides.

Finally it should be noted that some proteins are chemically combined with other classes of molecules such as lipids, sugars, and even inorganic metals. Haemoglobin is a well-known example of a protein that is complexed with a non-protein substance, the porphyrin haem ring containing iron. Such proteins are said to be examples of **conjugated proteins**; both parts of the molecule must be present for it to function.

All of the biological properties of proteins (e.g., the ability of some to act as catalysts—enzymes—in biochemical reactions) depend upon the precise maintenance of the tertiary or quaternary structure. Anything that even slightly disturbs this structure may have profound consequences for the biological activity. One common disturbing influence is temperature. Heating a protein will cause the tertiary or quaternary structure to break down or **denature**, usually in an irreversible manner. A common and obvious manifestation of the heat denaturation of proteins occurs on boiling an egg. This same example also provides an insight into the difficulty of renaturing (i.e., "unboiling") denatured proteins.

Proteins begin assuming their higher levels of structure as the primary structure is assembled one amino acid residue at a time, a process that will be examined in detail in a later lecture. To assist in the correct folding of this nascent polypeptide a group of protein molecules known as **chaperones** is enlisted. For example, one type of chaperone attaches to the hydrophobic portions of the nascent polypeptide to ensure that they are folded properly into the internal part of the molecule once its final form has taken shape. Meanwhile these chaperones also guard the hydrophobic sections against chance "misalliances" with other parts of the same or other peptides that would prevent their winding up in the proper internal location when the entire molecule is completed. There are several distinct types of chaperones, each with a specific role to play during the folding process. When their task is done, they exit from the scene, but why or how is not yet understood. Despite their importance in protein folding, however, it should be emphasized that chaperones do not dictate the shape of proteins but only facilitate the attainment of their thermodynamically most favorable and stable configuration.

DETERMINATION OF THE PRIMARY STRUCTURE

The first polypeptide whose primary structure became known was insulin,[3] an oligomeric structure composed of 21 amino acid residues in one strand and 30 in the other. The techniques pioneered

---

[*]One **Dalton (D)** is one molecular weight unit. It is equivalent to the mass of the Hydrogen atom.

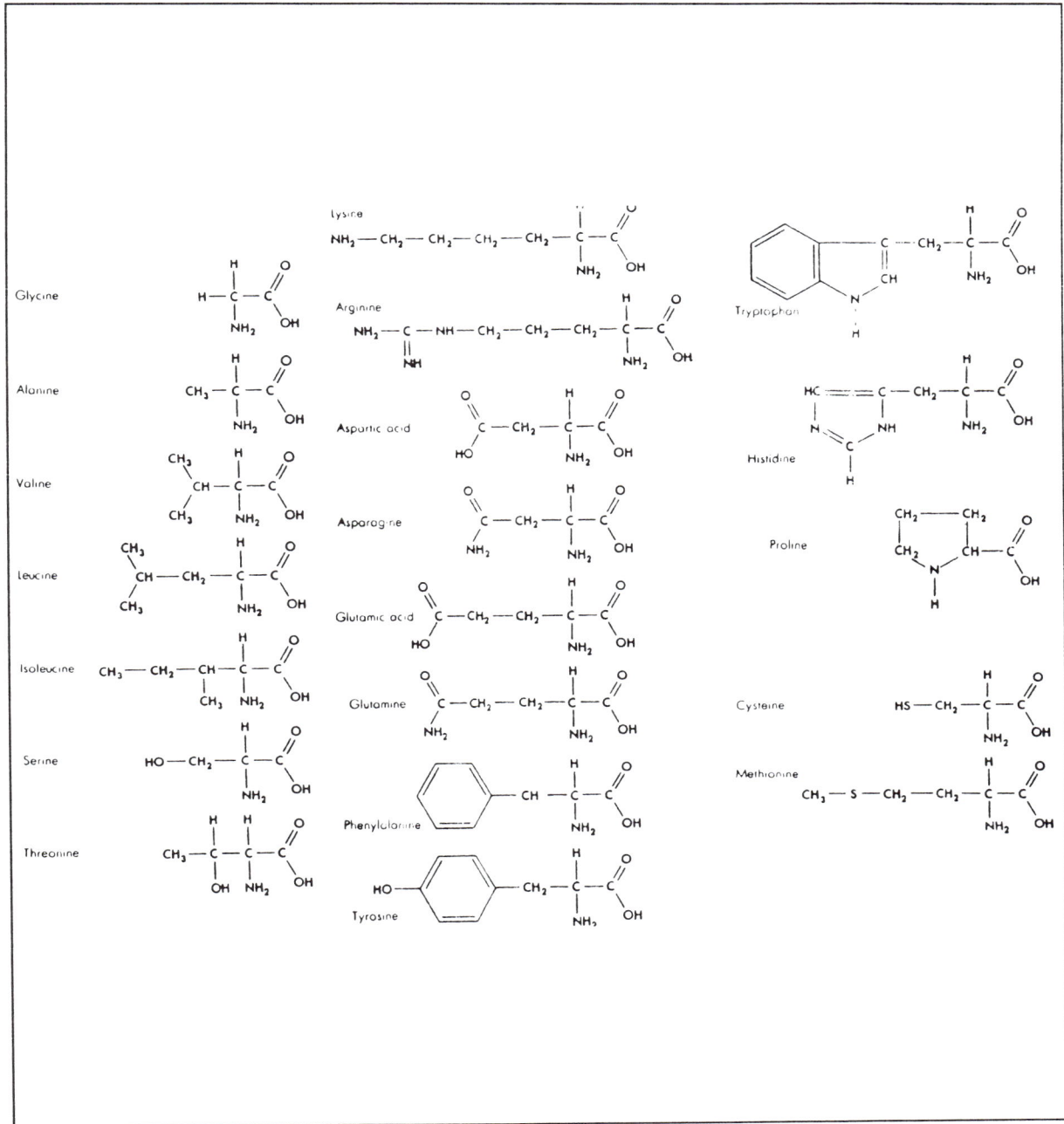

Figure 24.2 The twenty primary amino acids found in proteins. (Additional amino acids can also be found in proteins, but they are derived from these 20 by modification after their incorporation.)

by F. Sanger involve the splitting of each subunit by enzymes (proteases) that attack known specific peptide linkages, the separation of the resultant oligopeptide fragments by chromatography, and then the further analysis of these fragments by conventional biochemical techniques. Now it is possible almost routinely to analyze the sequence of very lengthy proteins, but given a primary structure a reliable prediction of its secondary and tertiary structure remains a task beyond the capability of our best brains and computers at present.

From the foregoing discussion it should be clear that the biological activity of a protein is dependent upon its precise three-dimensional shape and this in turn is mandated by its particular sequence of amino acid residues. To make this point concretely, we shall look briefly at two particular proteins, human haemoglobin and the one responsible for the genetic disease cystic fibrosis, and examine how their biological properties change with changes in their primary structure.

Note should be made of the common modern method of separating and purifying protein molecules for further analysis.. It goes by the name of **SDS-PAGE**, which stands for "sodium dodecylsulphate polyacrylamide gel electrophoresis". Fortunately is it a far less intimidating procedure than it sounds. SDS is a detergent; adding it to protein causes the protein to unfold and a large number of the negatively charged SDS molecules to adhere at fairly uniform points along the extended polypeptide chain. This imparts a negative charge to the polypeptide, the strength of which will be directly related to its length. A polyacrylamide gel contains a myriad of small, molecular-sized channels. It can be thought of as a "molecular sieve". When a mixture of SDS-treated proteins is placed at one end of a block of the gel and an electric current applied across the block, the charged polypeptide chains will be attracted to the positive pole and will migrate through the gel to reach it. Smaller polypeptides can travel faster through the maze than can longer ones; hence, the polypeptide mixture will become fractionated into bands that correspond to the length of the polypeptide, which in turn is very much related to its molecular weight. By using polypeptides of known molecular weight that are run simultaneously with a purified peptide of unknown molecular weight, the molecular weight of the latter can be estimated with great accuracy.

THE HAEMOGLOBIN STORY[4]

Haemoglobin A (HbA; the A stands for Adult) is a tetrameric conjugated protein composed of a pair of so-called alpha chains and a pair of beta chains. This is usually written in informal shorthand in the following manner: $\alpha_2 \beta_2$. The alpha and beta chains have been sequenced in their entirety; the former has 140 amino acid residues and the latter 146. Human fetuses have a somewhat different form of haemoglobin, HbF, that contains two gamma chains instead of the two beta chains of the adult form[*]. Beginning at birth, the HbF is replaced by HbA so that adults have 1% or less of their haemoglobin as HbF.

---

[*]In humans it is now known with certainty that two identical adjacent genes on the short arm of chromosome 16 code for the α chains and one gene on the short arm of chromosome 11 codes for the ß chains. The gamma chains of Hb F are of two types that differ in a single amino acid at position 136. The two genes for the gamma chains, one for each type of gamma chain, are also located on chromosome 11, very near the two identical genes for the beta chain.

A number of hereditary diseases affect the functioning of Hb, perhaps the best-known being **sickle-cell anaemia**, a disease which is inherited in a simple mendelian fashion as a homozygous recessive. A homozygote suffers severe loss of oxygen-carrying capacity and a greatly reduced life span. On microscopic examination the red blood cells are seen to have an abnormal sickle-like or crescent shape which leads to widespread clogging of capillaries and hence impaired organ function and early death. The molecules of HbS tend to bunch together, causing the usually disc-shaped erythrocyte to become distorted into a sickle.

In 1949 L. Pauling and H. Itano found that the haemoglobin they extracted from sickle cells had a mobility in an electric field slightly different from that of normal haemoglobin A. (This technique of separating molecules on the basis of differences in their intrinsic electrical charges at a given pH is the conventional type of electrophoresis.) When the techniques of detailed sequence analysis of polypeptides became applicable to lengthier chains, V. Ingram discovered that HbS (i.e., the form found in sickled cells) had two normal alpha chains, but both beta chains were slightly different from those found in HbA: each had a single specific substitution of one amino acid for another. The sixth amino acid in HbA beta chains is glutamic acid, but in HbS valine occurs at that position. In other words out of a total of 572 amino acid residues in the tetramer only two differ from the normal, yet this seemingly trivial difference is sufficient to condemn to premature death those whose only fault lies in position six of their haemoglobin beta chains.[*]

This case study of haemoglobin also uncovers two other very important facts. Since diseases of the haemoglobin molecule have a well-understood genetic basis, and since all higher orders of protein structure are dependent upon the primary structure of the molecule, genes, it would appear, need only specify the primary structure of polypeptides to control their higher orders of structure and hence their biological activity. The effects of a mutation, here the one conferring the sickle cell anaemia phenotype, are really manifested at the molecular level by a change in the amino acid sequence of a polypeptide. Genes, basically linear sequences of nucleotides along a DNA molecule, control the linear sequence of amino acids in the polypeptides they specify.[**]

---

[*] For unknown reasons the drug hydroxyurea seems to stimulate the production of HbF in some adults so that HbF can come to constitute 10 to 15% of their total haemoglobin. Since the defect in sickle cell anaemia resides with the beta chains of HbA, stimulation of HbF production in patients with this disease offers some measure of cure.

[**] Although it is true, as in the sickle cell anemia example cited here, that a single amino acid substitution within a protein can have severe and far-reaching consequences to the shape and hence the activity of a protein, it must not be concluded that *all* amino acid substitutions within proteins have equally devastating consequences. So far there have been discovered about 450 different variants of the human haemoglobin molecule, almost half of which are fully functional and therefore without effect on the phenotype. A change in the DNA which is not detectable by a change in the phenotype is called a "silent" mutation.

145

The second point is a very obvious one. Haemoglobin, although a genetically determined protein, is not an enzyme. Beadle and Tatum's generalization, "one gene, one enzyme", should then be read as "one gene, one polypeptide". Later we shall see that even this is too restrictive a definition of gene function.

THE CYSTIC FIBROSIS STORY

A more recent example of the exquisite relationship between primary structure and function in proteins comes from an analysis of the mutant gene causing **cystic fibrosis** in humans. This mutation is the most common serious autosomal recessive mutation in the Caucasian population. An estimated 5% of the population carries this mutation, leading to about one affected individual in every 2000 births. Affected individuals have a severe chronic pulmonary disorder, pancreatic malfunction, and other abnormalities. The actual protein responsible for the disease has not yet been isolated, but, astonishingly, the gene for the protein has been not only isolated (it resides on the long arm of chromosome 7) but also sequenced. Knowing the sequence of nucleotides in the gene and the genetic code (discussed in later lectures) permits investigators to write down the primary sequence of the sought-after cystic fibrosis protein. The nucleotide sequences from normal and affected individuals can then be compared, revealing that the disease is caused by a small deletion (three adjacent nucleotides are missing in the mutant gene) and this in turn causes a single amino acid (phenylalanine) to be missing from position 508 of the mutant polypeptide—just one amino acid missing out of a total of 1480 in the polypeptide.[5] (It is not yet known with certainty what role the normal protein plays in the cell, although it is thought to reside in the cell membrane and perhaps be involved in transport.)

REFERENCES

1. Doty, P., 1957. Proteins. Sci. Amer. (Sept.).

2. Doolittle, R., 1985. Proteins. Sci. Amer. 253: 88-96 (Oct.).

3. Thompson, E., 1955. The insulin molecule. Sci. Amer. (May).

4. Perutz, M., 1964. The haemoglobin molecule. Sci. Amer. (Nov.). This article also gives a simplified account of the principles of X-ray diffraction studies.

5. Riordin, J., et al. Identification of the cystic fibrosis gene: Cloning and characterization of complementary DNA. Science 245:1066-1073 (1989).

LECTURE 25

WHERE PROTEINS ARE MADE: 1

QUESTIONS OF SITE AND TEMPLATE

In the early studies on the mechanism of protein synthesis three questions explicitly or tacitly lay behind much of the work: What is the cellular *site* of protein synthesis? What is the *template* for protein synthesis? and Are the site and the template one and the same thing? Many years of experimentation went into answering these questions and others that were raised in the process. We can examine but a few of the actual experiments. The ones chosen for closer study in this account are thought (by this author) to be the most influential in shaping opinion or to provide the best illustration of techniques that have come to play a significant role in modern biology.

If genes specify the primary structure of proteins, one might easily conclude that these proteins are assembled directly on the genes that specify them. If this were true, eukaryotic organisms would carry on the bulk of protein synthesis in the nucleus. With some unicellular (but still eukaryotic) organisms it is possible to excise the nucleus from the cell and then to test for protein synthesis in the enucleate remains. The umbrella-shaped unicellular alga, *Acetabularia*, is such an organism. Its single nucleus is always found in the rhizoid at the bottom of its stalk, where it becomes vulnerable to anyone who wants to separate it from the rest of the cell. J. Hammerling and others found that enucleate *Acetabularia* continue for awhile to incorporate radioactive amino acids into protein at almost normal rates; the nucleus, therefore, cannot be the exclusive site for protein synthesis nor the genes the immediate template.

Since protein synthesis occurs primarily in the cytoplasm two questions immediately arise: In what part of the cytoplasm does it take place? and How do the genes, which of course are nuclear, exercise control over this synthesis? One undoubtedly significant datum was the long-standing observation, usually credited to J. Bracket and T. Caspersson, that those cells that most actively synthesize proteins are precisely those most rich in RNA. Since most of the cellular RNA is tied up in ribosomes, these latter became suspected of being the cytoplasmic site of protein synthesis. Since ribosomes are also found in the nucleus, perhaps they are also involved in the significant amount of nuclear protein synthesis that unequivocally does take place there.

IN VITRO PROTEIN SYNTHESIS

The suspicion became proof when in 1959 K. McQuillan, R. Roberts and R. Britten developed a good technique for isolating *E. coli* ribosomes *in vitro* and testing their ability to support protein synthesis outside the cell. They added radioactive S^{35} (in the form of the sulphate ion) to the *in vitro* mixture, which also contained most of the "soluble" cytoplasmic components of

147

the *E. coli*. They found that the S^{35} first became incorporated into amino acids (presumably methionine and cysteine) and that these radioactive amino acids then became attached to the ribosomes, where they were incorporated into nascent proteins. After a few minutes, these radioactive proteins were released from the ribosomes and could be found with the other non-particulate matter in solution.

The events involved in *in vitro* protein synthesis were examined more carefully the following year by A. Tissieres, D. Schlessinger and F. Gros.[1] Again using ribosomes extracted from *E. coli* but with pre-made radioactive amino acids, they confirmed the requirement for ribosomes, as well as a requirement for the *E. coli* supernatant (i.e., cytoplasm), ATP and magnesium ions. They also observed that protein synthesis *in vitro*, even under the best of conditions, was a very transitory phenomenon, having a "half-life" of only about four minutes. Even though the incubation mixture was still well-supplied with free amino acids, after a relatively short while no further protein synthesis occurred. Since adding DNAase to the mixture curtailed the active period of protein synthesis even further, the notion took hold that some essential feature of this synthesis, perhaps the template, i.e., the information molecule that provides the precise instructions for the amino acid sequence, is unstable and must be continuously replenished by the master set of instructions encoded by the nuclear DNA. As the ribosomes were clearly the site at which protein synthesis was occurring, it became the fashion to assume, not unreasonably, that the ribosomes also provided the exact instructions necessary for specifying the sequence for amino acid polymerization. Were this the case, the ribosomes would also have to serve as the unstable template for protein synthesis—Q.E.D.

THE ZALOKAR EXPERIMENT[2]

The likelihood that the ribosomes were "programmed", so to speak, at the time of their synthesis to produce thereafter specific proteins was considerably strengthened by some ingenious work by M. Zalokar with the fungus *Neurospora*. The hyphae of this mould are aseptate (i.e., lacking transverse cell walls) and multinucleate. If a single hypha is painstakingly fitted into a narrow glass needle and then centrifuged at high speed, the contents of the hypha will settle in the following top-to-bottom order: fats-vacuoles-cytoplasm-nuclei-mitochondria-ergastoplasm (i.e., membrane-bound ribosomes)-glycogen.

Having established that the various cell components are separable by centrifugation, Zalokar then undertook to determine the cytological origin of *Neurospora* ribosomes. Since ribosomes are the principal repository of RNA in the cell, he allowed growing *Neurospora* to incorporate H^3-labelled uridine for a few seconds before selecting a hypha for centrifuging. Examination of the layered cell contents revealed that virtually all the tritiated uridine was found in the nuclear fraction. If, in a follow-up experiment, he gave H^3-uridine for the same brief time interval and then allowed the *Neurospora* to continue growing in nonlabelled uridine for one hour before centrifugation, he discovered that the nuclei and what he called the ergastoplasm (i.e., the E.R.-bound cytoplasmic ribosomes) were both heavily labelled. The ineluctable conclusion from this work is that ribosomal RNA molecules (and probably whole ribosomes) are manufactured in the nucleus and later migrate to the cytoplasm, where they subsequently carry out the business of protein synthesis. The temptation was also to believe that at the time of their manufacture, the ribosomes were programmed to produce specific proteins and to be, therefore, the handmaidens of the genes in the cytoplasm.

RIBOSOMES[3] [4]

From the preceding account, it should be clear that no matter what their precise function, ribosomes are intimately involved in protein synthesis. We obviously need to know more about them.

Ribosomes in *E. coli* are free in the cytoplasm, whereas in eukaryotes they are found both free in the cytoplasm and also bound to the endoplasmic reticulum (or "ergastoplasm", in Zalokar's dated terminology). *E. coli* ribosomes, of which there are perhaps 15,000 per cell, measure 180Å in diameter, but the most common way of referring to them is by their "S" value of 70. "S" in this instance stands for **Svedberg unit** and is a measure of the velocity at which a particle or large molecule will pass through a standard solution under the influence of a standard centrifugal force. This velocity is determined largely by the molecular weight of the particle, but also significantly by its shape.

Since much of the work to follow has been carried out with bacterial ribosomes, these will be described in greater detail. They appear vaguely snowman-shaped, but under physiological concentrations of magnesium ions ribosomes are more likely to have separated into their two roughly spherical constituent subunits known familiarly as the 30S and 50S particles. (The obvious lack of additivity of the S-values—recall that the integrated particle is 70S—exemplifies the dependence of the sedimentation velocity on shape as well as on molecular weight.) Each subunit is composed of about 60 to 65 percent ribosomal RNA (rRNA); the remainder is protein. The 50S subunit consists of two distinct pieces of RNA (the 23S and 5S rRNAs) and about 35 different polypeptides; the 30S subunit has a single RNA piece (the 16S rRNA) and about 20 different polypeptides. Since low magnesium concentrations favor the disintegration of the 70S particle, ribosomes can be dissociated or reassociated at will by varying the Mg^{++} concentration of the preparation. Early *in vitro* studies of protein synthesis tended to be carried out fortuitously at high, unphysiological Mg^{++} concentrations. This resulted in a number of "lucky breaks" that originally simplified matters to a degree where it became possible to imagine we understood what was happening, but it also introduced its share of artifacts. We will on occasion have call to refer to both the opportunities and drawbacks of experimenting at relatively high Mg^{++} concentrations.

Cytoplasmic ribosomes in eukaryotes are somewhat larger then those in bacteria, having a diameter of 220Å and an S-value of 80. The dissociated subunits are 40S and 60S. The 40S subunit has a single 18S piece of RNA, whereas the 60S subunit has three: 5S, 5.8S, and 28S. By contrast, the ribosomes found in mitochondria and chloroplasts of eukaryotes tend to resemble those of bacterial cells.

SEARCH FOR THE TEMPLATE

If ribosomes are the site of protein synthesis, are they also the template? One argument against their being the template is the relative constancy of their RNA base ratios across bacterial species, even when the DNA base ratios of these species differ markedly. With such large variations in the make-up of the master templates, the genes, it would seem reasonable that this diversity be reflected *pari passu* in whatever carries out their instructions in the cytoplasm. All this assumes, of course, that it would be the RNA and not the protein component of ribosomes that serves as the functional template.

A second approach was to examine the stability of ribosomes during bacterial growth. Since the cytoplasmic template appeared to have a very short half-life—if (and at that point it was still a big if) the *in vitro* studies were an accurate reflection of events *in vivo*—the ribosomes, if they were the template for protein synthesis, should themselves be unstable. To test this idea, C. Davern and M. Meselson[5] (1960) grew *E. coli* for many generations on the heavy isotopes, C^{13} and N^{15}, to the point where all of the ribosomes acquired a pronounced density label. When the ribosomes were extracted and banded in a CsCl gradient, a single "heavy" band of ribosomes was obtained. This band was widely separated in the gradient from a reference band of added light (i.e., normal) ribosomes. The investigators then transferred a portion of the heavy *E. coli* culture to a medium containing only the light carbon and nitrogen isotopes, C^{12} and N^{14}. At various times during the subsequent growth of the bacteria a portion of the growing culture had its ribosomes extracted and banded in CsCl. A new band of fully-light ribosomes appeared in their gradients as soon as appreciable growth took place in the light medium, and as the bacteria continued their growth in light medium through 33, 66, 100 and 133 minutes (i.e., 0.7, 1.3, 2.2 and 3.3 generations, respectively), this new light band increased dramatically in size relative to the size of the heavy band. Nevertheless, the heavy band did not disappear nor did it decrease in absolute size, given that the number of bacterial cells was always increasing and that new light ribosomes were continuously being synthesized. More significantly, there was no evidence of a ribosomal band of hybrid density, which should have appeared if ribosomes were "turning over": that is, if old ribosomes were degraded and their component atoms being used in the synthesis of the new. Ribosomes seemed very stable indeed.[*]

---

[*]The Davern and Meselson experiment was fairly influential in turning people away from the idea that ribosomes were the template for protein synthesis, but within a year even better proofs were obtained from other labs. As it turned out, later work showed that the conclusions reached by Davern and Meselson, although correct, were based on a number of artifacts of their experimental procedures and hence were scientifically invalid. For one thing, their ribosomal extraction procedures resulted in the dissociation of the ribosomes before they were put into the cesium chloride gradient; this resulted in there being only a single band of ribosomes in the CsCl gradient whereas, as we shall see with later experiments, two would have been expected in retrospect. These two bands are usually called the "A" (more dense) and "B" (less dense) bands. Davern and Meselson obtained only A-band ribosomes. These A-band particles, unfortunately,

(continued...)

150

REFERENCES

1. Tissieres, A., et al., 1960. Proc. Nat. Acad. Sci. <u>46</u>:1450

2. Zalokar, M., 1959. Nature <u>183</u>: 1330.

3. Nomura, M., 1969. Ribosomes. Sci. Amer. <u>221</u>: 28 (Oct.).

4. Lake, J., 1981. The ribosome. Sci. Amer. <u>245</u>: 84-97 (Aug.).

5. Davern, C. and M. Meselson, 1960. J. Mol. Biol. <u>2</u>: 153.

---

[5](...continued)
are in a sense artifacts, since they do not represent anything that exists *in vivo*.

They are actually former 50S and 30S subunits which have lost, through exposure to CsCl, 20% of their protein; these **core particles** now have S-values of 42 and 23, respectively, and obviously cannot reassociate to form functional 70S ribosomes ever again. It would still be possible, therefore, for 20% of the ribosomal protein to have "turned over" during the time course of Davern and Meselson's experiment, and this would have escaped detection by their methods. As it happens, later work showed that this 20% does *not* turn over, and consequently their original conclusion is valid.

The B-band ribosomes, which Davern and Meselson did not find but which do show up in the work of other investigators, appear to be unchanged and probably still functional 70S ribosomes. Apparently the un-dissociated (70S) ribosome resists the action of CsCl in stripping away the proteins that are only loosely bound to the individual free subunits.

This lengthy footnote is by way of illustration of the fact, startling to some, that science—particularly molecular genetics—can often be usefully advanced by incorrect or artifactual results. We have occasion to witness many other instances of this phenomenon in this account.

LECTURE 26

WHERE PROTEINS ARE MADE: II

THE BRENNER JACOB & MESELSON EXPERIMENT (PART 1)

The "one gene, one ribosome, one protein" hypothesis, already looking shaky, was put to a critical test by S. Brenner, F. Jacob and M. Meselson[1] in 1961. Their choice of an experimental system that seemed almost perfectly suited to their needs and their clever exploitation of very simple techniques in order to achieve unambiguous solutions to very important questions are all features of what a scientist means when he calls an experiment "beautiful". Their work is presented here in two parts separated by some background information needed to follow the reasoning that led to their carrying out the second series of experiments.

From earlier sections of these notes we gained some insight into the events that take place inside an *E. coli* cell infected by a T-even phage. Besides monitoring the infected cell for the appearance of the first intact phage, one can also examine the cellular metabolism of various classes of compounds to see how they have been affected by the infection. Such studies reveal, for example, that global protein synthesis continues more or less unabated until near the end of the latent period. The specific types of protein produced do, however, undergo drastic change shortly after the phage invade. Bacteria-specific protein synthesis drops off very quickly; after about 10 minutes almost none of it still survives. Accompanying this fall-off in bacteria-specific protein synthesis there is a concomitant rise in the synthesis of phage-specific proteins—proteins which the cell has never before been called upon to produce. It is as if the protein-synthesizing machinery of the host cell has been hijacked by the bacteriophage and thereafter made to serve this new master. Since the specific instructions needed for the synthesis of these new proteins could not have been in the cell before the infection, Brenner and his colleagues first set about to settle two obvious questions: Are these phage proteins made on ribosomes, and, if so, did these ribosomes exist before the infection or were they made only afterward? If they were made before the infection, the ribosomes certainly could not have been "congenitally" programmed to make phage proteins and could not, therefore, be the template for phage-specific protein synthesis.

To establish the banding pattern of ordinary *E. coli* ribosomes, the investigators extracted them under conditions of fairly high Mg^{++} concentration and after density gradient centrifugation in CsCl found two bands, A and B. The denser A-band was composed of inactive subunits (or possibly core particles; see footnote in Lecture 25) whereas the less dense B-band contained the

70S ribosomes[*]. There was good separation achieved between these two bands. To determine whether either of these types of particles was active in phage-specific protein synthesis, *E. coli* were infected with T_4 phage and simultaneously transferred to a medium containing $(S^{35}O_4)^{-2}$ ion to label any proteins being synthesized. At the end of three minutes, the ribosomes from part of the culture were extracted and banded in CsCl; the remainder of the cells was given an eight-minute chase of cold $(SO_4)^{-2}$ and then these cells too had their ribosomes removed and banded.

Immediately at the end of the three minute period virtually all the S^{35} radioactivity associated with ribosomes was in the B-band (70S) particles. Eight minutes later much of this activity had been chased out of the ribosomes altogether and was found in the soluble cell fraction as completed, phage-specific protein. The synthesis of phage protein, therefore, takes place on 70S ribosomes, just as does the synthesis of any other protein in the cell.

These data, however, do not indicate whether these active ribosomes are made before or after infection. To do this, some means had to be found to distinguish pre-existing ribosomes from any that may be made after infection.

Cells were grown for many generations in C^{13} N^{15} P^{32} medium and their dense, radioactive ribosomes then extracted and banded in the same tube as a larger quantity of light, non-radioactive ribosomes. Two bands of radioactive particles were obtained (Heavy-A and Heavy-B) as well as two bands of normal particles (Light-A and Light-B). Because the density shift occasioned by the heavy isotopes was of the same order as that separating A and B bands of particles, the HB band appeared coincident with the LA band. Since the investigators had already proved (see above) that A-band ribosomes are not active in protein synthesis, any evidence they later gather of protein synthesis occurring on the LA band can be interpreted, therefore, as really occurring on the coincident HB band. The added unlabelled ribosomes simply serve as markers for the presence of the HB ribosomes, which are not present in sufficient quantity to be detected on their own by the U.V. absorption method employed here.

Now that a way could be devised to distinguish pre-infection ribosomes from any made after infection, the following experiment was performed. Cells were grown for many generations on the heavy isotopes C^{13} and N^{15}. When all their components had acquired the density label, the cells were switched to a light medium of C^{12} and N^{14}. Simultaneously they were infected with T4 phage, and S^{35} was added to the medium to label any protein synthesis taking place. At the end of three or four minutes, the ribosomes from the infected cells were extracted, mixed with a large quantity of normal *E. coli* ribosomes and banded in CsCl. All of the S^{35} label appeared associated with the reference LA band; that is, was actually associated with the HB band. Since these heavy 70S ribosomes could only have been made before infection—in other words, before

---

[*]Do not, incidentally, confuse density with molecular weight here. The A-band particles are lighter in weight but nonetheless more dense than the B-band particles. Cesium chloride gradients discriminate on the basis of density, not molecular weight.

the cell was called on to manufacture these alien phage proteins—the ribosomes, while serving as the site for the synthesis of phage proteins, could not possibly be also serving as their template.

If not the ribosomes, then what? Brenner, Jacob, and Meselson went on to answer this question in a second series of experiments reported in the same paper, but before we can follow this work, we must explore still other biochemical features of phage-infected *E. coli* cells.

VOLKIN-ASTRACHAN RNA

In the 1950s, S. Cohen had investigated the fate of RNA synthesis in phage-infected cells and found that net synthesis of RNA comes to a halt shortly after the phage enter. For several years, this observation was taken as proof that all RNA synthesis ceases on infection, a belief that would have been itself incompatible with the notion that ribosomes served as the newly-formed templates for phage protein synthesis. (Scientists, like the rest of mankind, are quite capable of holding contradictory views simultaneously.) In 1956, however, E. Volkin and L. Astrachan discovered that a small amount of RNA synthesis does survive phage infection; this RNA (for awhile called V-A RNA) is in a constant state of synthesis and degradation and does not, in consequence, ever accumulate in the cell. By careful measurements on the relative quantities of adenine, uracil, cytosine and guanine incorporated into this RNA, Volkin and Astrachan surmised correctly that these base ratios were more like those of the phage DNA than of the bacterial DNA (substituting, of course, thymine for uracil). Most likely, therefore, the V-A RNA is made under the direction of the interloping phage DNA[*].

The following year M. Nomura, B. Hall and S. Spiegelman succeeded in isolating V-A RNA and demonstrating that RNA of this composition could not be detected in uninfected bacteria. Their separation technique relied upon the fact that V-A RNA turns out to have an S-value slightly different from that of RNA in uninfected cells. This difference in sedimentation velocity could be detected if the RNA was spun through a gradient formed by layering in the centrifuge tube sucrose solutions of decreasing (bottom to top) concentrations and hence viscosities. Their entire procedure involves the following steps:

1. Giving a pulse of P^{32} to infected cells four minutes after infection. (This pulse will label any bacterial or phage-specific RNA being synthesized at that time.)

---

[*]At the time and for many years afterwards great significance was placed on the fact that this V-A RNA resembles in base composition the phage DNA and not the bacterial DNA. However, as our knowledge of the mechanism of phage infection increased, it became evident that this dissimilarity was merely fortuitous. The V-A RNA is copied from only a small segment of the phage DNA; consequently there is no reason why its base composition need reflect that of the entire phage genome or why, by chance, it could not even have the same base composition as bacterial DNA.

2. Giving, for the sake of comparison, a similar pulse to uninfected cells. (This will label all bacterial RNA being synthesized.)

3. Extracting the RNA from both cultures and comparing the sedimentation velocities of the P^{32} labelled RNA from each.

4. Isolating the P^{32} labelled RNA that appears in the infected culture but is absent in the uninfected one. This is the V-A RNA, by definition.

For an encore, Spiegelman's group the following year devised a technique to prove that V-A RNA was made from a template of phage, rather than bacterial, DNA. This technique, called DNA-RNA hybridization, is very similar to the one Marmur devised for reannealing complementary strands of DNA. The underlying assumption in these studies, and in all others that employ DNA-RNA hybridization methods, is that an RNA molecule transcribed from DNA will have a base sequence complementary to one of the strands of its template DNA. Therefore the finding that a single strand of RNA will under suitable circumstances form a hybrid, double-stranded molecule with a single strand of DNA becomes in itself proof that this DNA strand (or others of identical sequence) once served as the template for the synthesis of the RNA in question. The "suitable circumstances" are very similar to those required for the reannealing of DNA. The DNA is heat-denatured and slow-cooled in the presence of the RNA presumed to be complementary to one of its strands. Any DNA-RNA hybrids formed can be detected because their S-value, electrophoretic mobility or density may (again depending upon the precise circumstances of the experiment) differ from the equivalent values for unannealed RNA or for the single and double-stranded DNA molecules found in the cooled solution.[2]

Applying this technique to the problem of the origin of V-A RNA, Hall and Spiegelman readily demonstrated that V-A RNA annealed only with denatured phage DNA and definitely not with native or denatured DNA from uninfected *E. coli* nor with DNA from any other source.

THE BRENNER, JACOB & MESELSON EXPERIMENTS (PART 2)

What is the function of this newly synthesized RNA? Is it phage-specific ribosomal RNA or some as-yet unclassified type of RNA? These are the questions the second part of the Brenner, Jacob and Meselson experiment was designed to answer.

By adding a pulse of H^3-Uridine to a phage-infected *E. coli* culture, Brenner, Jacob and Meselson found that the V-A RNA synthesized between minutes three and five after infection became associated with the B-band, 70S ribosomes. A 16-minute chase of cold Uridine succeeded in eliminating most of this radioactivity from these ribosomes. Since, therefore, the V-A RNA is only transiently associated with the 70S ribosomes, it cannot be considered ribosomal RNA, which forms a permanent part of the ribosome structure. On the other hand, its unstable association with the very ribosomes active in the synthesis of the new phage-specific proteins make it a prime candidate for the long-sought template in protein synthesis.

To prove that the V-A RNA associates with pre-existing ribosomes, the investigators carried out one final experiment. Cells were grown for several generations in $C^{13}N^{15}$ medium, infected with T_4 phage and simultaneously transferred to a light medium. For the five minute interval between the second and seventh minute $(P^{32}O_4)^{-2}$ was added to label the V-A RNA. The ribosomes from the infected cells were then extracted, mixed with an excess of ordinary ribosomes, and banded in a CsCl gradient. Again, as with the S^{35} protein label discussed earlier, the P^{32} label was found only at the position of the Light-A band; that is, the V-A RNA had really become associated with the pre-existing, heavy, 70S ribosomes in the HB band. The conclusion seemed inescapable that the new proteins occasioned by the phage infection were synthesized on the pre-existing bacterial ribosomes and that these ribosomes were re-programmed to produce these proteins by V-A RNA transcribed from the DNA of the entering phage. In view of its function in communicating genetic information from the DNA to the ribosomes, VA-RNA is now given the more descriptive name of messenger RNA (m-RNA). The term was first suggested by F. Jacob and J. Monod in 1961.

The concept of a messenger RNA communicating between the DNA and the ribosomes is not limited to the admittedly atypical cellular events that follow a phage infection. Even uninfected bacteria continuously produce relatively small quantities of unstable m-RNA, the detection of which is usually complicated by the much larger quantities of r-RNA that are also continuously produced. Nor is the concept limited to bacterial cells. All cells—plant, animal and bacterial—transcribe m-RNA from a template of DNA; these RNA messengers then enable the unspecialized ribosomes to assemble the primary structure of various different proteins. As it turns out, however, most plant and animal messengers are considerably more stable than those typical of bacterial cells. It is a bit ironic that the instability of the template played such a significant role in its discovery when in fact the "typical" messenger RNA molecule is actually not unstable at all. *

REFERENCES

1. Brenner, S., et al., 1961. Nature 190: 576.

2. Spiegelman, S., 1964. Hybrid nucleic acids. Sci. Amer. 210: 48 (May).

---

*This account illustrates, in part, what the biochemist, V. de Vigneaud, had in mind when he wrote that "Nothing holds up the progress of science than the right idea at the wrong time". How do you think progress in discovering the proper roles of the messenger RNA and the ribosome in protein synthesis would have been affected had a more accurate understanding of mRNA stability been in hand at the time?

LECTURE 27

THE MECHANISM OF PROTEIN SYNTHESIS: 1

RNA POLYMERASE

In an earlier section the general properties of the RNA polymerase enzyme were outlined. Since all RNA in a normal cell (r-RNA, m-RNA, and, as we shall later learn, t-RNA) are transcribed by this enzyme, now is an appropriate time to review and extend our information about its activity.

RNA polymerase was first isolated independently in three different laboratories in about 1960. Using a double-stranded DNA template, the enzyme makes a single-stranded RNA product. To achieve any significant synthesis, all four ribonucleoside-5'-triphosphates must be present, as well as a DNA template. *In vitro* the enzyme will usually make a population of RNA molecules some of which, judged by annealing tests with the isolated single strands of the double-stranded template, are complementary to one DNA strand and some to the other. *In vivo*, however, and also in some *in vitro* preparations that cause least violence to the *in vivo* configurations, the polymerase copies only one of the two DNA strands for each segment of the template. It is possible, in other words, for the enzyme to copy half the template from the "Watson" strand and the remainder from the "Crick" strand, but not from both strands of a single segment.

Careful studies by A. Goldstein, J. Kirschbaum and A. Roman demonstrated that the nucleotides are polymerized into RNA beginning at the 5'-end; consequently the template DNA strand must be transcribed 3' to 5'. (Two enzymes copying adjacent DNA segments from different strands would be seen to converge.) The actual mechanism of transcription seems to involve a transitory DNA-RNA hybrid region—called a **transcription bubble**—surrounding the enzyme as it moves along the template. Since the RNA in a DNA-RNA hybrid is resistant to digestion by mild concentrations of RNAse, RNA in the growing region of the molecule would be protected from RNAse activity; but as the enzyme moves on, the completed RNA chain is released from the DNA template and thereupon becomes digestible by RNAse. Spiegelman's laboratory has done much of the elegant early work that established and confirmed this model.[1]

In order to serve as a template, the double-stranded RNA would presumably have to partly open up to accommodate the polymerase and to enable the nascent portion of the RNA to base-pair with one of the two strands. Exactly what is involved in forming this bubble within the DNA molecule is still a matter of speculation and debate. As the polymerase proceeds along the DNA, positive supercoils are introduced into the DNA ahead of the polymerase and negative ones behind. DNA gyrase removes the positive ones and **topoisomerase I** the negative.

The polymerase enzyme itself has a molecular weight of around 700,000D. Given that the "average" weight of a single amino acid is 120D, one can presume *a priori* that the enzyme is composed of several subunits. As it happens, RNA polymerase has been fractionated into four, three of which (α, ß, and ß') are essential for the actual polymerizing of ribonucleotides.

The fourth subunit, called **sigma** (σ), functions in the initiation of polymerization and presumably is responsible either for selecting the actual segment and strand of the DNA to be transcribed or for allosterically altering the over-all structure of the other three subunits (the **core polymerase**) so that they bind only to initiating sequences of the template DNA. After RNA synthesis has been initiated at a particular DNA site, the sigma subunit is known to be released from the enzyme and the remaining segments carry on with the business of elongating the RNA chain by themselves.

Besides RNA polymerase, transcription usually requires other proteins known as **transcription factors**. Their role is thought to involve facilitating the binding of the polymerase to the correct sites on the DNA. In at least one organism, the silkworm, an RNA factor is needed, as well.[2] The nature and precise role of this unusual RNA factor is unknown at present, nor is it known how widespread the phenomenon of RNA factors is in nature.

The situation in eukaryotes is somewhat more complicated, for there are three different RNA polymerases, each with a different role to play. RNA polymerase A exists in the nucleolus, where it transcribes the large molecules of r-RNA. RNA polymerase B is found elsewhere in the nucleus and transcribes the large RNA molecules that will later become the m-RNA (see Lecture 31 for the discussion of this m-RNA processing). Finally, RNA polymerase C, also found in the nucleus, transcribes the t-RNA (see below for a discussion of t-RNA) and the small (5S) r-RNA. Mitochondria and chloroplasts also have their own RNA polymerases.

INITIATING RNA TRANSCRIPTION

Since a succession of different m-RNA molecules is made at different times in the life cycle of a cell or infecting phage, the need for something to regulate the timing of this transcription and to select the appropriate DNA segments and strands to be transcribed appears obvious. Random transcription of DNA would, by contrast, flood the cell with incomplete, nonsensical or mis-timed genetic messages that would quickly spell disaster. Special sequences of base pairs in the DNA (called **promoters**—for more information about them, see Lecture 37 on gene regulation) serve as initiation signals for RNA synthesis, offering, as it were, an especially inviting bed to the RNA polymerase or at least to its sigma subunit. Where a timed succession of different RNA molecules is required, as for example, during phage development inside an infected *E. coli*, the initiation signals in the DNA for each succeeding segment to be transcribed may be somewhat different. As each new set of phage genes comes into play, a new and unique sigma factor capable of recognizing the appropriate initiation signals could be made. It is certainly the case that m-RNAs made late in the infective cycle differ significantly from those made immediately on entry of the phage—late messengers, for example, are not fully competitive with early messengers for binding sites on denatured phage DNA—and there is evidence that new phage-specified sigma factors appear during the course of the infection.

When discussing gene control in bacterial cells we shall have occasion to add a few more details to the problem of RNA initiation sites (promotors) on DNA.

The termination of transcription in bacteria is partially a consequence of the structure the growing m-RNA assumes at its 3' end. The natural ends of RNA molecules often have a run of about 8 bases followed by an interval of about five more bases, and then a further run of eight bases that are the complements of the first eight. This permits the end of the nascent RNA to assume a **hair-pin loop**. (The intervening 5 bases will form the bend.) The occurrence of the loop soméhow slows down the polymerase, and if it is followed by a run of Uracils (which, of course, make fewer H-bonds with the template DNA than do C and G), the polymerase will dissociate from the template and the RNA synthesized to that point will be released. There is also a release protein (called **rho**) which may at times be involved in termination, especially in those cases where the secondary structure of the RNA is not sufficient by itself to cause termination. Details of rho-dependent termination are still very sketchy.

TRANSFER RNA (t-RNA)

Before we can examine the synthesis of proteins in detail, there is one more general problem that must be faced: how are the individual amino acids conducted in the appropriate order to the ribosome-m-RNA complex? Since there is no apparent stearic relationship between particular amino acids and particular long or short runs of RNA bases, it seemed unlikely that amino acids would find their way to the ribosomes by themselves. In 1958, F. Crick made one of his famous "stabs in the dark" by predicting the existence of a class of "adaptor molecules", at least one for each amino acid, which would function to lead the amino acids, in the right order, to where they were being polymerized into protein. He further speculated that such adapters would be RNA molecules and hence capable of recognizing particular segments of the m-RNA by means of the same complementary base principle that he and Watson had hit upon earlier for DNA. In a sense, then, the amino acids would be passively carried to the m-RNA-ribosome complex as a consequence of the attraction between their adaptor molecules and the m-RNA.[*]

---

[*]The principal discoverer of amino acid activation and of the transfer of these activated amino acids to a species of RNA, M. Hoagland, was actually unaware of Crick's "adapter hypothesis" at the time he made the discovery. Soon afterward, when J. Watson apprised him of Crick's prescient speculation, Hoagland describes his reaction:

> I was bowled over by the ingenuity and beauty of Francis's idea and sensed that
> it had to be the explanation of our experimental findings. An image arose in my
> mind: we biochemical explorers were hacking our way through a dense jungle to
> discover a beautiful long-lost temple, while Francis Crick, flying gracefully

(continued...)

The year before, 1957, M. Hoagland, P. Zamecnik and M. Stevenson had discovered that the very first step in the preparation of an amino acid for polymerization into protein was its reaction with ATP. This resulted in the formation of a so-called **activated amino acid** (technically, an aminoacyl adenylic acid) and the release of inorganic pyrophosphate. As Crick predicted, this activated amino acid is indeed subsequently transferred to a small-molecular-weight RNA that had previously escaped detection in the cell. This RNA is now called **soluble RNA (s-RNA)**, or **transfer RNA (t-RNA)**, and it functions precisely as Crick had suggested.

Molecules of t-RNA, of which there are about 50 different ones in the cell, contain between 73 and 90 nucleotides in a single chain. Most chains contain about 75 nucleotides. This single chain is looped back on itself in complicated ways to produce a rough (very rough) L-shaped structure with some areas of intra-strand base pairing and other areas of unpaired turns or bulges. The molecule is about 90Å along its longest axis. Although the different t-RNA molecules have distinctive base sequences and may even have distinctive bases—there are altogether about 30 "unusual" nitrogenous bases found in the t-RNAs in addition to the standard adenine, guanine, uracil and cytosine—they do have some features in common. All of them, for example, have an unpaired C-C-A-OH 3'end to which the amino acid becomes attached, and most of them end in a paired 5'-guanine. They are all about 90Å long, no matter how many nucleotides they contain.[3]

Since there are more different t-RNA molecules than different amino acids to be transferred, it is obvious that at least some amino acids can be attached to several different t-RNA molecules. Serine is transferred to at least three different t-RNAs in *E. coli* and five in yeast; methionine, by contrast, attaches to only one type. The exact nucleotide sequence of most t-RNA molecules is known; the first one to be worked out was a yeast alanine t-RNA, an experimental tour-de-force requiring several years of effort by R. Holley and his co-workers at Cornell University.[4]

The enzymes that assemble the correct aminoacyl-t-RNAs from the free amino acids and t-RNAs in the cytoplasm are called **aminoacyl-t-RNA synthetases**. In general there is just one unique synthetase for each amino acid[*], which means that it must recognize all members of the family

---

[*](...continued)
> overhead on gossamer wings of theory, waited for us to see the goal he already was gazing down upon. [M. Hoagland, 1990. *Toward the Habit of Truth*. (W.W. Norton & Co., N.Y.) p.94]

In fairness, it should be noted that Crick did not originally publish his speculations when he first conceived the idea in 1955, but instead had circulated his manuscript privately among a small circle of friends, which at the time did not include Hoagland. However, by the time of Hoagland's publication (early 1957), Crick had discussed the idea in seminars and in the proceedings of a scientific society that did get published in 1956.

[*]Sometimes a different enzyme is employed for the protein synthesis occurring in mitochondria and chloroplasts.

of t-RNA molecules that carry that particular amino acid. It is crucial, of course, that the amino acid-t-RNA connection be error-free, despite the constant and possibly tempting presence of other t-RNAs of very similar appearance. If the wrong amino acid were to become attached, it would be passively led to the ribosome and incorporated into the protein in place of the one that should have occurred at that position. Because all synthetases carry out the same function, attaching amino acids to t-RNA, it was assumed that they would all share a strong structural similarity. This assumption was strengthened when the fist three synthetases whose structures became known were indeed very similar. However, when the structure of the fourth one, seryl-t-RNA synthetase, was published in late 1990, it surprised everyone by bearing almost no resemblance to the other three. If there are really two (or more) very different types of synthetases, it is difficult to understand how they evolved in concert. Some additional comments on the function of the synthetases and on the role of the t-RNA molecules will appear elsewhere in these lectures.

Once the amino acid has been transported to the m-RNA-ribosome complex and transferred to the growing end of the nascent polypeptide, the t-RNA is freed to pick up another molecule of the same amino acid and so be cyclically re-used in protein synthesis. The exact manner in which the transported amino acid becomes part of the growing polypeptide will be discussed later in this account.

POLYRIBOSOMES[5]

Shortly after the m-RNA concept became established, it became clear that a single m-RNA would be much longer than what could be conveniently attached to a single ribosome. Moreover, electron micrographs of ribosomes caught in the act of protein synthesis usually revealed closely associated clusters of ribosomes rather than single particles. Very high resolution micrographs further revealed a thread of RNA connecting these clustered ribosomes, suggestive of a lengthy message to which were attached several individual ribosomes. Such clusters, which we now know to be the active complex in protein synthesis, are called **polyribosomes**, now commonly syncopated to **polysomes**. The segment of the m-RNA that is in contact with any particular ribosome encompasses about 25-30 nucleotides; the individual ribosomes of a polysome are separated by about 30Å.

The ribosomes are threaded onto the m-RNA at a given point at or near the 5'-end, and as each ribosome proceeds along the m-RNA towards the 3'-terminus, the ribosome assembles, amino acid by amino acid, a nascent polypeptide. The attachment of the messenger in bacteria is thought to be aided by base-pairing between a short run of bases at the 5' end of the messenger and a complementary sequence at the 3' end of the 16S ribosomal RNA (the so-called **Shine-Dalgarno sequence**).[*] Each ribosome of a polysome is thus making a complete polypeptide, the length of which at any given moment will be exactly proportional to the length of m-RNA that the ribosome carrying it has already traversed. When the polypeptide is complete, that is, when

---

[*]In eukaryotes the special m-RNA "cap", discussed in later lectures, is thought to play an important role in the attachment of m-RNA to the ribosome.

the ribosome assembling that polypeptide has travelled the full length of the m-RNA segment encoding it, the polypeptide is released. Usually the higher orders of the polypeptide structure (i.e., the secondary and tertiary structures) had already been spontaneously assumed as assembly proceeded on the ribosome; hence the newly completed polypeptide will usually become functional as soon as it is released from the ribosome.*

THE DIRECTION IN WHICH PROTEINS ARE ASSEMBLED

Since polypeptides are assembled sequentially, it is of interest to know whether they grow in the N-terminal to C-terminal direction or the reverse. In 1961 H. Dintzis[6] answered this question by examining haemoglobin synthesis in immature rabbit reticulocytes (red blood cells). These cells synthesize almost exclusively large amounts of Hb alpha and beta chains. Dintzis first cooled a suspension of reticulocytes to slow down protein synthesis and then added to the culture a pulse of H^3-leucine, an amino acid known to be fairly regularly distributed throughout both the alpha and beta chains of rabbit haemoglobin. At the end of the pulse, he extracted the *completed* haemoglobin from the cells and observed that whenever H^3-leucine had been incorporated into these fully-complete Hb molecules, it occurred near the C-terminal end of the chain. These radioactive Hb molecules, a minority, of course, of the total amount of Hb present in the cell, must have been those that were completed and released during the pulse period. The fact that the H^3-leucine was found only near their C-terminal ends indicates that this must be the part of the molecule completed last. Protein synthesis must, therefore, proceed from the N-terminal to the C-terminal end of the polypeptide.

REFERENCES

1. Miller, O., 1973. The visualization of genes in action. Sci. Amer. (March).

2. Young, L.S., et al., 1991. Science 252: 542-46.

3. Rich, A. and S. Kim, 1978. The three-dimensional structure of t-RNA. Sci. Amer. (Jan.).

4. Holley, R., 1966. The nucleotide sequence of a nucleic acid. Sci. Amer. 214: 30 (Feb.).

5. Rich, A., 1963. Polyribosomes. Sci. Amer. (Dec.), p44.

6. Dintzis, H., 1961. Proc. Nat. Acad. Sci. 47: 247.

---

*Proteins with a quaternary or conjugated structure must, of course, await the cytoplasmic interaction of their subunits before becoming biologically active. Also, some monomeric proteins must be cleaved or otherwise modified by special enzymes in the cytoplasm before they can function.

LECTURE 28

THE MECHANISM OF PROTEIN SYNTHESIS: ll

THE GENETIC CODE: EARLY SPECULATIONS

In order to understand protein synthesis in further detail, a knowledge of the **genetic code** is essential. By "genetic code" is meant the precise sequence of nucleotides along a m-RNA molecule which specifies, or encodes, a particular amino acid. The unravelling of this code during the period from 1961 to about 1964 marks one of the greatest scientific achievements in biology since Watson and Crick's discovery of the structure of DNA a decade earlier. The first question to be settled is the so-called coding ratio, by which is meant the minimum number of nucleotides in m-RNA needed to specify just one amino acid. One nucleotide cannot specify a single amino acid since there are 20 amino acids to specify and only 4 different nucleotides which could be used. By a similar reasoning a coding ratio of two can be ruled out. A coding ratio of three would permit the specifying of 64 different amino acids and becomes somewhat of an embarrassment of riches. Almost all attempts to work out the genetic code, however, proceeded on the assumption that the code is triplet in nature.[1] This minimal coding unit is called a **codon**.   *codon — bitchin' triplet*

In 1954 G. Gamow proposed a triplet, comma-free overlapping code, in which the last two "letters" (bases) of one codon become the first two of the next. Gamow speculated that the amino acids aligned themselves directly on the DNA molecule and that different-shaped cavities along the molecule that arose as a consequence of a varying nucleotide sequence attracted the amino acids of a complementary shape. We now know that proteins are *not* assembled directly on the DNA molecule, but regardless of the actual mechanism of protein assembly, a triple, comma-free, overlapping code would, if true, have two interesting consequences:

    1. Only certain amino acids could follow others in a polypeptide, and

    2. A point mutation in the DNA affecting only one base in the m-RNA would possibly alter two or three amino acids in the corresponding polypeptide.

It was quickly shown by detailed sequence analysis of various polypeptides that any amino acid could follow any other and that many mutations affected only a single amino acid. So much for that theory.

F. Crick then suggested a triplet code which was non-overlapping, non-degenerate and comma-free. (A **degenerate code** has synonyms, two coding units specifying the same amino acid; a **non-degenerate code**, therefore, would have only one codeword for each amino acid. One with

163

commas would have "punctuation", something separating one coding unit from another. "Redundant" is another term for "degenerate".) Basically his proposal envisioned 20 "sense-words" (each coding for a different amino acid) and 44 "nonsense-words" (coding for nothing). The sense words were such that in which ever order they were to be written, only nonsense words would be generated if the reading of the message were begun accidentally at the second or third base in the sequence, instead of at the first.

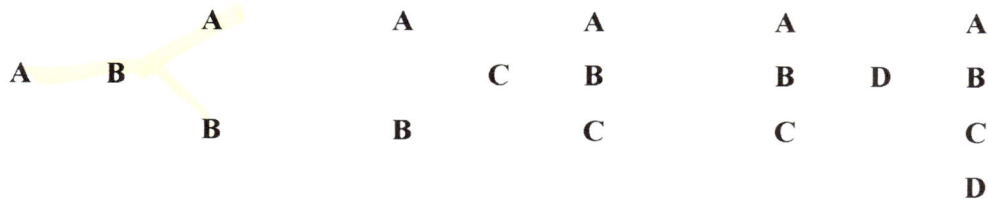

```
              A            A            A            A            A
  A     B                       C    B          B    D    B
        B            B           C          C              C
                                                           D
```

The 20 sense-words can be generated by a formula such as the one above, which yields the codons, ABA, ABB, ACA, ACB, ACC, BCA, etc. by selecting one letter from each row in each of the three groups.

It soon became apparent, however, that the code is extensively degenerate, that is, capable of employing more than a single unique codon to specify one amino acid. If, for the sake of argument, the code were non-degenerate, a great many mutations of sense words would convert them into nonsense words which, by definition, are un-translatable. Hence protein synthesis would halt at that point and the polypeptide, if released, would be a fragment quite unlike the finished product. It is clear, however, that most point mutations do not produce fragments, but instead full-length products only negligibly different in sequence from the wild-type—although often radically different in activity. The example of sickle cell anaemia given in an earlier lecture illustrates this point very clearly.

Two good tries, both in error.

THE WORK OF NIRENBERG & MATTHAEI[2]

The most promising approach came through the use of an enzyme, **polynucleotide phosphory-lase**, first characterized by S. Ochoa. Given the ribonucleoside *di*phosphates and some coaxing, this enzyme will make long-chain, single-stranded RNA molecules whose over-all base composition reflects the relative base composition of the reaction mixture. For example, given only one kind of nucleoside diphosphate the enzyme makes a homopolymer. Given a mixture of UDP and CDP in a 4:1 ratio, the enzyme makes a polymer in which, statistically speaking, about half (51.2%) of the triplet sequences within the polynucleotide product are UUU and 9.6% contain (in any order) two C's and a U. This enzyme was discussed earlier in a lecture on nucleic acid synthesis. At the time of its discovery, it was thought to be a synthetic enzyme, but was later shown likely to have a purely degradative function *in vivo*. (Its normal function in the living cell is still the subject of some debate.)

Interest in polynucleotide phosphorylase then quickly waned until 1961. It was then that M. Nirenberg and H. Matthaei reported taking a synthetic homopolymer, polyuridylic acid (i.e., containing only Uracil) prepared in the manner of Ochoa, using it to stimulate protein synthesis *in vitro*, and obtaining an unusual polypeptide product, polyphenylalanine.[*] Not too many assumptions must be made in order to identify UUU as the codeword for phenylalanine, the first amino acid whose codon became known. The method can be pushed somewhat further through the use of heteropolymers. For example, if the 4U : 1C synthetic polynucleotide referred to above were used for polypeptide synthesis, about half of all amino acids incorporated would be phenylalanine, but about 10% should be those amino acids whose codewords contain 2 C's and a U. By the use of this technique the base composition, if not the base sequence, of most codons soon became known.

Occasionally investigators would tinker with this method to squeeze out another datum or two. For example, through chemical manipulation of the completed phosphorylase product it was sometimes possible to attach a single discordant base to the 3'-end of a homopolymer. When this messenger (e.g., UUUU ... UA) was translated, leucine would be found at the C-terminal end of the polypeptide. Conclusion: UUA codes for leucine and translation proceeds in the 5' to 3' direction—assuming, of course, that peptide synthesis proceeds N-terminal to C-terminal.

Sometimes nucleases, RNA degradative enzymes that often contaminate the supernatant of the peptide-synthesizing system, would lop off the terminal nucleotide the chemists labored so hard to attach. Sigh.

The methods eventually devised to work out the genetic code in full rely on a chemical advance that enabled investigators to synthesize short polyribonucleotides of any given base sequence and then to make lengthy RNA molecules wherein this short sequence was a repeating unit. The exact base *sequence* of the synthetic RNA polymers was thus known, and when they were used to direct protein synthesis *in vitro*, the amino acid sequence of the polypeptide product could be precisely correlated with the base sequence of the template.[3] The exact details of these studies, while not at all difficult to follow, are best left to more advanced treatments.

The completed genetic code as we now know it is given in the accompanying table. (Later in this account it will be noted that mitochondria employ a slightly different code in translating their genome.)

---

[*]The *content* of the first synthetic genetic message (phe-phe-phe....) matches in banality only that of the first telephone message ("Mr. Watson—Come here—I want to see you."). In both cases, more inspired uses were soon to follow.

READING THE CODE

The process by which the genetic message is read is called **translation**. The individual ribosomes can be envisioned as travelling along the m-RNA and in so doing sequentially un-

Table 28.1 The nuclear genetic code.

Amino acid	**Codons**
Alanine (Ala)	GCU, GCC, GCA, GCG
Arginine (Arg)	CGU, CGC, CGA, CGG, AGA, AGG
Asparagine (Asn)	AAU, AAC
Aspartate (Asp)	GAU, GAC
Cysteine (Cys)	UGU, UGC
Glutamate (Glu)	GAA, GAG
Glutamine (Gln)	GAA, CAG
Glycine (Gly)	GGU, GGC, GGA, GGG
Histidine (His)	CAU, CAC
Isoleucine (Ile)	AUU, AUC, AUA
Leucine (Leu)	CUU, CUC, CUA, CUG, UUA, UUG
Lysine (Lys)	AAA, AAG
Methionine (Met)	AUG
Phenylalanine (Phe)	UUU, UUC
Proline (Pro)	CCU, CCC, CCA, CCG
Serine (Ser)	AGU, AGC, UCU, UCC, UCA, UCG
Threonine (Thr)	ACU, ACC, ACA, ACG
Tryptophan (Trp)	UGG
Tyrosine (Tyr)	UAU, UAC
Valine (Val)	GUU, GUC, GUA, GUG

Stop codons: UAA, UAG, UGA

covering—bringing into alignment at some point on the ribosomal surface—the various codons. This 'point of alignment' is usually called the **accepting site** or **A-site**, for it is to this interface of the m-RNA and ribosome that the appropriate aminoacyl-t-RNA is attracted. The appropriateness is determined solely by the specific base pairing possible between the codon at the A-site and a complementary triplet sequence carried by the aminoacyl-t-RNA. This t-RNA coding region is called an **anticodon** or, by a sort of anastrophe, **nodoc** ("codon" spelt backwards) and is found at one of the bends of the t-RNA structure.

When an anti-codon of an aminoacyl-t-RNA is properly paired with its complementary codon on the m-RNA, the continued forward motion of the ribosome pulls (or pushes) the t-RNA into a second ribosomal site called the **peptidyl site** or **P-site**.

Just before this shift takes place, however, the growing end of the polypeptide is switched from the P-site and becomes attached by a new peptide bond to the amino acid carried at the "stem-end" of the t-RNA standing at the A-site. The polypeptide has thus been lengthened by one amino acid residue, but that final amino acid is still attached to the t-RNA responsible for carrying it to the A-site. This polypeptide-t-RNA complex (called a **peptidyl-t-RNA**) now shifts over to the P-site, and in so doing it pushes out of the ribosome the "spent" t-RNA to which the nascent polypeptide had formerly been attached. The A-site, in summary, can be viewed as the place where the nascent polypeptide is transferred from its "old" t-RNA at the P-site to its "new", in-coming t-RNA, a process that lengthens the polypeptide by one residue and immediately results in its being shifted back into the P-site, but this time attached to a different t-RNA. The discharged t-RNA that is ejected will, of course, pick up another activated amino acid and be later re-used in protein synthesis.

The formation of the peptide bond is mediated by an enzyme, peptidyl transferase, located on the 50S ribosome subunit. It does not utilize an exogenous energy source, but the movement of the peptidyl-t-RNA from the A-site to the P-site requires the hydrolysis of one GTP molecule. Also involved in this rather complicated series of reactions are several cytoplasmic protein factors apparently needed for correct positioning of the charged t-RNA at the A-site and for hydrolysis of GTP. A discussion of their activities is best left for advanced courses.

One enzyme activity that does merit further comment is that of the peptidyl transferase, the enzyme responsible for

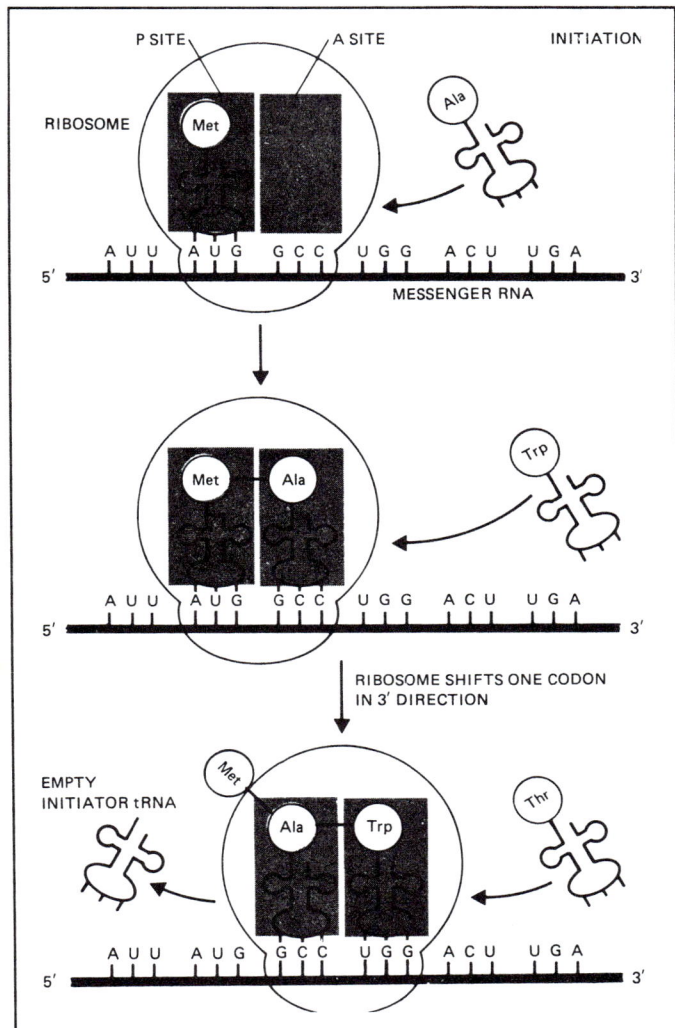

Figure 28.1 Translating the genetic code. See text for details. From CELL BIOLOGY, THIRD EDITION by John W. Kimball. Copyright © 1984 by John W. Kimball. Reprinted by permission of the author.

making the peptide bond. In searching for the protein that catalyzes this reaction, investigators used the approach of eliminating the various proteins, one by one, from the ribosome in order to determine which protein (or proteins) was essential for this reaction. To their surprise, the transferase activity survived the elimination of *all* the ribosomal proteins. Proteases (enzymes that destroy proteins) has no effect on the activity, whereas ribonuclease did. This crucial biological reaction, the synthesis of the peptide bond, is apparently mediated not by a conventional enzyme, but by a ribozyme, the 23S RNA of the ribosome![4]

CHARACTERISTICS OF THE CODE

As soon as the first workers began making tentative codon assignments their results began to suggest that the genetic code was extensively degenerate; that is, most amino acids could be specified by more than one codon. (Later, a certain amount of **ambiguity** was also noted when it was found that a single codon [AUG—see discussion of chain initiation] could sometimes specify two different amino acids.) One type of degeneracy that was particularly intriguing concerned the frequent inter-changeability of the two purines or the two pyrimidines at the third base position of some codons. It was observed, for example, that if a particular amino acid was specified by XYC, then XYU would also specify the same amino acid. Moreover, it seemed that this amino acid was being directed to either of its two codons by identical t-RNA molecules. That is to say, XYC and XYU must be able to pair with the same anticodon.

Using molecular models and some elegant physico-chemical calculations, Crick devised—some would say "divined"—his so-called **wobble hypothesis** to explain and predict this pattern of degeneracy in the genetic code. His model hinges around there being a certain "play" ("wobble") permissible in prescribed circumstances between the pairing of base number three of the codon and its pairing partner, base number one of the anti-codon. (To emphasize the anti-parallel character of the codon-anti-codon interaction, the numbering of the anti-codon bases is the reverse of that for the bases of the codons.) These permissible pairings are given below.

If base # 1 of anticodon is	then base # 3 of codon can be
U	A or G
C	G
A	U
G	C or U
I (Inosinic acid)	U or C or A

One obvious consequence of this degeneracy within the genetic code is the observation that not all changes (mutations) of the DNA lead to amino acid substitutions in the product. For the human haemoglobin molecule, for example, it has been calculated that there are 2,583 possible base substitutions in the DNA coding for the alpha and beta chains. Because of degeneracy, however, only 1,690 of these would actually result in amino acid substitutions in the product haemoglobin.

The genetic code was largely worked out by using protein synthesizing machinery (ribosomes, t-RNA, supernatant, etc.) derived from *E. coli*. Is this *E. coli* code the same as that used in other

Table 28.2 The codon assignments that differ between the nuclear and the mitochondrial DNA.

		MEANING IN MITOCHONDRION	
CODON	USUAL MEANING	MAMMALS	YEAST
UGA	stop	tryptophan	tryptophan
AUA	isoleucine	methionine	methionine
CUA	leucine	leucine	threonine
AGA	arginine	stop	arginine

cell systems? *A priori*, the least likely situation is each organism having a unique genetic code; again *a priori*, each organism having the same genetic code is probably the second-least likely possibility! Nonetheless, this latter seems to be the case. In 1967, R. Marshall, C. Caskey and M. Nirenberg reported[*] an experiment in which the codon recognition properties of *E. coli* t-RNA were compared with those of t-RNA from *Xenopus* (an amphibian) and guinea pig (a mammal). It was clearly demonstrated from the 50 codons tested in the three systems that a given codon gave an identical response (i.e., specified the same amino acid) whether that amino acid was carried by *E. coli*-, *Xenopus*-, or guinea pig-derived t-RNA. Differences were noted, however, in the relative strengths of the response among synonym codons in the different systems. This suggests that some classes of life have become accustomed to using one codon for a particular amino acid, whereas other classes use its synonym more commonly. A universal language with different dialects, in the metaphor of *The New York Times*.

EXCEPTIONS TO UNIVERSALITY

The belief that the genetic code is universal endured for many years, but as with many scientific beliefs, this one, too, has required modification in the light of subsequent investigations. In the lectures dealing with energy relationships within the cell, it was noted that mitochondria (and chloroplasts) have some DNA of their own and manufacture some proteins that are unique to these organelles. The ribosomes and the transfer RNA used in this synthesis also reside in the organelle. The chloroplast ribosomes resemble those of bacteria much more than those found in the cytoplasm of eukaryotes. Protein synthesis will proceed on chloroplast ribosomes using

---

[*] Their first reports of this discovery were made in *The New York Times* (Feb 22, p. 31), thus starting a trend in the reporting of "newsworthy" scientific discoveries that has been widely decried. Newspaper accounts rarely would provide sufficient details of the experimental procedures and the like to allow the claims to be competently evaluated by other scientists.

the chloroplast ribosome with the large subunit from *E. coli*, whereas this is not possible with a mix of cytoplasmic eukaryotic and prokaryotic ribosomes. As with bacteria, the initiation of protein synthesis in chloroplasts makes use of N-formylmethionine (see the later lecture on the initiation of protein synthesis for a discussion of the role of N-formylmethionine), but eukaryotes initiate nuclear protein synthesis with methionine. Many of these peculiarities also attach to mitochondrial protein synthesis, particularly the initiation of protein synthesis with N-formyl-methionine.

The genetic code used by mitochondria in particular has been the subject of recent investigations, and the results have been startlingly unexpected. As noted in the accompanying table, there are several codons that have a different meaning in mitochondrial DNA from what they have in nuclear DNA. Moreover, when the genetic code of yeast mitochondria was compared with that used by human mitochondria, it was seen that the code was not universal even among the mitochondria of this world. The meaning of this fact has not been settled yet.[5]

It turns out that even the code used by nuclear genes is not entirely universal. The codon CUG, which in everything else's nucleus specifies leucine, specifies serine in the yeast *Candida cylindracea*.[6] No one knows why. Even stranger is the observation that in a few genes of both bacteria and mammals the UGA codon, which normally is a stop codon, codes for an amino acid, selenocysteine, that is not one of the standard 20! (Selenocysteine has its own t-RNA in those bacterial species that code for it directly, but the situation in mammals has not been worked out in detail as of yet.)

THE "SECOND GENETIC CODE"

A final point about the genetic code concerns the region of the t-RNA recognized by the aminoacyl-t-RNA synthetases. Clearly the fidelity of translation depends just as much on accurate t-RNA recognition by the synthetase as it does on proper codon-nodoc pairing. What part of the t-RNA is recognized by the appropriate synthetase? The nodoc? Good evidence that it is *not* the nodoc comes from an experiment with seryl-t-RNA, of which there are three types:

$t\text{-RNA}^{ser}$ I , pairing with codons UCA and UCG

$t\text{-RNA}^{ser}$ IIa , pairing with codons AGU and AGC

$t\text{-RNA}^{ser}$ IIb , pairing with codons UCA and UCG

All three $t\text{-RNA}^{ser}$ species are charged to an identical extent by the same synthetase. Yet it is obvious, most so in the case of IIa and IIb, that these t-RNAs must carry different anti-codons.

The synthetase may be recognizing, therefore, some structural features of the t-RNA other than the anti-codon. This synthetase recognition whatever its basis, is sometimes called the **second genetic code**. In the *E. coli* t-RNA specifying alanine, this recognition site depends on a single

transferred to other t-RNA molecules by introducing this base-pair at an analogous location. For example, t-RNA^{phe} and t-RNA^{cys} can be transformed into alanine acceptors thereby.[7] The recognition site for the synthetase is becoming known as the **paracodon.**

REFERENCES

1. In the Lecture on "Frameshift Mutagens" the importance of Crick's early contributions to the question of the coding ratio are discussed. See also Crick, F., 1962. The genetic code. Sci Amer. (Oct.).

2. Nirenberg, M., 1963. The genetic code: ll. Sci. Amer. (March).

3. Crick, F., 1966. The genetic code: lll. Sci. Amer. (Oct.).

4. Noller, H., et al., 1992. Science 256: 1416-1419.

5. Grivell, L., 1983. Mitochondrial DNA. Sci. Amer. 248: 78-89 (March).

6. Kawaguchi, Y. et al., 1989. Nature 341:164-5.

7. Hou, Y., and P. Schimmel, 1988. Nature 333:140-145.

---

*Although t-RNA is single-stranded, because of the extensive secondary structure, many bases are paired.

LECTURE 29

THE MECHANISM OF PROTEIN SYNTHESIS: III

STARTING AND STOPPING PROTEIN SYNTHESIS

With this modicum of information about the genetic code, it is now possible to examine the last two topics in protein synthesis: initiation and termination.

STARTING[1]

In the early work on polypeptide synthesis directed by synthetic m-RNA templates, the initiation of synthesis was never viewed as a problem; synthesis began when the template, ribosomes, aminoacyl-t-RNAs etc. were present, and that was that. This synthesis, however, was occurring under conditions of very high magnesium ion concentrations, concentrations that would certainly not exist in a living cell. If the magnesium ion concentration was lowered to a level approximating that in a living cell—there is still some doubt as to exactly what that level is—polypeptide synthesis would usually stop. The significance of this was not understood at the time.

In the mid-1960s it was discovered that there existed in *E. coli* a t-RNA which had attached to it an amino acid, **N-formylmethionine** (F-met), which was never found in completed polypeptides. In fact, because this amino acid has a formyl group chemically "blocking" its terminal nitrogen, it would be impossible for N-formylmethionine to form a peptide bond with the carboxyl end of any other amino acid. If it were incorporated into protein, it would have to be the N-terminal amino acid residue—that is, be the first amino acid laid down in protein synthesis.

The t-RNA that carries N-formylmethionine (F-met-t-RNA$_F$) is not the same one that carries ordinary methionine (met-t-RNA$_M$), although both respond to the codon AUG. (The F-met-t-RNA$_F$ apparently also responds to GUG in *E. coli*, but not in eukaryotes.) Initially both t-RNAs pick up methionine, but afterwards a special cellular enzyme adds the formyl group only to the methionine attached to F-met-t-RNA$_F$. (In eukaryotes, the initiating t-RNA is also different from the regular methionine-carrying t-RNA, but the methionine it carries is *not* formylated; it is ordinary methionine.)

The very fact that F-met has a specific t-RNA to carry it strongly suggests that it does function in protein synthesis; its chemistry would demand that it serve as the first amino acid residue in the chain, since F-met cannot form a peptide bond through its formylated amino group. However, as noted above, F-met is not found in mature polypeptides, although more than half the proteins in *E. coli* do begin with methionine. A series of discoveries centering on the role

172

of F-met in protein synthesis then followed quickly on each other, and we came to see that the initiation of protein synthesis is a far more complicated business than was previously thought.

The first discovery concerned the product of *in vitro* protein synthesis when natural messengers were serving as the template. It is possible, for instance, to extract the RNA from an RNA phage, e.g., phage R17, and to use this genome to direct protein synthesis *in vitro*. (The genetic material of an RNA virus may serve as its own messenger RNA when the virus infects a cell.) When the products of this *in vitro* synthesis are examined—the phage RNA, as a genome, encodes only three polypeptides (coat protein, RNA synthetase, and maturation protein)—it was found that many of the nascent or completed polypeptide chains began with F-met despite the fact that none of these same proteins were found to have F-met attached when synthesized *in vivo*. Since the same m-RNA bearing the same genetic code was being used as a template in both instances, it seemed most probable that F-met was indeed the first amino acid positioned both *in vivo* and *in vitro*, but that a special cellular enzyme system, partly or wholly lost in the *in vitro* system, normally lopped off the initial F-met (or sometimes just the N-formyl group) from the polypeptide before its release from the ribosome.

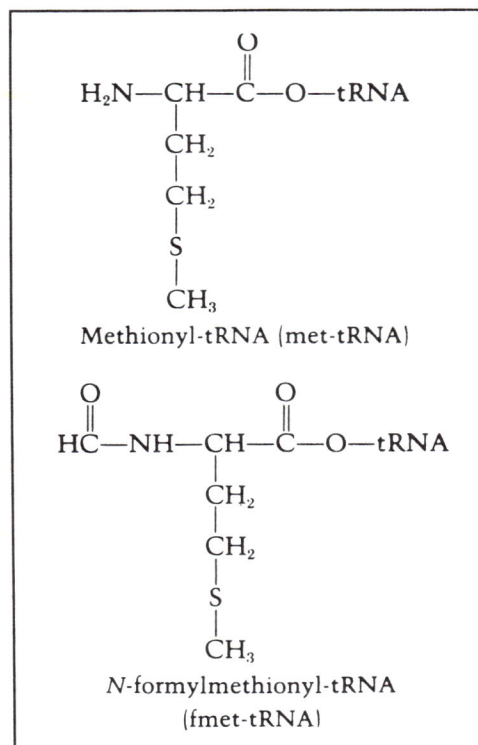

Figure 29.1 The two types of methionine discussed in the text.

The second discovery was that natural messengers, like R17 RNA, and synthetic messengers that began with the AUG codon would direct protein synthesis *in vitro* at much lower Mg^{++} concentrations than would be possible for synthetic messengers lacking an initial AUG. Moreover the AUG codon does not have to be precisely at the 5'-end of the synthetic polynucleotide; as long as it is somewhere near the 5'-end it will function to reduce the magnesium ion concentration needed to carry out protein synthesis. In these cases, however, the translation begins not with the first codon at the 5'-end, but only with the AUG codon itself. This codon, therefore, serves both to initiate protein synthesis and to establish the reading frame of the genetic message so that all codons beyond AUG towards the 3'-end are read (translated) in phase with the AUG codon.

It turns out that all bacterial protein synthesis is initiated by F-met, which is subsequently cleaved in whole or in part from the nascent polypeptide[*]. The positioning of the F-met-t-RNA[F] onto the m-RNA-ribosome complex, forming what is known as an **initiation complex**, is complicated and

---

[*]In eukaryotes, initiation also requires a special met-t-RNA; however, the methionine attached to this t-RNA is not formylated.

173

requires the participation of three specialized protein initiation factors and other goodies whose precise roles in the process took several years to elucidate. The most unusual feature has to do with the structure of the initiation complex itself, for it was found that it is only the 30S subunit of the dissociated ribosome that first complexes with the initiating AUG codon; only subsequently are the F-met-t-RNA^F and the 50S subunit attached.

Eukaryotic cells, as noted above, also employ a special t-RNA to initiate protein synthesis, but the methionine it carries is not formylated.

In over-view, and specifically omitting the roles of the various initiation factors, the process of beginning a polypeptide can be described as follows. A free 30S subunit uncovers an initiating AUG (or possibly, too, GUG) codon in the genetic message and forms a 30S-m-RNA complex to which is attracted an F-met-t-RNA$_F$ molecule. This t-RNA binds at the A-site, which is an integral part of the 30S ribosome. At this point the 50S subunit is added on, and the F-met-t-RNA$_F$ moves into the P-site, which is shared between the 30S and 50S subunits. Normally, of course, the P-site does not accept aminoacyl-t-RNA, but only peptidyl-t-RNA. Presumably in this case the N-formyl group functions, because of the blocked amino group, to mimic the structure of a peptide and hence to permit the F-met-t-RNA$_f$ free entry to the otherwise forbidden P-site. Once there, the F-met will be transferred to the next aminoacyl-t-RNA already waiting at the A-site, and chain elongation thereafter can occur in the normal way.

An interesting consequence of this method of chain initiation is that the ribosomal subunits are always changing partners with every new polypeptide they synthesize, and the newly-formed 70S ribosomes must dissociate into their subunits after each polypeptide they carry is completed. This "ribosome cycle" can, in fact, be detected experimentally if, for example, protein synthesis occurs *in vitro* with an equal mix of ribosomes derived from "heavy" and "light" bacteria. Sophisticated centrifugation experiments revealed that many of the ribosomes actually synthesizing proteins are composed of a heavy and light subunit.

A final puzzle, one not yet satisfactorily resolved, has to do with the AUG codon itself. Although it definitely serves to initiate polypeptide synthesis with N-formylmethionine, AUG is also the only codon specifying ordinary methionine. Since methionine is widely distributed in proteins, and is frequently an internal amino acid, how is the methionine AUG distinguished from the AUG initiator? Obviously a differentiation must be made, since chemically F-met cannot form a peptide bond through its amino group. The answer to this problem appears to be tied up with the problem of the actual secondary structure of m-RNA in solution, about which much remains to be learned. An initiating AUG, in this view, would be in an "exposed" part of the m-RNA, where it can be readily accessible to a 30S subunit. An "internal" AUG, then, would only encounter 70S ribosomes rolling down the line, and a 70S ribosome would read AUG as ordinary methionine. This idea gains support from observations with natural messengers, whose reading is sometimes significantly altered when pieces of the m-RNA are missing and hence their normal configuration becomes altered.

In eukaryotic cells, the ribosome must interact with the special 5' cap of the m-RNA before beginning translation further down the messenger (see below for details). Although most eukaryotic m-RNAs contain only a single gene, there are a few that contain two. In most of these cases, translation of the second gene must await the arrival of ribosomes that had travelled down from the 5' end, but there are occasional cases where the ribosome can apparently attach internally to the messenger, although the exact mechanism by which this is accomplished is still unknown.[2]

STOPPING

Endings, in human affairs as well as in molecular biology, can be as complicated as beginnings. The codon chart (see above) shows 61 out of the 64 possible triplets assigned to amino acids; the three left over, UAG, UAA, and UGA (called **amber***, **ocher** and—when imagination failed—**UGA**, although sometimes it is called **opal**) do not specify amino acids at all, but serve to terminate the synthesis of the polypeptide and to effect its release from the polypeptidyl-t-RNA holding it to the ribosome. Every unit of translation—a circumlocution for what in most instances would be the RNA equivalent of a single gene—begins with an initiator codon and ends in a terminating one. Ocher (UAA) seems to be the terminator commonly used in bacteria.

There is no t-RNA in the cell with an anti-codon complementary to a chain terminating codon. Although the absence of a t-RNA needed to translate the terminator will bring the elongation of a polypeptide to a halt at that point, this is not in itself sufficient to cause release of the polypeptide free from its C-terminal t-RNA. Release of the completed polypeptide is brought about by certain cytoplasmic proteins called **release factors** that act in conjunction with the terminator codons. Two are known in bacteria, one of which acts in conjunction with UAG and UAA and the other with UGA and UAA. There is also some evidence that the 16S r-RNA participates in the process by base-pairing with the stop codons and binding the release factors. In eukaryotes a single release factor interacts with all three stop codons, but the process is even less well understood than in prokaryotes. Presumably in both cases release of the polypeptide is correlated with the dissociation of the ribosome into its two subunits, but the exact sequence of events in both processes is still obscure.

Terminator codons can occur by mutation in the interior of the genetic message, in which case they are often called **nonsense mutations**. Their effect here is the same as when they occur naturally at the end of the gene: they trigger the immediate termination of synthesis and the release of the nascent polypeptide. This polypeptide, however, will be only a fragment of the normal gene product; its length will depend solely upon the distance the ribosomes travelled over the m-RNA before encountering the nonsense mutation.

---

* The names stem from an extended in-joke. Briefly, one of the discovers of stop codons named the first one after his mother, Amber. Since this is also a color, when a second one was discovered, the name of a similar color was chosen.

EUKARYOTIC MESSENGER RNA

Considerable controversy still surrounds the mechanisms by which eukaryotic m-RNA first is synthesized in the nucleus and subsequently is transported to the cytoplasm. Evidence has accumulated that eukaryotic messengers are not individually transcribed in their final translatable form, but instead comprise segments of much larger molecules known as **heterogeneous nuclear RNA (hn-RNA)**. For reasons not yet understood, much of this piece of RNA never leaves the nucleus and is rapidly degraded. After it is transcribed from the DNA, the part of the hn-RNA that will become the functional cytoplasmic messenger acquires a lengthy run (100 to 200 nucleotides) of adenylate nucleotides extending from its 3'-OH end. This **poly-A tail** is added by the activity of an enzyme other than RNA polymerase and is accomplished without the use of a template. The enzyme in question is known as **poly (A) polymerase**; the sequence AAUAAA, found 10 to 25 bases distal from the 3'-OH end where the tail will be appended, triggers the activity of the poly(A) polymerase.

The exact role of this poly-A is still in doubt. Some think it is needed in packaging the RNA to enable it to pass through the nuclear membrane;[*] others think it plays a role in translation. Some believe the poly-A tails become shorter as the messenger is repeatedly translated in the cytoplasm, and when the tail is completely removed, the messenger is degraded. Still others believe that poly-A can be added to m-RNA in the cytoplasm as well as in the nucleus. Since there are some messengers (those for the histone proteins) that lack poly-A tails altogether, clearly such tails are not essential in all cases. With all these conflicting views, the molecular biology of poly-A promises to remain a "growth area" in research for some time to come.

Eukaryotic m-RNA is also modified at the 5' end by the addition of a guanine in an unusual 5'-5' linkage, so that the 5' end of the m-RNA actually has the sequence 3'-G-5'ppp5'-N-3'p-5'-N-etc. ("N" can be any nucleotide). Moreover, the terminal Guanine is subsequently methylated at the N7 position of the ring, and methyl groups are usually added to the 2' positions of the penultimate and the antepenultimate nucleotides. This structure is known as the m-RNA **cap**. As noted above, this cap seems to play a necessary role in attaching the m-RNA to the small ribosome subunit.[**] It appears that eukaryotic messengers first attach at the 5' cap, and the small subunit then slides along the m-RNA until the initiating codon is encountered. At that point, the initiator t-RNA, various initiating factors, and the large ribosomal subunit complete the assembly of the functional 80S ribosome, and protein synthesis can begin.

Although the presence of the AUG codon establishes the reading frame for the translation of the subsequent codons, it is not unknown for there to be more than one "correct" reading frame for

---

[*]Mitochondrial messengers have tails, too, even though there is no membrane they have to pass through.

[**]The cap may also function to protect the m-RNA from digestion by cytoplasmic exonucleases.

the same RNA sequence. Certain small viruses, for example, have such overlapping genes, and so does the genome in the human mitochondrion. In such instances, there are two possible initiating codons and each, further down the m-RNA, has a terminator codon in phase with it. Such overlapping genes are apparently restricted to very small genomes where such economical use of the DNA is necessary.

Because of the rapid pace at which DNA is being sequenced in many organisms, it now common to "find" likely genes in a stretch of DNA long before the protein for which they code is discovered. These putative genes, begun with an initiating codon and ended by an in-phase terminator codon, are called **open reading frames**, awaiting the discovery of their function.

We shall see in a later lecture that prokaryotic m-RNA carries the transcripts of several genes on a single piece. Eukaryotic m-RNA, by contrast, is monogenic. One of the most unexpected discoveries made recently about eukaryotic m-RNA is the observation that it is much shorter than the length of DNA (one gene) from which it is transcribed! The original transcript is co-linear with its DNA template, but this m-RNA is then processed to form the functional messenger. In being processed, many segments (called **introns**) are excised and the remaining pieces (the **exons**) are mysteriously stitched together as the functional m-RNA. The total length of the introns often exceeds that of the remaining exons. What this means, of course, is that much of the DNA comprising a eukaryotic gene is seemingly "junk", that is, it is not transcribed into protein and serves no obvious function. Many investigations are currently under way to discover exactly how these introns can be so accurately removed from the initial RNA transcript and what it all means. These matters will be taken up again in the lecture on gene structure.

There has also appeared in the literature a hint about another possible kind of m-RNA processing, although little is known about it at present. In humans the protein glucose-6-phosphate dehydrogenase (G6PD) is composed of two peptides, one called the major and the other the minor one. A gene on the X-chromosome specifies the minor peptide. Part of the major peptide is also specified by a (different) gene on the X-chromosome, but the rest of this peptide is specified by a gene on chromosome 6! The mechanism that combines these two messengers (or combines the information from two messengers into one, or combines the information from two messengers into one peptide, or...) is not known. Whatever the underlying mechanism, G6PD is a striking example of "two genes, one polypeptide".[3]

Finally, mention should be made of protein synthesis in mitochondria and chloroplasts. These organelles contain their own ribosomes, t-RNAs, and synthetases. The ribosomes are smaller than their cytoplasmic counterparts, and the t-RNA are very different. Although the r-RNA for organelle ribosomes is synthesized in the organelle itself (and is smaller than the equivalents found in cytoplasmic ribosomes), the ribosomal proteins all appear to be coded in the nucleus and imported into the organelle after their synthesis in the cytoplasm. It has already been noted that organelles make use of a slightly different genetic code in translation. Like bacteria, but unlike eukaryotes, they initiate protein synthesis with N-formylmethionine.

REFERENCES

1. Clark, B. and K. Marker, 1968. How proteins start. Sci. Amer. <u>218</u>: 26 (Jan.).

2. Macejak, D., and P. Sarnow, 1991. Nature <u>353</u>:90-94.

3.Kanno, H. et al. Cell <u>58</u>:595-606 (1989)

LECTURE 30

GENES AND ALLELISM

HISTORICAL PERSPECTIVE

Classically the gene was thought to be:

    1) the ultimate unit of genetic function, and
    2) the ultimate unit of recombination, and
    3) the ultimate unit of mutation.

These can be viewed as operational definitions of the term **gene**, the operations obviously being function, recombination and mutation.

When people begin talking of "genes", it will not be long before the term **allele** crops up. In the earlier lectures on mendelian genetics, we have already learned that alleles are alternative functional states of the same gene. Here, we are going to push this concept a bit further in order to shed light on the basic structure of the gene.

Suppose one is dealing with a particular wild-type gene $\underline{a}^+$ which has mutated to $\underline{a}^1$. The genes $\underline{a}^+$ and $\underline{a}^1$ are said to be alleles and the superscript 1 both differentiates the mutant from the wild-type ($\underline{a}^+$) allele, and tells us that it is the first such mutant allele found.[*] (Sometimes the superscript of the mutant allele is a letter, perhaps the initials of the founder or some similar vanity.) During succeeding years, decades or centuries other alleles of $\underline{a}+$ may be found: $\underline{a}^2$, $\underline{a}^3$, $\underline{a}^{JLS}$, etc. This constitutes a "family of alleles"; all of its members are allelic to each other and to the unmutated $\underline{a}+$ gene. Usually these mutant alleles will manifest the same or very similar phenotype when they are homozygous; it generally would not be possible, therefore to tell by phenotypic inspection whether the mutant organism was genotypically homozygous $\underline{a}^2/\underline{a}^2$ or $\underline{a}^3/\underline{a}^3$, for example. (In this account, all the *mutant* alleles will, for the sake of simplicity, be recessive.)

---

[*]The phenotype expressed by each of the mutant genes in an allele family will usually be similar, but need not be identical. Moreover just because two recessive mutant genes may give rise to identical phenotypes when each is homozygous does not in itself prove that the mutants are alleles. White eyes in Drosophila, for example, can be caused by mutations at several distinct genes on different chromosomes even.

There did evolve, however, a set of agreed-upon criteria for deciding whether two or more independently-arising recessive mutations which manifest identical or closely similar phenotypes are alleles or not. One makes a heterozygote of the two mutations (for example, a and b) of the following sort: a + / + b. In a diploid organism this is called the **trans** arrangement of alleles and is distinguished from the other possible arrangement, (+ + / a b), which is called **cis**.

If the trans arrangement is wild-type, the two mutations are said to **complement** and are non-allelic; if the trans arrangement is mutant in phenotype, then the two mutations are said to be **non-complementary** and to be alleles. In effect what you have done in making a trans arrangement with actual alleles is to make a homozygote; that is, the two mutations which we had previously named a and b (and suggesting by that choice of nomenclature that they are non-allelic) should really have been named x^a and x^b, which clearly indicates that they are members of the same family of alleles.

Perhaps you can now tell why we would not have occasion to speak of the cis arrangement when dealing with true alleles.

THE DISCOVERY OF "PSEUDOALLELISM"

There would be no reason to belabor this point here were it not for the awkward consequences for classical genetic theory that followed from a discovery by Oliver in the early 1940s. He discovered that two different mutants of the lozenge gene in *Drosophila*, allelic by the criteria given above, would nevertheless recombine at low frequency. This result admits of two probable interpretations: the test for allelism is faulty, or intragenic recombination is occurring. Following this discovery similar untoward events were noticed by other investigators at other loci in *Drosophila* and in other organisms as well. Scientists who refused to believe that intra-genic recombination is possible (i.e. who continued to hold that the gene itself is also the ultimate unit of recombination) termed the phenomenon **pseudoallelism** and insisted that the observed recombination invalidated the presumed allelic status of the recombining units. Others, however, appeared eager to accept the possibility that the gene is far more mercurial than its accepted definitions made allowance for.

In order to resolve this dispute it became necessary to switch organisms and study the genetics of bacteriophage, the viruses that infect bacteria. Containing nothing more than DNA and several proteins, these bodies nonetheless have a genetic system (i.e., they reproduce, mutate, and recombine) that can be experimentally manipulated. A common phage type used in such work is T2 or T4.

RECOMBINATION IN BACTERIOPHAGE

As we have seen, when a phage particle infects a single bacterium growing in a confluent solid culture, it multiplies within the host cell, destroying it and liberating 50 to 200 progeny phage. These infect neighboring cells so that within a short time an area of clearing, known as a plaque, is observed. Since it is not practical to deal with phage individually, we often use characteristics like the size, shape and texture of the plaque to identify which phage stocks we are dealing with. This unusual procedure is possible because for each species of bacterial host the plaque characteristics in defined conditions are a constant hereditary property of the particular phage mutants used. (Other commonly used phage mutations may restrict or expand the host-range of the particle or affect its growth potential at elevated temperatures, for example).

Recombination can be observed in phage when a single bacterium is infected simultaneously with a pair of genetically differing phages. For example, if these parental phage each carry a distinct mutation (e.g., one parent carries a host-range restriction mutation and the other a gene causing a distinctive plaque morphology), it is not uncommon to discover among the progeny phage some that have inherited both mutations and some, the reciprocals, that carry neither. The parental phage are said to have "mated".

THE EXPERIMENTS OF BENZER[1]

The most famous mutant system of T-even phage is one that occasions an enlarged plaque during growth on *E. coli*. Such mutants are called r (for "rapid lysis"), and they occur at three distinct sites of the phage genome. Despite their having identical phenotypes, rI, rII and rIII mutations can be operationally distinguished by their plating characteristics on three strains of *E. coli*: B, S and K.

	B	S	K
rI	r	r	r
rII	r	+	-
rIII	r	+	+

It is important to note that rII mutations are (r) in phenotype on B, wild-type in appearance on S, and don't grow at all on K.

Although phage are haploid "organisms", it is possible to simulate a diploid condition by coinfecting a single bacterium with two parental phage and treating the infected cell as if it were a diploid phage. If this sounds contrived, consider the following example: Strain K is coinfected with wild-type and rII parents; twenty minutes later the cell ruptures releasing equal numbers of wild-type and rII phage. We know, however, that rII phage by themselves will not grow on K bacteria. The genome ("chromosome") contributed by the wild-type parent must have "covered up" for the rII defect in the mutant parental genome, with the result that both chromosomes replicated and eventually both types of progeny were released. Is not this masking of a recessive mutant gene by its wild-type allele much the same as what happens in a diploid heterozygote?

To follow on a bit further, what would happen if *E. coli* K were infected simultaneously by two parents each carrying an independently arising rII mutation? If you have been following the argument so far, you would say that no growth would occur, since different rII mutations, having been so defined by their lack of growth on K, would be true alleles, and in trans true alleles display a mutant phenotype. The actual experiment was carried out by S. Benzer in the

181

mid-1950s, and he discovered you would be right—about half the time. The other half the time (depending upon which particular pair of rII mutations were chosen) a normal infection followed and rII progeny phage released.

To summarize: no rII mutants by themselves will grow on K (by definition); all rII mutants will grow on K when coinfected with wild-type; rII mutants will grow on K when co-infected with particular other rII mutants. That is, some rII allelic pairs complement and others do not; that is, not all rII "alleles" belong to the same family. In fact, there are two rII genes, A and B, and all rII (point) mutations can be assigned to one or the other location. These separate genes are called by Benzer **cistrons**, the ultimate units of genetic function. The complementation test on K tells which rII mutations belong in which cistron.

Benzer then attempted to find out whether the family of rII mutants ascribed to a particular cistron (say B) could recombine. Since it is not possible to test for recombination of true rIIB alleles on K (Why?), the parents were first co-infected on bacterial strain B and the resultant progeny phage thrown onto strain K. Since both parents were rIIB and since no rII mutants can grow on K, any plaques on K must be the result of wild-type recombinants arising during parental mating on strain B. Such plaques were frequently observed, showing that true alleles will recombine, and consequently that the ultimate units of mutation and recombination are different from and smaller than the ultimate unit of function.

Considering the chemical structure of genes in terms of DNA, one can imagine the gene as a segment of several hundreds or thousands of base pairs. The function of the gene (or cistron) is altered by a change occurring at any of these different base-pair sites (**mutons**) and recombination can occur among any of them. The ultimate unit of recombination would be that occurring between adjacent base-pairs. This is called the **recon** and theoretically it would comprise two base-pairs of DNA.

It is now possible to begin relating the recombinational data we have accumulated about the gene to its biochemistry. As was noted in an earlier Lecture, linked genes are often described as being separated by so many "map units". This measure is derived from the frequency, in percent, with which the genes in question undergo recombination. If, for example, genes a and b on the second chromosome of *Drosophila* recombine at a frequency of 6%, then they are said to lie 6 map units apart. Because in diploids only two of the four chromatids in a tetrad are involved in each recombinational event, even if a and b were sufficiently far apart on the chromosome to recombine 100% of the time, the maximum frequency of recombinants among the progeny would be only 50%. In consequence when tested directly two genes cannot appear to be more than 50 map units apart. If, however, we examine three genes, a, b and c, known to lie in that order on a particular chromosome, and find that a and b are 30 map units apart and b and c are 35 map units apart, it becomes permissible to state that a and c are 65 map units apart, even though direct testing of a and c would reveal them to be separated by only 50 map units. Thus, by summing up small intervals that collectively span an entire chromosome, one can speak of its being, for example, 180 map units in length; and by summing the genetic lengths of all the chromosomes, one obtains the number of map units in the entire genome.

If allowances are made for the fact that phage are haploid and have a different mechanism of re-combination, one can, by a procedure analogous to that used in *Drosophila*, calculate map distances in phage, too. The phage genome, by this calculation, encompasses about 1500 map units. Since the most distal mutons of the rIIA cistron recombine with a frequency of 6%, the cistron itself occupies, therefore, 6/1500 of the phage genome. The entire phage genome, we know from biochemical calculations, contains about 2×10^5 base pairs; the rIIA cistron, therefore, contains about 800 base pairs.

The smallest non-zero intra-genic recombination value for rIIA is 0.02%. By a calculation similar to that carried out above, we obtain a figure of $.02/1500 \times 2 \times 10^5$, or about 3, for the number of base pairs in a recon. This figure is very close to the theoretical value of 2 and lends strong support to Benzer's concept of gene structure.

All of these calculations, it should be noted, assume that recombination is an equally probably event at any point along the DNA molecule. There are some indications in the literature that this is not always the case.

Before you generalize too freely from the genes of micro-organisms to those of higher organisms, I would like to register a few *caveats*. Genes of higher organisms (e.g. *Drosophila*) do not give evidence of having a large number of recombinable sites (mutons) like those of phage, bacteria and fungi. Moreover, there is often a great deal of apparent intra-genic complementation which in turn makes it awkward to use the term cistron. These embarrassments may simply be the result of emerging differences in the mechanism of recombination as one mounts the evolutionary ladder, or may be attributable to funny cytoplasmic phenomena that obscure chromosomal events. Nevertheless, I avert my eyes when confronted with attempts to clothe classical organisms in the terminology of the "newer" ones. The terms have little substance when removed from their original models.

REFERENCES

1. Benzer, S., 1962. The fine structure of the gene. Sci. Amer. (Jan.).

LECTURE 31

SPLIT GENES IN EUKARYOTES

INTRONS AND EXONS

One very striking (and before 1977, completely unsuspected) difference between genes in microorganisms and their counterparts in higher organisms has already been alluded to. Genes in higher eukaryotes are extensively split into coding (**exons**) and non-coding (**introns**) sections.[1] In some genes the introns can constitute most of the gene. An extreme case, perhaps, is the gene in chickens for collagen where more than 50 introns are found, representing about 90% of the DNA. Being split into introns and exons is typical of most protein-specifying genes in vertebrates. The introns can assume a wide variety of lengths and sequences and can even divide the triplet sequence (a codon) specifying a single amino acid. However, as one descends the evolutionary ladder, introns become less common and their lengths shorter when they do occur. For example, the gene for cytochrome C in humans is split, whereas it lacks introns in yeast; similarly, the Drosophila alcohol dehydrogenase gene has introns, the yeast equivalent does not.

But not all protein-specifying genes are divided, even in the higher eukaryotes. Those that specify histones, for example, are rarely split (histones, remember, form the structural components of nucleosomes), nor are those specifying interferons (which are proteins functioning in the immune system). There seems no obvious regulatory or other reason that determines whether a gene is split or not.

Moreover, it is not only protein-specifying genes that contain introns. Many of the nuclear genes that specify t-RNA also contain introns (generally just one each), which range in length from 14 to 60 nucleotide pairs. The cases involving introns in r-RNA are from lower eukaryotes and occur in the gene for the large RNA molecule in the ribosome.

Removing introns must obviously be an extraordinary precise business. This is especially apparent in the case mentioned above where an intron occurs right in the middle of a codon, but the potential for causing frame-shift mutations (see the later lecture on gene mutation) requires that exons be removed with absolute accuracy no matter where they occur within the gene. There have evolved several mechanisms for removing introns, depending upon whether they occur in m-RNA, t-RNA or r-RNA.[2] This account will focus on the mechanism used for splicing pre-messenger RNA. (The cap and poly-A tails would already have been added to the messenger before splicing begins.)

The ability to isolate, clone and sequence particular stretches of nucleic acid that recombinant DNA techniques now routinely permit (see the later lecture on recombinant DNA techniques) has resulted in the sequencing of a large number of introns (and their bordering exons) from many different genes. When these sequences were compared, it became clear that a common sequence, called a **consensus sequence,** existed at the beginning (5' end) and end (3' end) of all introns, regardless of how many bases comprised the entire intron. Specifically, all introns began with GU.... and ended with ....AG. It also seems likely that there are strong similarities in the sequences found at the ends of the bordering exons. The situation can be represented in the following manner, where the bases in bold are those of the intron:

(5')......(C/A)AG**GU(A/G)AGU**..........(C/U)A**GG**...... (3')

This is commonly known as the **GU-AG rule.**[*]

The actual excision of the intron is done by a **spliceosome,** the active components of which are five major (and at least that number of minor) riboproteins known as **snRNPs** (small nuclear ribonucleoproteins) or, more commonly, as "snurps". Each snurp contains a molecule of RNA (one of the five snurps contains two) and several proteins; all five major snurps seem to have a role in the spliceosome, although what it is has not been precisely defined in each case.[**] It is known, however, that a cut is made first at the 5' end of the intron (this seems to be the function of the snRNP called U1) and this 5' end is then covalently bonded (through a phosphodiester bond) to the 2' position of a nucleotide 20 to 55 bases "upstream" (i.e., towards the 5' end) of the 3' end of the intron. The U2 snRNP seems to play a role in facilitating this bonding. The 3' end of the intron appears to be recognized in a complex manner by one or more "auxiliary proteins"; following the attachment of them plus the U1 and U2 snRNPs mentioned earlier, a complex containing the remaining snRNPs attaches and completes the splice. The unusual 2'-5' phospho-diester bond creates a **branchpoint** in the RNA; when the intron is then cut at its 3' end and the two ends of the adjacent exons joined, the intron is released as a sigma-shaped (σ) structure known to those innocent of Greek as a **lariat.** This lariat is soon degraded. Only after all introns have been removed is the m-RNA transported to the cytoplasm.[3]

Although the splicing of introns is highly accurate, sometimes an intron within a single gene can be spliced under one set of circumstances and not under others. This **alternative splicing** yields different functional messengers from the same DNA segment. Guiding the choice of splice sites

---

[*]The necessity for specific sequences at splice junctions suggests the possibility of mutations affecting the splicing process. Indeed, one is already known in humans. One of the proteins necessary for the proper clotting of blood is known as factor IX; its absence or functional impairment causes the disease, hemophilia B. Although factor IX can be missing or impaired for a variety of reasons, in one well-characterized instance the cause is a splicing error in the primary transcript of factor IX m-RNA. In this case a mutation from GU to UU alters the obligatory 5' consensus sequence of an intron and in consequence the transcript is improperly spliced.

[**] One of the RNA molecules, the one known as U6, has been extremely well conserved through evolution, being very similar in sequence in organisms as divergent as yeast and humans. This has led to the suggestion that it plays a key role in the splicing process.

by the snRNPs are additional protein factors whose precise roles (or even how many there are) have not been worked out. It is clear, however, the presence of some of these factors results in a different splicing pattern from what is obtained in their absence. For example, in one gene studied, there are two possible 5' splice sites preceding a single 3' site. When a particular splicing factor is absent (or present in only small quantities), only the first splice site is used. As the concentration of this factor is increased, however, both sites are used, the second one preferentially over the first. Alternative splicing may be an important regulatory mechanism in development and tissue-specific gene expression.

The introns of some genes sometimes themselves nest functional genes. In humans, for example, the gene on the X-chromosome that specifies Factor lll (one of the proteins which, when missing or misfunctional, causes hemophilia) houses an entirely different gene as one of its introns. An even more striking case is the gene for human neurofibromatosis or Von Ricklinghausen disease.* In this case, no fewer than three additional genes have been identified within introns of the principal gene. Moreover, the neurofibromatosis gene itself is of huge proportions, comprising somewhere between 500,000 and 2,000,000 base-pairs on chromosome 17.[4]

Some introns have been found to have introns of their own! In the unicellular alga, *Euglena gracilis*, one of the chloroplast genes has an intron within it that is excised before the "host intron" is itself spliced out of the gene. Why should it be necessary to remove separately the internal intron, given that the host intron will be eliminated in its entirety, anyway? Why, indeed.

Finally, it has recently been found that the sequence of exons in the functional messenger RNA is not always an accurate reflection of their order in the gene or in the pre-spliced m-RNA.[5] The mechanism that accounts for this **exon shuffling** is unknown at present, nor is it known how widespread the phenomenon is.

One can't help wondering why eukaryotic genes have introns, since at first blush it would seem far more "reasonable", if only because it would be far simpler, for them to be patterned on the prokaryotic model. This assumes, of course, that such complexities in the genome arose concomitant with or following the origin of eukaryotes. But there is another view, widely held, that sees introns and exons as very early features of the genome, predating the diversion of plants and animals and predating even the incorporation of mitochondria and chloroplasts as endosymbionts. Modern prokaryotes, this argument goes, probably have lost this complexity instead of never having had it. This theory gains some support from the recent finding of an intron in the gene coding for the leucyl-t-RNA in the cyanobacteria, which are the modern

---

* Neurofibromatosis affects all races at a frequency of about 1 in 3500 live births. As such, it is one of the most frequent human autosomal dominant disorders. It causes "cafe au lait" spots, numerous benign tumors known as neurofibromas, and learning disabilities. Sometimes malignant tumors result, as well. There is no effective treatment for the disease. It was once thought that the so-called "elephant man" syndrome, the subject of a well-known movie of the title, was neurofibromatosis, but that diagnosis is no longer believed to be correct.

descendants of formerly free-living bacteria that invaded plant cells to become the chloroplasts. Moreover, modern chloroplasts have an intron in the same t-RNA gene, and in both cases it is the self-splicing type (see below). It could be argued, therefore, that this particular intron has existed since before the origin of chloroplasts, more than a billion years ago. If so, the question becomes ever more puzzling: just what purpose do introns and exons serve?

One popular theory is that exons represent specific "domains" of the polypeptides they encode. By domains is meant functional, folding or structural regions of the protein. Proteins, this view holds, are built on the modular plan by the combination and recombination of particular exons that specify particular necessary attributes of the required protein. Proteins would evolve, therefore, by a sort of **exon shuffling**[*] and not primarily by the testing of amino acids one at a time. This in turn suggests that there may be a finite number of exons that are shuffled in evolution. One attempt to estimate this number has been made by W. Gilbert and colleagues, who surveyed data banks for the nucleotide sequences of all exons that had been determined by 1990. After eliminating obvious sources of redundancy, such as exons that derive from basically the same gene in different species or from genes that seem clearly to have arisen by duplication of other known genes, they then translated the remaining exons into their peptide sequences and compared the resulting sequences with each other. Their surprising conclusion is that the "world of exons" is actually very small, from one to seven thousand.[6] Whether indeed all biological proteins have been derived by assembling a relatively few exons in a wide variety of ways is highly speculative and controversial at this time, but the idea is certainly intriguing.

SELF-SPLICING RNA[7]

Another surprising discovery to come out of this work is self-splicing RNA. In the ciliate, *Tetrahymena*, it was first learned that the primary transcripts of r-RNA genes require no additional proteins or RNA in order to correctly excise their introns; later it was found that the same is true for certain primary transcripts of both m-RNA and r-RNA in yeast. In these cases, and perhaps in many others yet to be found, the primary transcript RNA is autocatalytic. Such **ribozymes** contradict the long-held belief that only proteins can function as biological catalysts.

RNA EDITING

An even more recent discovery is the phenomenon of **RNA editing**, in which *untemplated* nucleotides are added to the coding portions of the m-RNA after it is transcribed. In the slime mould (*Physarum polycephalum*) it has been discovered that numerous Cytidine residues are added to the mitochondrial m-RNA for a particular protein (one of the subunits for ATP synthetase)[8]. The Cytidines are added by insertion *de novo* into the m-RNA at an average, but very variable, spacing of about 26 nucleotides. The insertions appear to be non-random, for there is a seeming preference for insertion at the third position of codons and for being 3' to a purine-

---

[*] Note that the term "exon shuffling" is currently used to denote two different phenomena. It is unlikely this confusion will last long.

pyrimidine dinucleotide. It is not known at this time whether the positioning of these insertions is determined by the nature of the m-RNA itself (e.g., by some aspect of its secondary structure) or by some sort of facilitating RNAs that are complementary to parts of the m-RNA where insertion is to occur. How widespread this phenomenon is also remains to be seen.

REFERENCES

1. Chambon, P., 1981. Split genes. Sci. Amer. 244: 60-74.

2. Darnell, J., 1983. The processing of RNA. Sci. Amer. 249: 89-99.

3. Seitz, J., 1988. Snurps. Sci. Amer. (June).

4. Wallace, M., et al., 1990. Science 249: 181-186.

5. Nigro, J., et al., 1991. Cell 64:607-613.

6. Dorit, R., L. Schoenbach, and W. Gilbert, 1990. Science 250: 1377-1382.

7. Cech, T., 1986. RNA as an enzyme. Sci. Amer. 255: 76.

8. Mahendran, R., et al., 1991. Nature 349: 434-438.

LECTURE 32

MUTATION: RANDOM OR ADAPTIVE?

INTRODUCTION

Before we go on in the next lecture to examine how chemically induced mutations can arise, there are some important aspects of spontaneous mutation that deserve prior comment. One has to do with the nature of spontaneous mutation itself. By labelling a given mutation as "spontaneous" in origin, we are in effect pleading ignorance of its true cause. Spontaneous mutations originate, as it were, in the interstices of our accumulated knowledge of induced mutations. This ignorance has often permitted a number of false notions to flourish, perhaps the most hardy having to do with the relationship between a mutation and the environment in which it arose. On one hand, there are those who believe that some or all mutations are elicited in an adaptive, or directed, manner by the environment so that the organism (or rather its descendants) would be better able to cope with the particular environmental circumstances that allegedly evoked the mutation in the first place. On the other hand, there are those who believe that all mutations are nondirected, or random, in their effects and that the sole role of the environment is to select in favor of those that happen to have beneficial effects and against those whose effects are less beneficial. (The definition of "beneficial", now usually ascribed to greater reproductive capacity, need not concern us here). The former view usually goes under the name neo-Lamarckism; the latter, neo-Darwinism.

THE LURIA-DELBRUCK EXPERIMENT[1]

At first thought it might seem fairly easy to determine whether a particular environment is actually directing mutation or merely selecting randomly-produced ones, but in practice it is exceedingly difficult to devise convincing experiments to disprove either view. The first convincing proof of the random, non-directed origin of spontaneous mutations was offered by S. Luria and M. Delbruck in 1943. It is commonly known as the **fluctuation test** and owes its inspiration to Luria's observations of the manner in which Las Vegas slot machines dispense their rewards. (Those interested in following his ratiocination can consult his highly readable autobiography, *A Slot Machine, A Broken Test Tube*, Harper and Row, N.Y. 1984.) The scientific underpinnings to their experiment can be outlined as follows.

Wild-type *E. coli* are sensitive to infection by T_1 phage; genotypically they can be called $\underline{Ton}^S$ ("T-one sensitive"). We know that mutations to T_1 resistance occur somehow; such cells are called $\underline{Ton}^R$ mutants. The question is: are $\underline{Ton}^R$ mutants induced by T_1 phage and occur, therefore, only in the physical presence of T_1 phage, or, alternatively, do such mutations occur spontaneously and non-directedly, and therefore can occur even when T_1 phage are not present. The difficulty in answering this question lies in the fact that in order to know whether $\underline{Ton}^R$

mutants are present, T_1 phage have to be added to the bacteria in order to kill off the (non-mutant) sensitives so that only the resistant mutants will be left to form colonies, which can then be counted. But having added the phage, how do you know it wasn't then and only then that the mutation to phage resistance occurred? In other words, how do you know that these phage that selected for the presence of the mutation to phage resistance didn't induce it as well?

To answer this question, Luria and Delbruck posed the following scenario. Suppose a liquid culture of bacteria were established from a very few wild-type ($\underline{Ton}^S$) cells—ideally, from just one cell, although this is often impractical. These cells are permitted to multiply in the absence of T_1 phage until the culture consisted of a very large number of cells, let us say 100 million. If a mutation to $\underline{Ton}^R$ were to occur during the growth of this culture, a certain fraction of the final 100 million cells would be $\underline{Ton}^R$. The size of this fraction would depend upon exactly when during the growth phase the mutation occurred. If it had happened early in the growth period, a large fraction of the final 100 million cells would be $\underline{Ton}^R$; if it had happened much later, only a small fraction of the total would be $\underline{Ton}^R$. If you had established not one, but many such cultures, each would likely have a different fraction of the 100 million being $\underline{Ton}^R$, since in different cultures the mutation to $\underline{Ton}^R$ would occur at different times during the growth period. Some cultures may not have any $\underline{Ton}^R$ cells at all, since 100 million cells isn't that large a number when dealing with an event, the mutation to $\underline{Ton}^R$, that is known to be very rare in any case. To put this situation in another way, if you had several independent bacterial cultures each established in the manner described, there would likely be a significant fluctuation in the number of $\underline{Ton}^R$ cells when the numbers of $\underline{Ton}^R$ mutants in the final generation of each culture are compared.

All this assumes, of course, that mutations to $\underline{Ton}^R$ do occur in the absence of T_1 phage. Suppose, now, that they don't. In this case, all cultures would still be homogeneously $\underline{Ton}^S$ at the beginning and at the end of their growth, since at no time during this growth was the putative "mutagen", the phage, present. So, in summary, the "Darwinian" view would hold that there would already be $\underline{Ton}^R$ mutants in many of these cultures, but some may have none and some may have a large number; the "Lamarckian" view is that none of the cultures have any $\underline{Ton}^R$ cells at this point.

Now suppose that the bacteria from each liquid culture were plated on a solid culture plate that was laced with T_1 phage. Only $\underline{Ton}^R$ cells would grow into colonies under these circumstances. In the "Darwinian" interpretation, these $\underline{Ton}^R$ cells already existed in the cultures; whereas in the "Lamarckian" interpretation, they would be induced at this point. Lets us look at the consequences of this latter view with respect to the number of $\underline{Ton}^R$ cells that would be induced in a series of independent cultures set up as described previously. Since each culture contains 100 million cells, we are exposing the same number of cells to the "mutagen" in each case. Hence, we would expect about the same number of mutants to be induced in each culture. There should, in other words, be little fluctuation in the numbers from one culture to another. This doesn't mean there should be no variation at all, of course, since "sampling error" will make some variation inevitable. If, for example, the phage were inducing mutations at a frequency of 1 in 10 million—an average of ten per culture—then some variation around this figure would be expected from culture to culture, in accordance with well-known statistical laws. But the extent of this variation in the number of Ton^R cells from culture to culture would be considerably smaller than that expected from a "Darwinian" interpretation of their origin.

The actual experiment undertaken by Luria and Delbruck was very straightforward, given the above background information. Twenty small (0.2 ml) cultures of nutrient broth were seeded with a very small number of Ton^S E. coli. These cultures were allowed to grow for about 17 generations, by which time the number of cells in each tube had reached about 10^8 per ml. The T_1 phage were now added to each culture and the mixture immediately plated on solid nutrient agar plates. After incubating these plates until the Ton^R colonies could be counted, the investigators could compare the number growing on each of the plates. The data are given below.

Tube Number	Number of Ton^r Colonies
1.	1
2.	0
3.	3
4.	0
5.	0
6.	5
7.	0
8.	5
9.	0
10.	6
11.	107
12.	0
13.	0
14.	0
15.	1
16.	0
17.	0
18.	64
19.	0
20.	35

It is obvious that there is considerable variation in the number of Ton^R mutants that arose from one culture to the next. Before one concludes that these data settle the issue in favor of a "Darwinian" interpretation, however, there is a further question that must be addressed: is this fluctuation greater than what would be expected due to sampling error? Can one argue, for example, that the observed variation between the extremes of 0 and 107 is the normal range expected for a mean frequency of about 50 Ton^R mutants induced per cells? To answer this objection, the investigators initiated a single 10 ml. culture with a small number of Ton^S E. coli and grew in until it was composed of about the same density of cells that had been growing in the individual 0.2 ml. tubes. Ten separate 0.2 ml. samples of this "mass culture" were then mixed with T_1 phage, plated and incubated on solid nutrient agar. Note that each of these samplings from the mass culture contained about the same number of bacteria as did each of the 20 individual cultures described above. Hence, in both cases, the same number of cells are being exposed to the putative "mutagen", the T_1 phage. The only difference is that in this second case, the cells all originated from a single large culture, not 20 individual small ones. The results on this second set of plates are given below.

Tube Number	Number of $\underline{Ton}^R$ Colonies
1.	14
2.	15
3.	13
4.	21
5.	15
6.	14
7.	26
8.	16
9.	20
10.	13

There is a striking difference in the variation observed on these two sets of plates. The variation due to sampling error is seen to be very low—meaning, therefore, that the much greater variation observed within the 20 individual cultures was due to something else altogether. It was due, actually, to the fact that the $\underline{Ton}^R$ mutants arose at different times during the growth of the cultures—at times when there were no T_1 phage present. The phage could not, therefore, have induced the mutation to phage resistance; the phage only selected for survival the $\underline{Ton}^R$ that already existed in these cultures.

(Luria and Delbruck also rigorously analyzed mathematically the experimental results presented here. The statistical analysis supports the conclusions I have derived in this account "by inspection", as it were.)

THE NEWCOMBE EXPERIMENT[2]

For those bereft of mathematical intuition, a more penetrable proof was offered six years later by H. Newcombe. Newcombe also made use of the fact that wild-type *E. coli* are sensitive to attack and lysis by T_1 phage, but that mutations can, very rarely, arise in *E. coli* to make a cell and its descendants resistant to T_1. The question Newcombe set out to answer can be simply put: do $\underline{TON}^R$ mutations arise within a $\underline{TON}^S$ population because T_1 phage induce them? or, alternatively, do they arise randomly but become identifiable because T_1 phage subsequently select them? If it can be shown that the $\underline{TON}^R$ mutations actually arise before $\underline{TON}^S$ bacteria are

192

exposed to T_1 phage, the question can be unambiguously settled in favor of a neo-Darwinist interpretation.

To answer this question, Newcombe plated about 50,000 $\underline{TON}^S$ cells on each of 12 petri dishes and allowed each plate to incubate for about five hours. During this interval each of the 50,000 cells per plate would become a micro-colony of about 5,000 cells. Since every cell in a micro--colony is derived from a single parental cell, these micro-colonies each constitutes a clone.

Newcombe then took six of the twelve petri dishes and spread the cells in a uniform layer over the surface of the agar; the remaining six petri dishes he left untouched. Note that spreading the micro-colonies over the surface of the agar does not change the number of bacterial cells on the plate. The spread and unspread dishes contain the same number of cells, but their spacial orientation is different in the two sets.

Immediately after spreading the 6 petri dishes, he sprayed all 12 (i.e., 6 spread and 6 unspread) dishes with a fine mist containing concentrated T_1 phage and incubated the lot overnight. The next day the 6 spread dishes displayed a total of 353 $\underline{TON}^R$ colonies growing on them, the 6 unspread ones had only 28. From these data, Newcombe concluded that the T_1 phage merely selected but did not induce the mutation to $\underline{TON}^R$. This argument runs as follows:

- The mutation from $\underline{TON}^S$ to $\underline{TON}^R$ we know to be an exceedingly rare event, one that occurs under the best of conditions at a frequency of less than 10^{-6}.

- A micro-colony of 5,000 $\underline{TON}^S$ cells, therefore, could never contain more than one $\underline{TON}^R$ mutant, if the mutation were induced by the T_1 phage sprayed on   at the end of the 5-hour incubation period.

- Both spread and unspread plates contain equal numbers of bacteria at the end of the first incubation, and hence equal numbers of cells were exposed to T_1 in each case.

- If the phage were inducing $\underline{TON}^R$ mutations, equal numbers of the rare $\underline{TON}^R$ would be induced on each set, since their frequency would be proportional only to the number of cells exposed to the "mutagen", supposedly the T_1 phage.

- If $\underline{TON}^R$ mutants arose spontaneously before T_1 phage were added, a micro-colony would probably contain more than one $\underline{TON}^R$ cell, since once the $\underline{TON}^R$ mutation arises in one cell, all its descendants within the micro-colony inherit the change. Spreading such a clone permits each $\underline{TON}^R$ cell to grow into a clone in its own right, during the overnight incubation.

-The fact that spreading did greatly increase the number of $\underline{TON}^R$ colonies compared to the number on the unspread plates demonstrates that the mutation to T_1-resistance must have occurred before the T_1 phage were added to the bacteria;

193

the role of the T_1 phage, therefore, its limited to one of selection: selection in favor of those cells which had already acquired T_1 resistance, and against those which had not.

THE LEDERBERGS' EXPERIMENT[3]

Three years later an even simpler disproof of the neo-Lamarckian interpretation of mutation was published by J. and E. Lederberg. Their technique also used $\underline{TON}^S$ (wild-type) *E. coli* and selected for rare $\underline{TON}^R$ mutations arising within the $\underline{TON}^S$ population. A cloned culture of $\underline{TON}^S$ cells was densely plated onto a petri dish and incubated until a confluent lawn of bacteria resulted. In addition, six other petri plates were prepared with an appropriate *E. coli* nutrient medium. These plates were left without bacteria, but were instead sprayed with a concentrated preparation of T_1 phage. At this point, then, the materials for the experiment consisted of a $\underline{TON}^S$ master plate, and six additional bacteria-free plates whose nutrient agar had been sprayed with T_1 phage.

The Lederbergs then obtained a cylinder whose diameter closely matched that of the petri dishes and covered one end of it with a tight-fitting piece of velvet cloth. The velvet surface of the cylinder was first carefully pressed against the bacterial surface of the master plate and then sequentially applied lightly to the surface of the six so-called "replica plates". The orientation of the velvet surface relative to the surface of the petri dish was noted when the velvet was first pressed onto the master plate, and it was kept the same when pressed in turn to the phage-agar surface of each replica plate. The six replica plates were then incubated overnight in order to allow for the growth of any $\underline{TON}^R$ colonies on the selecting medium.

The next day it was observed that all six replica plates had a few $\underline{TON}^R$ colonies growing up overnight. More significantly, all the replica plates had the same number of $\underline{TON}^R$ colonies, and the spatial orientation of those colonies, moreover, was identical in all six. These $\underline{TON}^R$ colonies must have been transferred to each successive replica plate by the velvet, which had picked them up originally from the master plate. But the master plate, although it evidently contained $\underline{TON}^R$ cells, had never itself been exposed to T_1 phage. The mutation from $\underline{TON}^S$ to $\underline{TON}^R$ arises, therefore, in the absence of T_1 phage.

At present we know of no agent that will produce specific, predetermined changes in gene function.

REFERENCES

1. Luria, S. and M. Delbruck, 1943. Genetics $\underline{28}$: 491

2. Newcombe, H., 1949. Nature $\underline{164}$: 150.

3. Lederberg, J. and E. Lederberg, 1952. J. Bacteriol. $\underline{63}$: 399.

LECTURE 33

MUTATION: SPONTANEOUS AND INDUCED

INTRODUCTION

The first chemical mutagen known, **nitrogen mustard**, was discovered through the work of Charlotte Auerbach in 1944. Like many other mutagens found since, there is still little known about the precise way it causes heritable changes in the DNA. In this section we shall discuss exclusively those chemicals for which we have some experimental basis to propose a mutagenic mechanism. Broadly speaking, these chemicals fall into two categories: base analogues or base analogue-type mutagens, on one hand, and acridine-type mutagens, on the other. Before either type of mutagen can be studied, however, some basic definitions and concepts must be understood.

TYPES OF MUTAGENIC EVENTS WITHIN GENES

A hereditary change in a wild-type gene leading to a mutation (e.g., a^+ to $\underline{a}$) is termed a **forward mutation** or "a mutation in the forward direction". Conversely, a hereditary change that converts a mutant gene into a wild-type (e.g., $\underline{a}$ to $\underline{a}^+$) is called a **reverse mutation**. Since any number of changes within a wild-type gene can lead to a perceptible change in its function (i.e., to a forward mutation), but only one or a very limited number of changes can reverse the alteration which led to the mutation in the first place, reverse mutations will occur much less frequently than forward ones.

In going in the forward direction, the intra-genic changes may affect one or more of the base-pairs composing the gene. If only one base-pair is affected (e.g., by being changed to another base-pair or by being deleted), the change is known as a **point** or **single-site mutation.** Because only one base-pair site is affected, point mutations can usually revert spontaneously or by induction with a mutagen. (**Mutagens**—agents that change genes—will change the gene in both the forward and the reverse direction). If the mutation involves more than one base-pair, it is said to be **multi-site**, the most common of which are deletions of two or more contiguous base-pairs from the gene. Such mutations rarely revert to wild-type spontaneously or by induction, since the likelihood of randomly restoring exactly the precise sequence that was originally altered has a vanishingly small probability. Hence, a mutation that does not revert to wild-type is usually considered to be multi-site in origin, and one that does revert is usually considered to be single-site—although we shall presently see some interesting exceptions to this rule.

Spontaneous mutations, as argued in the previous lecture, are those that arise without apparent cause. Nothing, however, happens without *some* cause, and there have been some heuristically interesting attempts to link spontaneous mutation to chemical or physical events that could plausibly occur by chance in DNA. The best-known of these are Watson and Crick's speculations on the results of spontaneous tautomerization of the normal DNA bases.

SPONTANEOUS TAUTOMERIZATION OF BASES

Watson and Crick postulated that a base of DNA may on occasion undergo a **tautomeric shift** that would alter its pairing properties. If, for example, thymine was tautomerized to its enol form or adenine to its imino configuration, and if this shift happened at a time when that base was seeking a complement during DNA replication, then an enol-thymine could pair satisfactorily with guanine, or an imino-adenine could pair with cytosine. But since enol-thymine and imino-adenine are extremely unstable molecules, it is more than likely that the next round of DNA replications would find them back in their normal configuration and pairing with their steady partners, A or T, respectively. The G and C, which were their previous partners, go on to pair with their usual complements, C and G. Now let's look in summary at what has happened, focusing on the consequences of a tautomeric shift in the thymine of a particular A/T base-pair somewhere in a DNA molecule:

1. Thymine tautomerizes to the enol configuration: A/T → A/T*

2. The DNA strand replicates to form two daughter molecules: A/T and G/T*

3. The enol-thymine reverts to its more stable form: G/T

4. This molecule of DNA replications to form two daughter molecules: G/C and A/T.

5. What was, therefore, an A/T site in the starting DNA molecule has become a G/C site—a mutation—in one of its grand-daughter molecules.

Watson and Crick's explanation, in summary, envisioned spontaneous mutations arising as a consequence of well-understood physical-chemical events that might occur during semi-conservative replication of a double-stranded DNA molecule.

Later in this section we shall see that this seemingly apodictic explanation can at most account for about 15% of the spontaneous mutations that occur in phage and probably in other organisms too. Nevertheless, the theory vaticinates many of the ideas we have now come to accept in explanation of chemically induced mutation.

A more recent study by Tobal and Fresco takes Watson and Crick's speculations a bit further. These investigators propose that enol-keto or amino-imino tautomerization and anti-syn tautomerization (rotation of the base around its glycosidic bond from its normal *anti* position through 180° to an uncommon *syn* position) enable a double helix to accommodate the following "complementary mispairs" without intolerable distortion:

A = C imino
G = T enol
A imino = C
G enol = T
A imino = A syn
G enol = G syn
G imino = G syn
A imino = G syn
G enol = A syn
G imino = A syn

Note that purine-purine pairs are possible, but not pyrimidine-pyrimidine ones. It is possible to calculate the frequency with which these complementary mispairings occur under physiological conditions. It turns out to be surprisingly high: one in 10^4-10^5 for the enol or imino forms and one in 10-20 for the syn forms. Of course, this is much higher than the known rate of spontaneous mutation: one in 10^9-10^{12} base pairs replicated. The authors than propose that the number of errors is reduced by the editing mechanism that is discussed elsewhere in these lectures. Thermodynamic considerations also yield an estimate of the fraction of complementary mispairs that would survive the error-correcting process. This final figure closely resembles the known frequency of spontaneous mutations.

THE DISTRIBUTION OF SPONTANEOUS MUTATIONS WITHIN GENES

There is yet another aspect of spontaneous mutations that should not go unnoticed. It is possible to collect hundreds of rIIA or B mutons, for example, and to map their location within the cistron. One quickly discovers that they are not randomly distributed; some base-pair sites are extremely mutable whereas others are seemingly very resistant to mutation. The former are termed **hot spots** and the latter, **silent areas**. Since there are only four base-pairs that can occur anywhere in DNA (i.e., GC, CG, AT, TA) and since there are only one or a few hot spots per cistron, it is difficult to know how to account for them in molecular terms. It would be absurd to think that a hot spot represented the unique occurrence of one of the four base-pair combinations. We usually ascribe hot spots and silent areas to "neighborhood influences", which isn't saying much.

A more quantifiable explanation for hot spots is that they are sites of 5-methyl cytosine (see the Lecture on "DNA Repair"), which indeed many of them are. Silent areas are likely areas of the gene that code for parts of a polypeptide wherein a certain amount of variation is possible without alteration in function.

BASE ANALOGUE MUTAGENESIS

The compounds that have been most extensively studied as mutagens fall into the category of **base analogues**, compounds that mimic to some degree the behavior of the natural bases of DNA. **5-Bromouracil** (more correctly, 5-bromo-deoxy-uridine, and more conveniently, 5-BU) provides a good illustration of base analogue mutagenesis. When this compound is added to an *in vivo* system where DNA replication is occurring (e.g. to phage-infected bacteria), the 5-BU may be "mistaken" (assuming non-sentient things can make mistakes) for thymine (a.k.a. 5-methyluracil) and incorporated into replicating DNA opposite adenine. As long as this incorporated 5-BU continues to behave chemically in a manner indistinguishable from thymine, there is no particular problem. However, 5-BU is more likely to tautomerize to the enol form than is thymine, and 5-BU in the enol configuration will pair with guanine, not adenine. Consequently, as the DNA continues to replicate, the incorporated 5-BU pairs with guanine, which, on further replication pairs, of course, with cytosine. That is, what started as an AT pair at a particular site in the DNA molecule, becomes first an A-BU site, then a G-BU site. This would take at minimum *three* rounds of DNA replication, although in practice many rounds of replication may intervene between the time 5-BU was first incorporated and when it "misbehaves".

The change from AT to GC is in effect the substitution of one purine for another (G for A) in one strand, and in the other the substitution of one pyrimidine for another (C for T). This is called a **transition mutation**. And because it comes about through the mispairing during replication of an *already incorporated* base analogue, it is called an **error of replication**. In actual fact, errors of replication with 5-BU are not very common, since the polymerization of the base analogue tends to reduce its likelihood of tautomerizing.

A more common occurrence would be the initial "misincorporation" of 5-BU (in its less likely enol form) opposite G. During a subsequent replication the 5-BU tautomerizes to its more usual form and pairs with A. This A in a subsequent replication will pair, of course, with T. The effect has been to change an initial GC site to AT. This, too, is a transition, but is not an error of replication. Rather, it is an **error or incorporation**, since 5-BU was *initially incorporated* opposite its less likely pairing partner.

Since 5-BU engineers more GC to AT changes (errors of incorporation), it is said to have a preferred direction of mutagenic action. Nonetheless, it would still have to be classified as a two-way transitional agent, since some errors of replication do occur with 5-BU.

The base analogue **2-amino purine** (2-AP) is also a two-way transitional agent, pairing normally with T, but frequently with C, too.

Nitrous acid (HNO_2) is not itself a base analogue. It acts on resting DNA to deaminate cytosine to uracil, adenine to hypoxanthine, and guanine to xanthine. In subsequent replication of the DNA, the uracil will obviously pair with adenine, and at the next replication, the A will pair with T. Consequently, HNO_2 will cause GC to AT transitions. It will also cause AT to GC transitions, since hypoxanthine pairs with C, not T.

All agents discussed so far cause to some degree two-way transitions. One base analogue-type mutagen which causes unique transitions is HA— **hydroxylamine** (NH_2OH). It attacks cytosine in resting DNA to initiate an obscure chain of breakdown products, one of which pairs with A if replication were to take place subsequently. Hence HA will cause only GC to AT transitions.

The theoretical framework outlined above contains the basics of what is called eponymously Freeze's mutation theory. If it is correct, it allows some speculation on the nature of individual mutons in DNA. If a given mutation reverts to wild-type, it can first be classified as a point mutation, i.e., originating only at a given base-pair site. If it is induced to revert to normal at a higher than spontaneous frequency by 2AP or NA, it is a transition. If it is sluggish to revert to normal under the influence of 5-BU, it may be AT at the mutant site. If HA does not induce it to revert, this identification becomes more likely. If on the other hand, it reverts well in response both to 5-BU and HA, it seems likely that the mutant site is GC.

The theory derives some support from what may be termed internal evidence. On the basis of the theory, one would not expect mutants originally induced in the forward direction (i.e. from wild-type to mutant) by HA to be reverted to normal by HA or at a high rate by 5-BU. But they should revert fairly well by 2-AP or NA. This is usually the case, and other predictions of this sort also work out.

There are objections, however, to these types of studies. Chief among them is the frequent lack of proof that the change which reverts a given mutation to normal is in fact the exact reversal of the change that gave rise to the mutation in the first place. Other interpretations are certainly possible (e.g. a second transition closely linked to the first, which restores function to the protein product. Nevertheless, for certain types of agents, Freeze's theory remains quite enticing. (Needless to say, the advances in direct DNA sequencing detailed in earlier Lectures have made these types if indirect methods entirely obsolete.)

TRANSVERSIONS

There are, though, many mutations that cannot be induced to revert through the action of any known base analogue. Included in this category is the vast majority of spontaneous mutations. Originally these recalcitrant mutations (which do nonetheless revert spontaneously) were called **transversions** by Freeze, meaning changes that altered the purine or pyrimidine content of each DNA strand. Although some spontaneous mutations may indeed be transversions, we shall see that many of them are not. Nor are they transitions. They are single or multiple base-pair additions or deletions. All that can be said for transversions at this point in history is that they may exist, but we have discovered no agent that will specifically induce or revert them.

One feature of the Tobal and Fresco hypothesis (*vide supra*) which has relevance here has to do with the mutagenic consequences of their proposed "complementary mispairs". Working out the steps that follow such mispairing, one notes that only the purine-pyrimidine mispairs lead to transitions and only the purine-purine mispairs to transversions, classically defined. The mutagen, 5-BU, a pyrimidine, can only form purine-pyrimidine complements, all of which lead to transitions. The purine, 2-AP, might be thought to form some purine-purine pairs and hence to yield transversions. Detailed analysis of just what these particular 2-AP-purine complements could be however, shows that none of them, in fact are structurally possible. This mutagen remains, therefore, exclusively a transition agent, as Freeze originally proposed.

OTHER TYPES OF MUTATIONS

Besides events that involve base substitutions, additions, and deletions, there are many other events happening at the level of the DNA that can occasion profound phenotypic effects. One of the most bizarre is the recently discovered phenomenon of **triplet repeat mutations** in humans. Of these, the best characterized so far is the mutation causing **fragile-X syndrome**. This disease involves moderate to severe mental retardation (actually one of the most common types, with a frequency of 1 in 1250 males and 1 in 2500 females), large ears, head and testicles, and a long face. The name "fragile-X" derives from the fact that the X-chromosome from affected individuals tends to break in a specific location in tissue culture. Molecular studies of this break site shows it to comprise a vast repetition of the sequence CGG—some 250 to 4000 times. Even normal people have some repetition of this sequence, usually about 6 to 54 copies, with an average of about 29. Males usually transmit to their offspring about the same number of repeats that they themselves inherited, but in some females the copy number may increase during meiosis. Individuals with copy numbers somewhat outside the normal range are usually still unaffected by the syndrome, but as the copy number increases, the stage seems set for them in turn (if female) to have affected offspring with the very large copy numbers characteristic of the disease. The severity of the disease seems linked to the copy number, although precise correlations have not been done as yet. (This disease is also discussed as an example of gene imprinting in Lecture 17.)

LECTURE 34

FRAMESHIFT MUTATIONS

PROFLAVIN AS A MUTAGEN

Although few spontaneous mutations can be reverted by base analogues, there are agents that will revert them. Of particular significance in this respect is **proflavin**, one of the family of acridine dyes. Proflavin is a mutagen capable of inducing both forward and reverse mutations in microorganisms. Proflavin will revert, in addition to many spontaneous mutations, proflavin-induced mutations. Most proflavin-induced mutations will also revert spontaneously, but very few of them will be induced to revert by any base-analogue type mutagen. Moreover proflavin will revert almost none of the mutants induced by base-analogues. In other words, mutations induced by proflavin have much in common with those that occur spontaneously, but little in common with those of base-analogue origin.

Proflavin has other interesting effects too, as first pointed out by Crick and others in 1961.[1] In an experiment with T4 to test the relative effectiveness of 5-BU and proflavin in inducing rII and host-range (h) mutations, it was found that both agents were capable of inducing rII mutations, but only 5-BU would induce recoverable h mutations with any efficiency. A clue to understanding this difference can be found by analyzing the likely difference between the r and h gene products. The r locus makes a dispensable product (if the phage knows enough to avoid K bacteria), whereas the h gene specifies an essential protein which can be modified slightly but yet cannot be too greatly altered and cannot be eliminated. If true, this suggests that proflavin was having considerably greater effect on the functioning of a gene than does 5-BU. The logic here is close, though.

FRAME-SHIFT MUTATIONS

From this observation evolved the notion that proflavin adds or deletes base-pairs to the DNA. When the gene is transcribed into m-RNA, this message presumably has a base missing or in excess, with the result that in translation the reading frame of the message is thrown off for all triplets beyond the insertion or deletion, and the amino acid sequences of the nascent peptide become completely in error beyond that point. Whereas, therefore, the actual change in the DNA itself is relatively slight and confined to a particular base-pair site, the ramifications of this change are drastic and far-reaching in the polypeptide product of that gene. On the face of it, one would expect that proflavin-induced mutations (which we will presumptuously call **frame-shift mutations** from now on) would be resistant to reversal, since the chances of inserting or deleting just that base-pair which was initially deleted or inserted seem exceedingly remote. The facts, however, are otherwise, since frame-shift mutations, like many spontaneous ones, do revert with fair frequency. But they, again like spontaneous mutations, can only be reduced to revert by other frameshift agents, not by base analogue-type mutagens. A close analysis of these revertants, moreover, reveals that they are not genuine revertants at all, but rather double mutants that nonetheless have most of the characteristics of the wild-type. In other words, the specter of a second mutation, closely linked to the primary mutant site and suppressing its effects—a theoretical consideration we raised in connection with Freeze's scheme—has come back to haunt us again. This time with considerably more substances: the second mutation can actually be isolated (by recombination) free from the first, and is found *by itself* to be a mutation in the same gene and to sport a mutant phenotype in no way different from that of the mutation it reversed.

The explanation offered by Crick is quite simple. Supposing, the reasoning goes, the first mutation were a single base-pair addition (a "plus sign" mutation), then the triplets (in the transcribed m-RNA) read to the right of this change would be thrown out of register by one base, and the polypeptide product would obviously be inactive. A closely linked deletion mutation (a "minus sign" mutation) to either side of it would correct (or be corrected by) the first change, so that only the interval between the two lesions would be misread. As long as this region were not essential to the functioning of the translated polypeptide, or the misread region did not uncover nonsense triplets, the restored product could have close to normal biological activity.

In consequence, it should be possible to collect families of intra-cistronic mutations (mutons) having the same sign, to put them together on the same chromosome, and to predict the results. For example, any mutation that suppresses a given reference mutation—the reference arbitrarily defined here as a "plus"—should be a minus; any that suppresses this suppressor should also be a "plus", and so on. Two plus mutations together should yield a mutant phenotype, and similarly for two minus mutations. On the other hand, three pluses or three minuses should yield a pseudowild phenotype. By and large the system does work as outlined here, as demonstrated first and most elegantly by Crick for families of frameshift mutations of the rII cistrons.

Crick observed another effect of proflavin which substantially enhanced the plausibility of his theory. Normally the cistrons rIIA and rIIB, each having its own initiator and terminator, behave independently during translation into their respective polypeptides—whatever they may be. There exists, however, a well-characterized deletion mutation, rII 1589, which excises the right-hand end portion of rIIA and the beginning part of rIIB, such that during translation of the rII 1589 messenger RNA a single run-on polypeptide is produced from the rII region. This polypeptide has no remaining rIIA function, but does, strangely, retain full rIIB function, the initial part of rIIB (the so-called B1 region of this cistron) being apparently unnecessary for full rIIB activity and the extraneous rIIA polypeptide fragment tacked on apparently not interfering with it.

Transition or transversion mutations induced in rIIA of rII 1589 phage would not, of course, interfere in any way with the rIIB activity of the run-on polypeptide (unless such point mutations result in chain terminating codons: see later sections of these notes). In consequence, most base analogue-induced mutations in rIIA of such phage do not result in loss of rIIB activity. Proflavin-induced mutants in rIIA, by contrast, usually do obliterate rIIB activity. The best explanation for this observation is, again, that proflavin mutants throw off the reading frame of the m-RNA, and this misreading carries through to the rIIB cistron when rII 1589 converts the whole rII region into a single unit of translation.

(Parenthetically, it can also be observed that many spontaneous deletions of rIIA, presumably those that contain a non-integral number of codons, will also obliterate rIIB activity in rII 1589 phage).

DEDUCING THE BASE SEQUENCE OF DNA

Crick's frame-shift theory can also be convincingly verified through biochemical analyses that reveal the actual amino acid sequence of mutant and pseudowild peptides resulting from proflavin treatment. The gene (e) that specifies the enzyme lysozyme in T4 is especially useful for these studies since, unlike the rII products, lysozyme can be readily isolated and sequenced. The results of Terzaghi's study of this system can best be summarized as follows:

```
   e+ ... thr lys ser pro ser leu asn ala ...
              AGU CCA UCA CUU AAU
              GUC CAU CAC UUA AUG
double ... thr lys val his his leu met ala ...
mutant
```

Note that a unique base sequence for this part of the lysozyme messenger can be inferred, and that the double mutant arose as a result of a deletion of the underlined A and an addition of the underlined G. This is the only sequence capable of linking the normal, singly mutant, and restored (pseudo-wild) amino acid sequences.

Later studies of the lysozyme system showed that proflavin can at times also occasion double base-pair additions and deletions. (And, of course, later technical developments that permitted the ready direct sequencing of DNA rendered such indirect methods of sequence analysis outlined here completely obsolete.)

Finally, the study of frameshift mutations provided one clue to the mystery surrounding the origin of certain mutational hot spots, discussed earlier. Here, however, the problem is cast in somewhat different terms, as the site in question, ejDII of the lysozyme gene, reverts spontaneously to wild-type at a frequency a hundred times or more that of any other muton in the cistron. (Our earlier discussion of hot spots, it will be recalled, focused on forward mutations). Moreover, these reversions are occurring at the ejDII site itself and lead to an exact restoration of the wild-type amino acid sequence in the lysozyme. This clearly differentiates this phenomenon from that observed at other frame-shift sites, where reversions are second frame-shift mutations closely linked to, not distinct from, the initial mutant sites.

Through analysis of the mutant and restored amino acid sequence governed by eJDII and its surrounding base-pairs, it was determined that eJDII represents the insertion of an A/T pair into a region of DNA that already contains five A/T pairs in sequence. The addition of the sixth A/T apparently introduces a replicative instability which makes it increasingly probable that the inter-loping A/T will be lost.

It is obvious, of course, that this observation is in no sense an explanation of hot spots, since it leaves unanswered the question of why the replicating enzymes should become jittery over a sequence of six A/Ts. It does, however, begin the definition of inherently unstable DNA sequences, and that, presumably, is the first step necessary for any explanation.

At this point we are faced with two complementary mutation theories: one for base analogue-type mutagens and one for proflavin-type mutagens. Undoubtedly, there is room for others, too, since not all mutagenic agents fit neatly into one category or the other.

INTENTIONAL FRAME-SHIFTING DURING TRANSLATION

On occasion, organisms make intentional use of the alternative coding possibilities afforded by frame-shifting during translation. In *E. coli*, for example, it is known that a particular release factor (RF2), which is required for normal termination of translation, will reprogram about one-third of the ribosomes to change to the +1 reading frame on encountering particular UGA codons in the messenger RNA. The result, of course, is a continuation of translation, rather than termination. (Other signals, such as an upstream sequence in the m-RNA, are also required for this frame shift to occur.) A somewhat similar case is now also known for a gene in mammals. Certain retroviruses undergo a -1 frame shift at a specific 6-nucleotide sequence in the m-RNA, and this, too, is propitiated by sequences at other points in the m-RNA being translated. Although the full extent of such **recoding** events is still not known, it is clear that frame-shifting has greater significance in the genetic message than just as a mutagenic event.

TESTING FOR MUTAGENS[2]

The ability of almost all carcinogenic (cancer-producing) compounds to induce gene mutations as well has led to an efficient bacterial test for carcinogenicity. The bacteria are a specially manipulated strain of *Salmonella* which lack the ability to produce their own histidine, one of the amino acids found in most proteins. The cause of the deficiency is a mutation in one of the genes governing histidine biosynthesis. Besides this defect, the *Salmonella* are defective in some of the genes involved in DNA repair and demonstrate an enhanced error rate during SOS repair. (For a discussion of DNA repair, see the following Lecture.) Finally, their cell membranes are defective in order to make the DNA accessible to a much greater variety of externally applied compounds than would normally succeed in gaining entry to the cell.

Compounds whose mutagenicity is being tested are mixed with the special *Salmonella* and the cells are plated on a nutrient agar plate that lacks histidine. The only cells capable of forming colonies under these circumstances are those that are back-mutated in the his⁻ gene by the added compound. In other words, if the cells to which the suspected mutagen, had been added form significantly more colonies than a control group without the test compound, then that substance is a mutagen—and most likely a carcinogen, too.

When this test—called the **Ames test** after its inventor, B. Ames—was first used to screen suspected mutagens, a few substances known from animal studies to be highly mutagenic or carcinogenic were strangely ineffective in reverting the his- mutation in bacteria. The reason was soon discovered to have more to do with the test animal than the compound being tested, for it happens that many substances, harmless in themselves, are converted into mutagens by enzymes in the mammalian liver. Our livers would do the same thing. This problem is obviated by adding a preparation derived from rat liver to the nutrient broth for the bacteria. If these enzymes convert the test compound into an active mutagen, as would happen in our own bodies, the compound will now register as a mutagen in the bacterial system.

REFERENCES

1. Crick, F., 1962. The genetic code. Sci. Amer. (Oct.)

2. Devoret, R., 1979. Bacterial tests for potential carcinogens. Sci. Amer. (Aug.)

LECTURE 35

DNA REPAIR

INTRODUCTION

Although the physical-chemical complementarily of the two pairs of DNA bases assures reasonably accurate replication of the molecule, this principle alone cannot account for the very low mutation rates that characterize biological inheritance. It has been calculated, for example, that about one gene in ten would experience an alteration in its base-pair sequence—in effect, a mutation—each cell generation if organisms had not evolved mechanisms to correct the inevitable errors such a replicative process would engender. The actual observed rate of mutation is several orders of magnitude lower than that. The mechanisms that keep the mutation rate within acceptable bounds are many and correspond to the diverse ways in which alterations in the DNA sequence may arise.[1]

EDITING REPAIR OR PROOFREADING

The most straightforward error-correcting mechanism is known as 3' to 5' **editing** or **proofreading** and rests with a special property of both the DNA polymerizing enzymes, pol I and pol III. Both these enzymes have, besides their obvious 5' to 3' polymerizing activity, a 3' to 5' exonuclease activity, as well. This seemingly antagonistic function only comes into play when the polymerase has mistakenly attached an "improper" (i.e., non-complementary) base to the growing 3' end of the chain. The resultant distortion of the growing end is detected by the polymerase and the newly-made phosphodiester bond is immediately cleaved by this same polymerase, using its 3' to 5' exonuclease activity. In effect, the polymerase always gives itself another chance to get it right, continuously checking the accuracy of its base selections before adding a subsequent one. Wrong bases are quickly excised, correct ones added in their place, and the polymerase moves on.[*]

---

[*]The necessity for an editing function to ensure fidelity of replication may explain why a 3' to 5' polymerizing function has never evolved. The energy for the formation of the new phosphodiester bond comes from splitting the beta and gamma phosphates from the incoming nucleoside-5'-triphosphate. If growth were in the 3' to 5' direction, the beta and gamma phosphates released would have to be those at the 5' end of the growing chain, not those at the 5' end of the incoming nucleotide. Now, if this newly added nucleotide had been incorrectly added and subsequently cleaved (by a 5' to 3' exonuclease of some sort) the resultant 5' end would no longer be a triphosphate, but at best a monophosphate. The 5' end would no longer be able to accept a new nucleotide and chain elongation would cease.

MISMATCH REPAIR

Even so, some mismatches do slip by the editing processes operating at the replicating fork, causing occasional "odd couples" within the newly made double helix behind it. The repair of these falls to a series of enzymes that constitutes the **mismatch repair system**. These enzymes constantly monitor the DNA to detect mismatches by the distortion in the helix they cause. But the mismatch repair system confronts an immediate and obvious dilemma—which is the correct base and which the error? Should a mismatched TG pair, for example, be repaired to TA or to CG?

Mismatches, as noted above, usually occur as a misincorporation that passed undetected through the editing processes at the replicating fork. In consequence, the incorrect base will be the one in the newly made strand and the correct one in the parental strand of the double helix. Is there a way these two strands can still be discriminated once the fork has moved well beyond? There is. The DNA from most natural sources usually is extensively methylated. Cells employ special enzymes—methylases—to add methyl groups to particular bases, when these bases occur in specific sequences within the molecule. It is the adenine of the sequence GATC, for example, that is frequently methylated in *E. coli*. The activity of these methylases lags well behind the replicating fork, however, and in consequence for awhile the newly made strand will be under methylated in comparison with the parental strand (which, of course, was made at least one cell generation earlier). This difference in methylation can be detected by the mismatch repair enzymes; repair, therefore, will always occur on the undermethylated strand.[*]

The steps leading to the repair itself were once thought to be fairly straightforward and common to several of the repair systems we shall be examining (see "short patch excision repair" below). However, recent success in getting mismatch repair to occur *in vitro*[2] has shown that it actually is accomplished in the following manner. A protein (the product of a gene called MutH) puts a single cut in the phosphodiester backbone of the strand being repaired. This cut occurs on the 5' side of the mismatch, but may be 1000 or more base-pairs distal to it. The MutH protein recognizes the sequence GATC and incises the DNA at the beginning of the *unmethylated* GATC sequence. This is the key decision that then directs the subsequent steps in the repair process to the correct strand. Although all the proteins needed for these subsequent steps have been

---

[*]Methylation of DNA in bacteria is the basis for the phenomenon of host restriction and the explanation for why restriction endonucleases (see the section on Recombinant DNA) do not attack the DNA of the host cell. In a particular strain of bacteria the target sequences for the restriction enzymes it carries are methylated at specific base-pair sites to render them unrecognizable to the enzymes. Foreign DNA entering the cell will have a different methylation pattern from the host. The target sequences are not likely to be appropriately methylated, therefore, and this DNA can be degraded by the vigilant restriction enzymes of the host.

identified,[*] the exact means by which the large segment of DNA between the mismatch (and, of course, containing it) and the cut at the unmethylated GATC sequence is removed and correctly resynthesized has yet to be understood in detail. It is clear, however, that the synthesizing enzyme must be pol III and that the final phosphodiester bond that seals the DNA is catalyzed by DNA ligase. Moreover, it is likely that helicase II is unwinding the cut DNA as pol III progresses and that the discarded fragment is digested by exonuclease I.

Of the eight possible mismatches (G/T, G/G, T/T, A/A, A/C, C/T, A/G, and C/C), only C/C seems refractory to this **methyl-directed repair** mechanism, at least *in vitro*.

dUTP REPAIR

The biochemical pathways that yield the deoxyribonucleoside-5'-triphosphates necessary for DNA synthesis occasionally result in the synthesis of dUTP, as well. The polymerizing enzymes cannot readily distinguish TTP from dUTP, as both make effective hydrogen bonds with adenine. There are two mechanisms that prevent the stable incorporation of uracil into DNA. The first of these operates before misincorporation can even occur; so it is not, strictly speaking, a repair process at all. Most cells contain an enzyme, **dUTPase,** whose function it is to cleave the terminal two phosphates from dUTP, preventing its incorporation into DNA.

Some dUTP will inevitably escape castration by dUTPase, however, and the DNA polymerase will on occasion incorporate it into DNA, in place of thymine. For reasons that will become obvious later on in this account, the presence of uracil in DNA can be highly mischievous; it must be removed. To do so, an enzyme known as **Uracil-N-glycosylase** (UNG) is employed. This enzyme constantly monitors DNA for the presence of uracil, and when it finds it, the unwanted base is removed. It does so by cleaving the glycosidic bond between the base and the sugar; the free uracil is released, but the sugar-phosphate "backbone" of the strand remains intact. The result is a gap in the nucleotide sequence of the strand, but none in the backbone.

Another enzyme, **AP endonuclease**[**], detects this lacuna in the sequence and puts the necessary cut in the backbone, on the 5' side of the phantom uracil. This cut creates, of course, a new 3' and a new 5' end. The cut is not directly beside the missing base but about 12 bases away. Another enzyme with a 5' to 3' exonuclease activity, probably pol I, then begins to chew away a section of the strand until the missing base—and four or five "good" bases on its 3' side, as well—are eliminated. (Or, alternatively, the 12- to 13-nucleotide strip is removed and degraded

---

[*]They are the two protein products of the genes MutL and MutS; DNA helicase II; SSB proteins; DNA pol III; DNA ligase; ATP; and the four deoxynucleotide triphosphates.

[**]AP endonuclease stands for "apurinic acid endonuclease". Yes, I know uracil is a pyrimidine, not a purine. The reason for this name is historical and will become clear later in this account.

later by an enzyme acting on single-stranded DNA.) To fill the gap, pol I then adds the correct bases by lengthening the new 3' end created by the endonuclease. (Pol I could do both the excision and the synthesis simultaneously by **nick translation**, in effect moving the nick right through the lesion and thereby correcting it.) The rejoining of the backbone is done by ligase, as in the final joining of the Okazaki fragments during DNA replication. This excision of a flawed segment and the resynthesis of it in the correct manner is known as **short patch excision repair**. It was until recently thought that this short-patch excision repair was the only type that existed, but recently there has been found another, termed **long patch excision repair.** In this instance, the excised patch is about 1000 nucleotides long and the gap is filled by pol lll, not pol l. Long patch excision repair is very similar to the mismatch repair system discussed above, except that it need not be methyl-directed. When excision repair is more carefully scrutinized in the future, it too may turn out to be even more complicated than outlined here.

There is another way in which uracil can find itself illegitimately in DNA: by the spontaneous deamination of cytosine. If this uracil were allowed to remain, it would have very pernicious effects when the DNA containing it

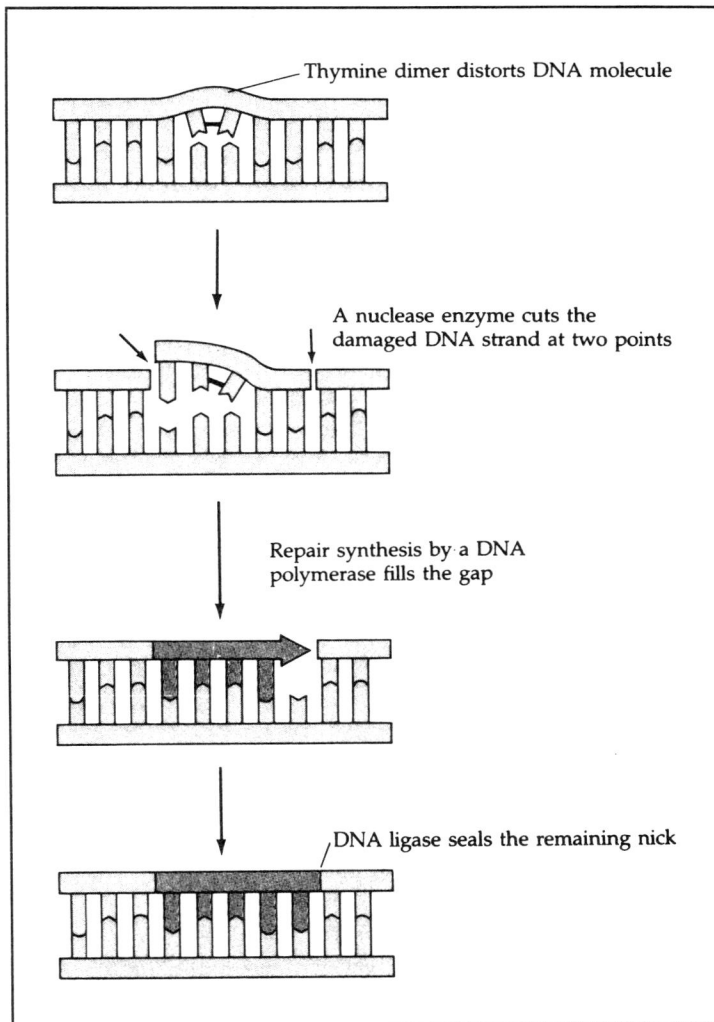

Thymine dimer distorts DNA molecule

A nuclease enzyme cuts the damaged DNA strand at two points

Repair synthesis by a DNA polymerase fills the gap

DNA ligase seals the remaining nick

Figure 35.1 Excision repair of DNA. From Biology, Second Edition, by Neil Campbell (Redwood City, CA: Benjamin/Cummings Publishing Company, 1990) p. 319. Reprinted by permission.

subsequently replicated, since uracil pairs not with guanine—its current partner, since it arose from cytosine—but with adenine. Hence an original GC site in the parental DNA would become an AU site in the next molecular generation and an AT site thereafter. The repair of the GU site proceeds in the same manner as for an AU site arising from misincorporation: uracil-N-glycolsylase and AP endonuclease activity, followed by excision repair.

The ability to repair GU sites arising from the deamination of cytosine probably explains why DNA contains thymine instead of uracil in the first place. If uracil were a normal constituent of DNA, there would be no way of differentiating between a GU pair that arose through the misincorporation of G (and warranting the repair of the G-containing strand) and one that arose through the deamination of cytosine (and warranting the repair of the dU-containing strand). When thymine is the normal constituent, all uracil in the molecule is illegitimate and the strand that contains it is the defective one; the molecule can then be repaired unambiguously.

In many organisms some of the cytosine bases have a methyl group at position 5 of the pyrimidine ring. If this 5-methyl cytosine is deaminated, the resultant base, 5-methyl uracil, is one more familiar by its common name of "thymine". The GT base-pair site that now exists in the DNA cannot be faultlessly repaired, as there is no way of ascertaining which of the two strands contains the error. Many of the DNA sites known to have very high mutation rates—"hot spots", they are called—have been shown to be 5MC/G sites in the wild-type molecule.

FALSE OKAZAKI FRAGMENTS

Unravelling the events involved in uracil repair also solved a puzzle surrounding the early work of Okazaki and the discovery of discontinuous synthesis of the lagging DNA strand *in vivo*. In his original experiments the most recently synthesized DNA was labelled with tritium under conditions of slow synthesis, and the DNA was recovered from the *E. coli* under alkaline conditions. This resulted in the denaturation of the DNA and the recovery of the newly synthesized, tritiated DNA as short segments of single-stranded DNA—the celebrated Okazaki fragments. However, *all* the newly synthesized DNA was in fragments, whereas we now know that since one strand is continuously synthesized, only about half the newly synthesized DNA should be fragmented. The remainder, that synthesized along the leading strand, should be in much larger pieces and only their ends should be radioactive. The reason for this anomaly can now be appreciated. The action of uracil-N-glycosylase creates nucleotide gaps in the DNA sequence (and so does the spontaneous depurination of DNA—see later in this account). DNA containing such gaps is readily susceptible to alkaline hydrolysis at that point in the sequence. Wherever gaps exist in the leading strand (and, of course, in the true Okazaki fragments along the lagging strand, too), breaks will occur during the extraction procedure. Hence all new DNA will be to some extent fragmented when extracted in this manner.

This point can be illustrated by the use of *E. coli* carrying certain mutations of the uracil repair process. Strains can be obtained which are deficient in either dUTPase activity (dut⁻) or the uracil-N-glycosylase (ung⁻) activity. Mutants carrying the dut⁻ gene should have elevated levels of uracil incorporation into the DNA and therefore yield shorter fragments when the DNA is extracted under alkaline conditions. All the new DNA should be fragmented. Mutants carrying the ung⁻ gene, by contrast, should have true Okazaki fragments of about normal lengths, but only about half the newly made DNA should be fragmented. Experiments confirm these predictions.

Gaps can occur in the DNA sequence by other means than by the express action of uracil-N-glycosylase. A common occurrence is the spontaneous depurination of DNA as a consequence of the inherent instability of the glycosidic bond between the sugar and a purine base at physiological temperatures and pH. About one in every 300 purines is lost in this manner every day. This process is considerably enhanced if alkyl groups have been attached to the purines. This can happen when "alkylating agents" gain access to the DNA and add an alkyl group to the N7 of purines. This weakens the glycosidic bond and aids in the loss of the purine. The repair of such gaps employs the same mechanism that follows from the activity of uracil-N-glycosylase. (This is why apurinic acid endonuclease is so named.)

REPAIR OF ULTRA-VIOLET LIGHT DAMAGE

The most common UV-induced damage to DNA is the induction of **thymine dimers**. These occur when adjacent thymine bases become bonded together, the two carbon 5s and the two carbon 6s joining to make a cyclobutane ring. The dimer cannot hydrogen bond with any bases on the complementary strand and must, therefore, be removed before the strand containing it can be used as a template for DNA or RNA synthesis. To do so, several different mechanisms have evolved.[*]

The most straightforward is by means of the **photoreactivation enzyme**, sometimes called "photolyase". Somehow—the exact chemical details are unknown—this enzyme uses visible light to cleave the 5-5 and 6-6 bonds of the thymine dimer, regenerating the two independent thymines. The enzyme is also active against the occasional cytosine-cytosine and cytosine-thymine dimers that are also caused by UV light; these are much less frequent than thymine-thymine pairs.

The best studied repair mechanism is that of "excision repair" in a manner similar to the "short patch" route discussed earlier in this account. For the repair of thymine dimers, an enzyme, UV endonuclease, puts *two* cuts in the DNA. One of these is eight bases to the 5' side of the dimer, and the other, four or five bases away on the 3' side. The 12-base segment is then excised from the DNA and degraded. The gap is filled in the usual manner by pol I followed by ligase.

A third method of UV damage repair is called **SOS repair**. It involves several proteins that are only found in UV damaged cells and are likely, therefore, induced by such damage. The key protein in SOS repair is a product of the RecA gene, which is involved in recombination; the binding if the RecA protein at the site of the dimer is thought to be the signal for the induction of the other repair proteins. The RecA protein (likely acting in concert with the induced SOS proteins) then permits the polymerase to pass beyond the dimer site by adding two bases to the growing strand despite the lack of a template at the dimer site. The two added bases are often

---

[*] The well-known mutagenic and carcinogenic properties of sunlight are caused by the so-called "ultraviolet B" portion of the spectrum, and it is ultraviolet B that induces thymine dimers in DNA. The dimerization properties of UV B have been used as an inexpensive measure of UV B intensity. A strand of DNA is enclosed in a quartz vial; when UV B passes through, the number of induced pyrimidine dimers is proportional to the UV B dose.

adenines, in which case the defect has been successfully by-passed, if not actually repaired. On occasion, however, the bases are not adenines, and a mutation results. In fact, this is the commonest origin of UV-induced mutations. It is not known how the SOS proteins permit the by-pass of the dimer, but it is assumed that part of the answer lies in the inactivation of the editing 3' to 5' exonuclease function of the polymerase. Because SOS repair frequently leads to inaccurate repair—the trade-off for cell survival—it is also called by some investigations **error-prone repair**.

REFERENCES

1. Radman, M. and R. Wagner, 1988. The high fidelity of DNA duplication. Sci. Amer. (Aug.).

2. Lahue R, Au K, and Modrich P., 1989. DNA mismatch correction in a defined system. Science 245:160-164.

LECTURE 36

BASIC BACTERIAL GENETICS

THE DISCOVERY OF BACTERIAL SEX[1]

Initially it was thought that bacteria could not undergo genetic recombination because they did not mate. True, bacteria could be pictured under the microscope in suspicious carriage, but the suggestion that these cells were mating was dismissed as mere anthropocentrism or simple dirty-mindedness on the part of the investigator. By 1946 J. Lederberg and E. Tatum had begun to question seriously the conventional view-point and set up a genetic experiment to demonstrate recombination in *E. coli*. They took two strains of *E. coli* each containing a number of nutritional deficiency mutations not present in the other. After mixing these parental strains, they plated them on a minimal medium that would support the growth only of wild-type cells. Not much to their surprise, but much to everyone else's, a number of wild-type colonies were observed growing on the minimal medium.

Before this can be interpreted as mating, however, a number of alternative explanations must be ruled out. Transformation as a possibility comes immediately to mind, although too many genes were acquired by one or the other strain to make this a fully plausible explanation. This interpretation can be further discredited by showing that actual cell-to-cell contact is necessary for such a massive recombination. Each parental strain was placed separately in one arm of a glass U-tube, the arms of which were separated by a filter at the bend that would not allow whole bacteria through. Air pressure was applied alternately to one side and then the other in order to facilitate complete mixing of the media supporting the two parental strains. After complete mixing of the media had been achieved, samples of the multiply auxotrophic parents were taken from each side of the U-tube and plated on minimal medium. No colonies were obtained. Transformation—the acquisition and subsequent incorporation into one cell of DNA fragments carrying genes from another cell—could have been effected under these circumstances, but still no wild-type colonies occurred. It is clear that Lederberg and Tatum had uncovered a new mode of bacterial recombination, one requiring cell-to-cell contact. (Parenthetically it can here be noted that this same U-tube experiment, thought up by Davis, also rules out transduction as an explanation for these results. Transduction, not known at the time, is a phenomenon we shall discuss peripherally later in these lectures).

It soon became established that any bacterial gene could be involved in recombination, but that it was not a frequent phenomenon for any. Furthermore, crosses could only take place between certain strains of *E. coli*, whereas other strains of the same species remained infertile. Moreover when recombination did occur, it was rare to find many genes recombining simultaneously. When simultaneous recombination did occur, it was taken as evidence that the genes involved were closely linked on the putative bacterial chromosome.

213

BACTERIAL SEXES

At this point it becomes necessary to distinguish between two possible mechanisms underlying the origins of recombinant genotypes in bacteria. We can, for example, imagine an earthworm-like model where each parent fertilizes the other and both give rise to recombinants. Here it would be possible to kill either parent after mating, and recombinants would still occur as long as just one parent survived. Alternatively, we can imagine an arachnoid (spiders, and certain other species) model where there are distinct sexes and after fertilization the survival of one (the female) but not the other is necessary for the obtaining of recombinant offspring. By adding a streptomycin-resistant gene first to one and then, in a separate experiment, to just the other parental strain, it is possible to kill selectively either parent—the one still susceptible to the drug—after mating has occurred. It was quickly found that indeed the survival of one parent, but not the other, was absolutely necessary for the fertility of the cross. Bacteria, it seems, have not only sex, but sexes too. When the appropriate cells are in contact, genes, in the form of chromosome fragments, are transferred unidirectionally from what is called the **donor** ("male", or F^+) cell to the **recipient** ("female", or F^-) cell. Some or all of the transferred genes are subsequently integrated into "her" genome in place of alleles already there. However, the sex of the recombinants is almost always male, for reasons which will become clear later.

The difference between donor and the more prevalent recipient bacteria is attributable to cytoplasmic, autonomous, virus-like particles—the **sex factors** or F-factors—found in donors and absent in recipients. These sex factors are highly infectious, and the commonest result of mixing F^+ and F^- strains is not mating and recombination, but the conversion of the F^- into F^+ by simple infection. All cells then become the same sex, and mating becomes even less likely. Presumably early in the development of the recombinant genotypes inside the recipient, these recipient cells also acquired a sex factor by infection, and that is why, even though the recombinants originated in an F^- cell, they are F^+ by the time they are isolated as a clone derived from this recombinant F^- cell. Sex factors have now been isolated free of the bacterial cells and are used extensively in recombinant DNA work (*vide infra*).

HFR DONORS

Research in bacterial genetic was speeded up, so to speak, by the discovery by W. Hayes (1953) of a donor capable of transferring its chromosome 1000 times more frequently than does an F^+. These **Hfr males**, however, transfer only some genes with high frequency, although different Hfr strains transfer different genes at this elevated rate. Moreover, for each particular strain there is a definite gradient of transmission noted for the genes which are transferred, such that some genes are transferred highly efficiently, others with somewhat lessened efficiency, and others hardly at all. The efficiency of transfer is also correlated with the time it takes for a particular gene to be transferred. Some are transferred right away, and they are the ones that are transferred most often. Others get transferred only after about 90 minutes of continuous mating, and these are the ones that get transferred least often in an Hfr cross. In between there is a continuous gradient of times and correlated frequencies.

214

It should be obvious that these observations add up to a way of mapping bacterial genes. A large number of parallel Hfr x F⁻ crosses of the same strains can be established under as tranquil culture conditions as possible. Then at various intervals afterwards the mating parents can be separated by a blender treatment and plated to determine which genes have had the opportunity of being transferred before the parents were separated. In this way one determines the position, measured in minutes, of each gene on the Hfr chromosome.[2]

As a matter of interest, it is possible to calculate the rate of DNA transfer. We know, by direct radioautographic measurements, that the bacterial chromosome (a "genophore") is 1,100 microns long. We know, too, that bases in DNA are 3.4 A apart. Consequently, there must be about 3.2×10^6 base-pairs in the entire chromosome. Since it takes 90 minutes for complete transfer (at 37°C), about 3.5×10^4 base-pairs are transferred per minute. This amount of DNA would presumably correspond to 30 to 50 genes.

THE SEX PLASMID

Hfr x F⁻ crosses also differ from F⁺ x F⁻ ones in that the former give rise to F⁻ recombinants and the latter to F⁺. It turns out that Hfr cells, although chromosome donors, do not have any free sex factors with which to infect F⁻ strains. This fact is causally related to their becoming Hfr's in the first place, since the creation of an Hfr involves the integration of a sex factor with the bacterial chromosome and its being carried along thereafter as a regular piece of the bacterial chromosome in no way distinguishable from any other segment. Simultaneously, any autonomous sex factors already present or infecting in future are frozen (repressed) and cannot replicate in that cytoplasm. Within a few cell divisions, the inactive sex factors are diluted out so that by the time an Hfr "mutant" is isolated in pure culture, no sex factors remain in the cytoplasm.

The sex factor, therefore, represents a class of genetic elements known as **episomes**. Episomes can lead an autonomous existence in the cytoplasm of a cell or can integrate into the chromosome and behave as just another section of DNA. Other examples of episomes are certain viruses of bacteria, plants and mammals. In the case of the F-factor, its DNA molecule is circular, and so is that of the host *E. coli* cell. Integration is accomplished by a single act of recombination between the two circles, which has the affect of inserting the smaller circle as a linear segment of the larger one.

Every Hfr strain transfers a different set of genes with high frequency. The order of transfer (and consequently the efficiency of transfer) is determined by the chromosomal location of the sex factor. Before the chromosome can be transferred, it must be converted from its normal circular configuration to a linear structure. The break occurs at the sex factor attachment site, wherever that happens to be for that strain. The end of the chromosome having the sex factor is transferred *last*; hence no bacterial gene should be transferred less frequently than the sex factor. Those very few recombinants which actually acquire the sex factor become Hfr instead of F⁻.

This transfer is accomplished by a modification of the "rolling circle" mode of DNA replication. A break occurs in *one* strand of the circular DNA, thus creating a 5' and a 3' end. New nucleotides are added to the 3' end, displacing the 5' end from the duplex as a single strand. This lengthening single-stranded 5' end is then threaded into the recipient cell, where some, all or none of it is integrated into the genome of the recipient. The details of this integration process are still obscure.

When the sex factor is free in the cytoplasm and is transferred independently to an F⁻ cell, it is transferred in the same manner—that is, by modified rolling circle replication. In fact, the transfer of the Hfr chromosome to an F⁻ recipient can best be imagined as simply the consequence of transferring an abnormal sex factor—abnormal in that this sex factor has the entire bacterial chromosome integrated into it. (We usually think of the smaller object as being integrated into the larger, but there is no reason it cannot equally well be viewed as the larger being integrated into the smaller.) Because the free sex factor is so much smaller than the bacterial chromosome, there is usually sufficient time for it to be transferred in its entirety when it infects by itself; however, when it is "carrying" the whole of the chromosome, there is usually only time for some of the chromosomal genes to get transferred before mating is interrupted. Because the break-point initiating rolling circle replication occurs within the DNA of the sex factor (although towards one end), the entire sex factor is not transferred in an Hfr cross unless the whole of the bacterial chromosome is. Hence, the recipients usually remain F⁻.

Having made so much of the differences between Hfr and F⁺ crosses, I'm now going to ask you to believe they are actually much the same, despite the superficial dissimilarities. The *mating* F⁺ cell (as opposed to the vast majority which are not mating—indeed, cannot mate) is not an F⁺ at all, but an Hfr mutation that just arose within the F⁺ population. Since the sex factor could have settled anywhere on the chromosome to form this Hfr, any gene has, statistically speaking, an equal chance of being transferred. Finally, since the recombinant cells are surrounded by F⁺ donors, they become infected by sex factors and become, therefore, F⁺. And so it is that F⁺ x F⁻ crosses are just special cases of Hfr x F⁻ crosses.

TYPE I DONORS

In 1959 Adelberg discovered a donor strain of *E. coli* that had properties intermediate between those of F⁺ and Hfr strains. He called it **Type I** (for "intermediate"). It transferred its chromosome with one-tenth the Hfr frequency, but otherwise with the same characteristics as an Hfr strain. It also transferred sex factors to F⁻ recipients, which in consequence became not F⁺, but Type I. Furthermore, these new Type I's (i.e., the former F⁻ cells that became Type I by infection) transferred their chromosome with the same polarity as did the donating strain from which they received the sex factor. That is to say, when this Type I sex factor comes to settle on the recipient's chromosome (enabling it thereby to transfer that chromosome), it does not do so at a random location, but selects the same location where it or its forefathers/mothers were originally lodged. Like salmon, it retains a "memory" of its chromosomal home, even though (unlike salmon) it may be only a distant descendant of the sex factor that first shook loose from the chromosome.

The interpretation given to Type I is as follows. They are derived from F^+'s in the same way as are Hfr's except that the association between the chromosome and the sex factor is an unstable one. In coming loose from the chromosome, however, the sex factor takes, as a consequence of some aberrant recombination-like event, a small piece of bacterial chromosome with it and leaves an even smaller piece of its own DNA behind. This means that the Type I sex factor is not the same as the F^+ factor. Radiological evidence suggests it is twice as large as a normal factor. To distinguish it from the wild-type, it is usually written F'.

The F' spends about 10% of its time on the chromosome, during which time the chromosome can be transferred but any autonomous sex factors (F' or F) cannot replicate. But since the sex factors are on the loose in the cytoplasm 90% of the time, they are free to replicate during this time and do not become diluted out. When transferred to F^-, the piece of bacterial chromosome the F' carries directs it to the homologous site on the recipient's chromosome, and hence its "memory".

Besides being a highly entertaining phenomenon, Type I's are of great practical worth—to geneticists at any rate. Consider the consequences of the following experiment. An Hfr strain was selected that transferred a particular bacterial gene *last*, i.e., at about 90 minutes. After just 10 minutes, however, the mating strains were artificially separated and plated on a medium that selected specifically for those recombinants which might have received that particular gene. A few were actually found. On analysis they proved to be Type I donors with the expected polarity, but more interestingly, had acquired the ability to transfer the gene in question with great efficiency and speed to F^- recipients. It seems as if the sex factor had, as its "memory" piece, picked up the selected gene, the one that was adjacent to the sex factor when it had a chromosomal existence. This process, called **sexduction** (a special case of **episomal transduction**), can be used to transfer any given gene—sometimes several genes simultaneously—to another cell. All one has to do is find a Hfr that transfers last the genes you are interested in, and then carry out the appropriate selection experiment.

The sexduced (recipient) cell becomes a **heterogenote** for the bacterial region in question. Sometimes after many cell divisions the bacterial genes on the **exogenotic fragment** may integrate on to the bacterial chromosome and replace the alleles of the endogenote. Usually, however, that particular region remains functionally heterozygous for many cell generations.

REFERENCES

1. Jacob, F. and E. Wollman, 1956. Sexuality in bacteria. Sci. Amer. (July).

2. Cairns, J., 1966. The bacterial chromosome. Sci. Amer. (Jan.).

LECTURE 37

CONTROL OF GENE ACTIVITY IN BACTERIA

INTRODUCTION

It is obvious that some mechanism must exist for the control of gene activity, since the products of most genes are not always in evidence. Although all cells of the body must contain the same set of genetic instructions, brain cells, for example, do not produce the specific proteins found in liver cells, nor do liver cells produce specifically brain proteins. In each cell type some mechanism must exist to switch some genes on and to switch others off.

THE BETA-GALACTOSIDASE STORY[1]

An instructive illustration is presented by analyzing the metabolism of the sugar lactose in *E. coli*. This disaccharide requires as the first step in its catabolism an enzyme, **ß-galactosidase**, to split it into two simple sugars. When cells are growing in the absence of lactose, only the minutest trace of this enzyme is found in their cytoplasm. Transfer these cells to a medium where lactose is the sole carbon source, however, and these same cells start manifesting within minutes ludicrous quantities of ß-galactosidase, enough to account under some circumstances for a few percent of the entire protein synthesis of the cell.

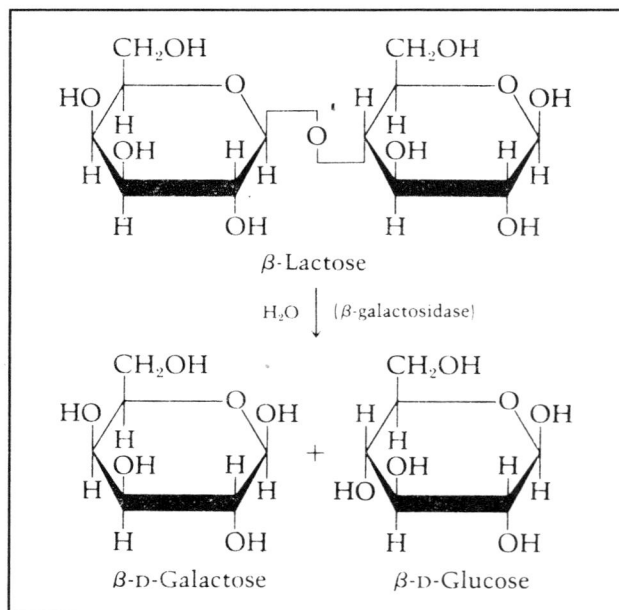

Figure 37.1 The activity of beta-galactosidase

Is this a *de novo* synthesis in response to a particular environmental stimulus (lactose), or is it merely the activation of a pre-existing dormant enzyme or pro- enzyme? A simple experiment employing radiotracers can distinguish between these two possibilities. A batch of *E. coli* was grown on a medium containing simple sugars. The only sulphur available was the radioactive S^{35} isotope, and it was present in very limited quantities to prevent pools of stored radioactive sulphur from accumulating within the cells. The cells were then transferred to a medium where lactose was the only carbon source and where cold sulphur replaced the hot. Large quantities of ß-galactosidase were produced which, when analyzed, were found to contain only normal (cold) sulphur. This protein could not have been synthesized before the transfer or it would have been radioactive. Later experiments of a different type demonstrated that even the m-RNA for the ß-galactosidase was synthesized after the transfer, thus confirming the activation of a particular gene by a specific environmental agent.

Concomitantly with the induction of ß-galactosidase, at least two other proteins which play a role in lactose utilization also appear: a **membrane protein** (termed **M-protein**) and **thiogalactoside transacetylase**. These latter two proteins are in some way involved in the active transport of lactose through the cell membrane, but their exact role remains somewhat obscure. The bacterial genes specifying these products—z, y and a for ß-galactosidase, M-protein, and thiogalactoside transacetylase, respectively—are closely linked on the bacterial chromosome and lie in the order given. They are nonetheless independent genes, each with its own initiator and terminator codons.

The induction of z, y and a activity is accomplished fairly specifically by lactose. Some lactose analogues, however, will also induce the system. For example, some thiogalactoside sugars will induce but cannot be metabolized by the enzymes produced. These are called **gratuitous inducers**. Some compounds cannot induce, but are satisfactory enzyme substrates. Phenylgalactosides fall into this category.

Strains of bacteria that produce the lactose proteins only when called upon to do so are called **inducible**. Certain other strains, by contrast, are found to produce these proteins at all times. These cells are called **constitutive** producers. They differ from inducible strains by a single gene mutation. This mutation occurs not in any of the three structural genes mentioned above, but in a fourth gene, i, closely linked to z and to its left. A mutation in gene i appears, therefore, to affect the control that governs the activity of the structural genes; gene i is an example of a new class of **regulatory genes** which function to control structural genes. Since it is the constitutive strain that carries the mutant i gene, the ability to regulate the synthesis of the z, y and a proteins should be regarded as the normal or wild-type phenotype. Consistent with this interpretation is the fact that almost all *E. coli* randomly tested in nature are inducible.

How can one gene, the **regulator**, serve to control the functioning of other genes that specify polypeptides? One might at first imagine the i gene to exert a direct physical effect on the z, y, and a genes, since they lie in a row almost directly beside the regulator. Alternatively, one could envision the regulator producing a cytoplasmic product which is capable of gumming up the functioning of the three structural genes, z, y, and a. In fact, i does produce a cytoplasmic product—called R, for **Repressor**—which, when present in even minute quantities in the cytoplasm, somehow prevents the structural genes of the lactose system from being expressed.

So, before this system becomes even more complicated, let us tie our disparate observations together. Inducible strains produce R at all times, and R prevents the structural genes of the lactose system from functioning. We know, however, that lactose will stimulate z, y and a activity, and consequently lactose must antagonize (inactivate) R. Constitutive strains lack active R, either because none is produced or, more likely, because the R that is produced does not function properly.

If the regulator gene i^+ makes active R and $\underline{i}$ inactive R, it should be possible to find still other types of i mutations, as, for example, one whose product retains its ability to repress the structural genes but loses its ability to be blocked by lactose. Cells carrying such a mutation would be permanently repressed, that is, unable to produce the $\underline{z}$, $\underline{y}$ and $\underline{a}$ proteins under any circumstances. Such a mutation, termed $\underline{i}^s$, has been found and confirmed to be an alteration of the regulator rather than of the structural genes.

How does R control the activity of the structural genes? It was earlier mentioned that induction involves not just the *de novo* coordinate synthesis of the three lactose proteins, but also of the m-RNA that encodes them. The propinquity of the lactose structural genes led to the suggestion that all three proteins are encoded on the same messenger. Regulation would then be fundamentally a question of controlling m-RNA transcription off the segment of the DNA that encodes the genes for lactose utilization. It would also be reasonable to assume that the synthesis of this messenger has a definite starting point on the DNA, and that R need only block this starting point to prevent appearance of the entire m-RNA and hence all of the lactose proteins. The existence of this controlling site, called the $\underline{o}$ or **operator** region, has been confirmed and located by recombination studies between the $\underline{i}$ and $\underline{z}$ genes.

The strongest evidence for an operator gene also came from mutation studies. For example, one could predict the existence of $\underline{o}^c$ mutations of the operator which have lost their ability to be blocked by R. Strains containing such lesions should be constitutive, and therefore resemble $\underline{i}$- mutants. They would not map, however, to the $\underline{i}$ gene, but to a region (the operator) much closer to the $\underline{z}$ structural gene. Such mutations have indeed been discovered.

THE CLASSICAL OPERON

The scheme outlined so far, encompassing the two types of regulator genes ($\underline{i}$ and $\underline{o}$) and the various structural genes under their control, is known as the classical **operon concept** and was first proposed by F. Jacob and J. Monod in 1961. It has since undergone some modification; chief among them is the discovery of a third type of controlling gene, the **promotor**, discussed later on. The lactose operon is about the simplest of the many operons investigated in the last decade. Obviously central to this operon concept is the presumed existence of a cytoplasmic repressor, R. Initially it was thought to be an RNA molecule, since this would neatly explain its ability to complex specifically with the DNA segment known as the operator. Later this notion became untenable—the synthesis of R was blocked by the usual experimental inhibitors of *protein* synthesis—and we came to accept the fact it was a protein. A very elusive protein. Because it is produced in very small quantities (about ten molecules per cell), it was years before investigators succeeded in isolating it from *E. coli*. During this interval a number of molecular geneticists lost the faith and proposed alternative mechanisms that obviated the need for a repressor molecule. However, in the late 1960s M. Ptashne rekindled our belief in the existence of the repressor molecule when he succeeded in isolating from bacteria infected with phage lambda a repressor molecule that functions in a way analogous to the Lac repressor. Shortly thereafter, W. Gilbert and B. Muller-Hill found a way to isolate the Lac repressor itself. These investigators made use of the fact that the repressor must bind to a gratuitous inducer, such as

220

isopropylthiogalactoside (IPTG). Using radioactive IPTG, they observed an alteration in its sedimentation rate when the repressor was bound, thus serving as a means to identify the presence of the repressor. It was then a simple matter to isolate the repressor free from the IPTG and further characterize it functionally and biochemically. As predicted, the repressor bound only to DNA carrying the normal Lac operon; also as predicted, it did not bind to operons carrying o^c mutations.[2]

ENZYME REPRESSION

Up to now we have looked at the operon model only as an explanation for gene induction. There is also a parallel phenomenon, enzyme **repression**, which is even more common. Repressible cells cease manufacturing a particular substance whenever that substance is present in the growth medium. This is the case for most of the amino acids elaborated by bacteria. Repression can be looked on as an economy measure which prevents cells from expending energy making a product already available to them. As with the lactose system, many enzymes may be coordinately regulated, usually just those unique to the synthetic pathway being repressed. The mechanism of repression bears a striking similarity to that governing induction. Here the repressor molecule, R, is thought to be inactive when it is produced by the equivalent of the i gene. When, however, the *external* repressor compound is added to the medium and is taken up in small quantities by the cell, it interacts with R to form an active *internal* repressor which can now complex with the o gene of the operon to block messenger RNA synthesis from that region.

THE PROMOTOR

The penultimate wrinkle I want to add to the operon concept has to do with the control of the quantitative output of an operon. Some operons, once they get revved up, produce enormous quantities of their protein products, whereas others under the best of circumstances produce only exiguous amounts. It was also noted that single gene mutations occasionally altered the quantitative control of the operon, increasing or decreasing its output markedly, These mutations occur in a region closely linked to the operator, and to its left. This region has become known as the **promotor** and appears to be the initial site of RNA polymerase attachment prior to transcription. Presumably, more efficient promotors offer a more accommodating bed for the polymerase. The nature of these promotor sequences of DNA are now well understood, thanks largely to the ability to clone and sequence long stretches of DNA.

In bacteria, about five to ten bases to the left ("**upstream**") of where m-RNA transcription actually begins, there is a region on the DNA template strand known as a **Pribnow box** having the sequence TATAAT. The RNA polymerase binds to this sequence, making it, in effect, the promotor. There is, however, another sequence further upstream, the **-35 sequence** (so called because it is 35 bases to the left of where transcription begins) that is also important in many operons for the correct positioning of the RNA polymerase.[*]

In the Lac operon, the repressor has been shown to bind to the DNA in such a way as to prevent the attachment of RNA polymerase to its promotor region. The first 21 or so bases that are transcribed by the polymerase are covered by the repressor molecule; these are some of the same bases that are covered by the RNA polymerase when it binds to the promotor. In some systems (e.g., lambda phage) the operator and promotor regions are actually fully coincident, as originally proposed by Jacob and Monod.

In some operons, including the Lac operon, it has been noted that certain other regions of the DNA that lie hundreds of base-pairs away from the promotor-

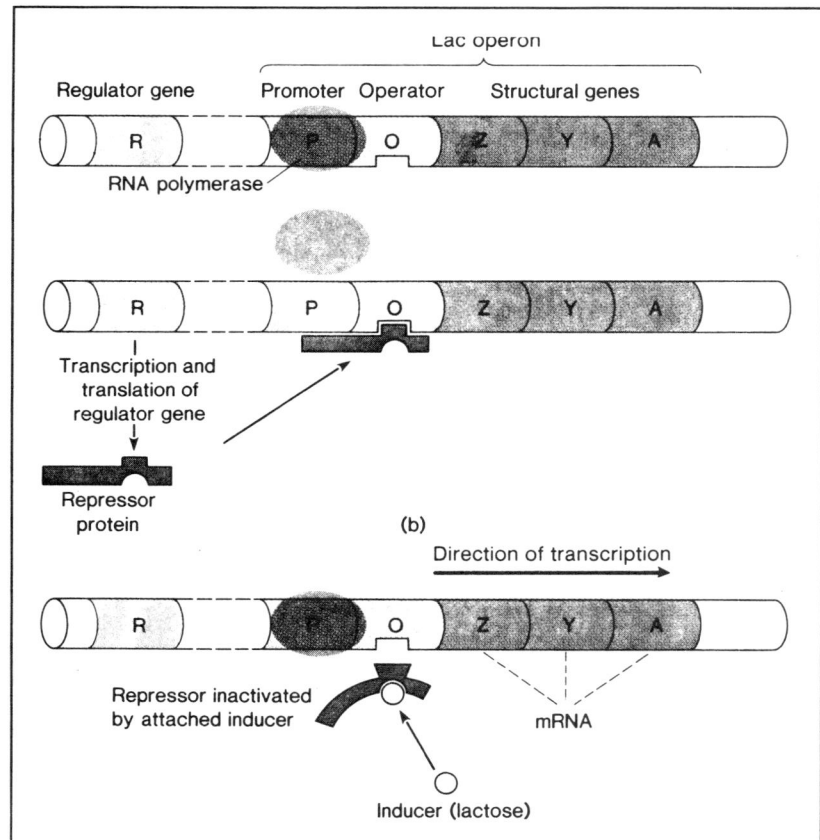

Figure 37.2 The lactose operon. From Leland G. Johnson, Biology, 2nd ed. Copyright © 1987 Wm. C. Brown Communications, Inc., Dubuque, Iowa. All rights reserved. Reprinted by permission.

proximal operator also seem capable of binding the repressor and do play a part in repressing transcription. The Lac operon, for example, has such sites within both the i and the z genes. This seeming "action-at-a-distance" has now been explained as a ligating effect of the repressor protein. Repressor molecules bound at these disparate sites bind together, causing the DNA to

---

[*]The situation in eukaryotic organisms is somewhat different. In eukaryotes there is a sequence known as the **TATA box** located about 25 bases upstream from where transcription begins. The TATA box has the usual sequence TATAAAT. There is also another, lengthier sequence located in the -75 region that is very important. The actual binding of the RNA polymerase II (the one that transcribes m-RNA in eukaryotes) involves the mandatory participation of six **transcription factors**, termed TFIIB, -D, -E, -F, -H, -J. TFIID (which is itself composed of several proteins) binds first to the TATA box, and is then joined by TFIIB. At this point, the RNA pol II joins the gang, followed in an ordered way by the E, H, and J factors. TFIIH is thought to be especially important in this process, since it apparently has the ability to phosphorylate a repeated sequence of amino acids at the C-terminal end of the polymerase, thus converting it into an active enzyme for elongation.

222

form loops held together by the repressor. Apparently this looped DNA is much more resistant to translation than would the linear molecule be. Widely separated sites of common function are a common feature of regulatory systems in eukaryotes; perhaps looping of DNA tied together in this way will explain some of the puzzling results obtained from studies with higher organisms.

The i gene has its own promotor which may mutate to increase the quantities of R produced. Some of these mutants (e.g. i^{SQ}) produce incredible quantities of R and enabled the latter eventually to be isolated in almost gram amounts. This in turn led to the ready biochemical characterization of the Lac as a tetramer whose complete amino acid sequence has been determined.

OTHER EFFECTOR MOLECULES

Two other molecules are also necessary for the initiation of transcription at the promotor site. One of these, cyclic AMP (cAMP), is found ubiquitously in living systems and assumes a role in many different regulatory processes. The other is a protein known still be a variety of names, from which I shall arbitrarily choose **CRP—catabolic repression protein**. These two molecules interact to *enhance* the binding of RNA polymerase to its Lac promotor (and to the promotors of several other operons, as well). Neither molecule by itself is effective in enhancing this binding. The level of cAMP in the cell is itself controlled by an unknown product of glucose catabolism. When glucose is being catabolized, this mysterious "catabolite" inhibits the production of cAMP. In consequence, the cAMP-CRP complex is in short supply and RNA polymerase is not efficiently bound to the Lac promotor. This accounts for the long-standing puzzling observation that *E. coli* growing in the presence of both glucose and lactose (together) choose to catabolize glucose and seemingly to repress the catabolism of lactose. But why glucose is preferred as a carbon source over lactose is unknown. (This inhibition of lactose utilization by glucose is called **catabolic repression**, and from whence derives the name "catabolic repression protein".)

One strong clue to the puzzling ability of the CRP-cAMP complex to influence translation that actually begins some distance away (between 41 and 103 base-pairs distant), is the recent discovery that CRP puts a 90 degree bend in the DNA at the point where it binds.[3] This could perhaps bring the initiation site and the CRP binding sites into greater proximity.*

Knowing something about gene control in micro-organisms, you must not now assume you know anything at all about this control in higher forms. Here the process is apparently quite different.

---

*Lately several molecules that bend DNA have been discovered, and with them a growing awareness that such molecules may play a very important part in the regulation of gene activity in general. As one example, **testis determining factor**, the protein that changes the course of fetal development from the "default" female to the male direction, is a potent bender of DNA. In this manner it is thought that regulatory proteins, inactive by themselves and bound at distal regions of the DNA molecule, could be brought into direct physical contact, an interaction that would result in DNA transcription.

It has been exceedingly difficult to date to prove the existence of a single classical operon in higher organisms, perhaps because they are not present.

There is one well-characterized regulation in eukaryotes worth noting here, that involving the immune protein **interferon** α, a molecule involved in fighting off viral infections. When interferon interacts with a cell, that cell is stimulated to produce several proteins by *de novo* transcription of their respective genes. Each of these genes carries the same specific DNA sequence at their beginning which is recognized by a particular transcription factor. Normally, however, this factor is not associated with the leader sequences—unless the factor is complexed with three other proteins that normally occur not in the nucleus, but in the cytoplasm. Moreover, these three cytoplasmic proteins have no inherent ability to associate with each other or to enter the nucleus. The membrane of the appropriate cell harbors a receptor protein for interferon α, and this receptor protein is linked inside the cell to a particular enzyme, **tyrosine kinase**. These are the *dramatis personae*; actual co-activation of the nuclear genes comes about as follows: interferon α, circulating in the blood stream, contacts its receptor protein on a cell membrane; this activates the associated tyrosine kinase to add phosphorus groups to certain tyrosine residues of the three cytoplasmic proteins that interact with the transcription factor; these three proteins, now somewhat modified, acquire the ability to associate, and do so; the associated protein complex now can pass through the nuclear membrane, and inside the nucleus encounters the transcription fact, with which it binds; the whole regulator complex now binds to the regulator sequences of the genes in question, thus initiating transcription.[4] How similar this "house-that-Jack-built" sequence is to the workings of a bacterial operon is left to the reader to puzzle over.

REFERENCES

1. Maniatis, T. and M. Ptashne, 1976. A DNA operator-repressor system. Sci. Amer. (Jan.).

2. Ptashne, M. and W. Gilbert, 1970. Genetic repressors. Sci. Amer. (June).

3. Schultz, S., et al., 1991. Science 253: 1001-1007.

4. Schindler, C., et al., 1992. Science 257: 809-813.
   David, M. and A. Larner, 1992. Science 257: 813-815.

LECTURE 38

RECOMBINANT DNA (OR "DESIGNER GENES"): 1

INTRODUCTION

Undoubtedly the most important practical discoveries to come out of the biological sciences during the past decade have been in the general area of recombinant DNA. These discoveries promise to revolutionize the fields of medicine and agriculture as much as the invention of the transistor has influenced the field of electronics. The birth of this technology, however, has not been without controversy. In this account we shall confine ourselves to the bare scientific questions, although this should not be taken as an expression of the author's lack of concern about the social implications of this work. In order to keep in mind the importance of considering the social significance of these and other potentially dangerous discoveries within molecular biology, one should mull over the nuances of this exchange between R. Oppenheimer, the developer of the atomic bomb, and his teacher, M. Born, a Nobel laureate in physics:

"I was inspired by the deepest interest in my
science, but I was unaware of the relation between
man and his society."
Oppenheimer

"It is nice to have praised such clever and
competent pupils, but still I wished that they
were more wise and less cleaver. It is certainly
an error on my part if they learned only the
methods of research from me and nothing else."
Born

None of this is to suggest that there is anything inherently scrofulous about recombinant DNA techniques or any other aspect of modern molecular biology—only that one should always try to peer as far as possible into the future before taking the next step in that direction.[1]

BACTERIAL PLASMIDS

In an earlier section of this account, the nature of bacterial sexuality was discussed and the *E. coli* sex factor, the F plasmid, was introduced. The F plasmid is only one of many different types that are known to exist in the cytoplasm of bacterial cells. Recently plasmids of various types have been found in plant, insect, and animal cells, as well. Many of the plasmids that exist in bacteria, and especially those that are most useful in gene splicing experiments, carry genes that confer resistance to particular drugs on the cell that harbors them. It is because such plasmids can be passed readily to other bacteria, making them in turn resistant to the drug in question, that bacteria that are resistant to antibiotics can spring up and spread so rapidly in modern hospitals.

Although plasmids can be acquired naturally only by cell-to-cell contact, particular plasmids can be experimentally extracted from bacteria and isolated free of contaminating "chromosomal" DNA and other types of plasmids. They can also be re-introduced into cells, but only after the membrane of the cell has been chemically somewhat damaged in order to permit the plasmids to pass through. Treatment of the cells with calcium is a common way to render them somewhat permeable to plasmids *in vitro*. Once the plasmid has regained entry to a cell, it will replicate and establish itself as a permanent cytoplasmic resident thereafter, assuming, of course, that it had not been damaged in being extracted from its original host cell. One key aspect of a plasmid that cannot be altered without destroying its ability to replicate inside a cell is its circularity.

RESTRICTION ENDONUCLEASES[2]

Besides plasmids, another essential component of gene splicing technology is a class of enzymes known as **restriction endonucleases**. Indeed, it was the discovery of these enzymes and how they worked that actually gave rise to this technology in the first place. Restriction enzymes cut double stranded DNA. They do not do so randomly, however, but act only at given sites on the DNA molecule. Since DNA is only a long sequence of base pairs, "sites" in this context can only mean particular sequences of base pairs along the DNA molecule. There are many different restriction enzymes, and in general each restriction enzyme recognizes and cuts at only one particular site. That is to say, RE-X will cut only at sequence X in DNA, and RE-Y will cut only at sequence Y. These enzymes can be isolated, purified, and made to function *in vitro*.

What is the nature of these **restriction sites** in DNA? In general, they comprise about a dozen base pairs in length, but both longer and shorter ones are known, too. Often, but not always, they have a peculiar characteristic that distinguishes them from most other DNA sequences of similar length: they constitute a kind of **palindrome** within the DNA molecule of which they form part. Palindromes, in English prose, are sequences of letters that read the same way in two directions. The sentence, "Are we not drawn onwards, we Jews, drawn onward to new era?", can serve as an example. Similarly, a kind of palindrome can exist in DNA, as the following example illustrates:

5'...A C T G A C G T C A G T ....
...T G A C T G C A G T C A ....5'

Reading along the top right to left we generate the same sequence as along the bottom when reading in the opposite direction.

When a given restriction enzyme cuts DNA at its particular restriction site, the palindrome which it recognizes, it does so in a characteristic manner. It may, depending upon exactly which enzyme we are dealing with, cut the DNA at a particular location within the palindrome so that the breaks in the two strands are precisely opposite each other. These restriction enzymes are the less interesting for our purposes here. The ones that are by far the more useful in recombinant DNA work are those that make staggered cuts in the DNA to produce DNA molecules that have projecting single stranded ends that are complementary. In the example given above, a particular restriction enzyme that recognizes this site might cut each strand between the 3rd and 4th nucleotides in from the 5' end of the palindrome. This would produce cut DNA whose ends would look like this:

```
5'...A C T         G A C G T C A G T.....
   ...T G A C T G C A G         T C A.....5'
```

These projecting ends are, of course complementary, as they used to be opposite each other in the double helix before it was cut by the restriction enzyme. Because they are complementary, they can also reanneal with each other to reform the complete palindrome. All one needs to do, then, to completely undo the action of the restriction enzyme is to allow the cut pieces to associate under annealing conditions and then add the enzyme, ligase, which is capable of forming a phosphodiester bond between adjacent nucleotides that are correctly base-paired with their complements on the opposite strand. (Recall that ligase performs just this function in joining up Okazaki fragments during DNA synthesis. It can also be made to perform this joining function *in vitro*, as in the example just described.) These projecting complementary single strands generated by the action of restriction endonuclease are often called **sticky ends**.

One very important corollary of what we have just described should not go unnoticed. Because a given restriction endonuclease only cuts one particular site and always cuts it in a characteristic way, the sticky ends produced by its action will always be complementary regardless of whether the DNA is from the same source or not. That is to say, restriction endonuclease-X will cut site X whether that site exists in DNA from bacteria, toads, or humans, and in all cases the sticky ends produced by its action will be the same. The bacterial DNA, then, can be annealed with human DNA and the part bacterial, part human (i.e., "recombinant") DNA molecule so formed sealed by ligase, which also does not care about the source of the DNA it is joining. In this way DNA molecules from very different sources can efficiently be joined.[*]

Because restriction enzymes recognize fairly short sequences of DNA (typically 4 to 8 nucleotide pairs in length), it is likely that there would occur many such vulnerable sequences in a genome. In the human genome, which consists of about three billion base pairs, there would be tens of thousands of such sequences. Hence, any attempt to chop this amount of DNA with even the most selective restriction enzyme would produce such a welter of fragments that it would be impossible to separate and analyze them. For large stretches of DNA, more selective means of cutting must be found. One approach that seems promising is the use of sequences capable of

---

[*]The two scientists who originated DNA splicing, S. Cohen and H. Boyer, sketched out the seminal experiments in a delicatessen on Waikiki Beach (Hawaii). Alas, this historic site goes unmarked. In fact, it has been obliterated and replaced by a shopping center.

forming triple-stranded sections within normal duplex DNA. In Chapter 7 the existence of triple-stranded DNA and its characteristic bonding were noted. This third strand can be manufactured so as to bind to any given sequence of natural DNA. It is also technically possible to attach an iron-containing oxidizing agent to the ends of the synthetic strand so that the natural DNA is cut wherever the third strand binds. In this way, sequences that occur only once or twice in billions of base pairs can be attacked.[3]

The cutting of DNA with a variety of restriction enzymes or by other means has become an important feature of modern biology because the technique has applicability that goes well beyond genetic engineering *per se*. It is worth pausing, therefore, to take a closer look at how the fragments generated by restriction enzymes can be identified and manipulated.

SEPARATING AND IDENTIFYING RESTRICTION FRAGMENTS

It is often necessary, for example, to know whether a particular lengthy sequence of DNA is found among the fragments produced by cutting up whole genomes of DNA with a restriction endonuclease. A case in point could be the genome from *E. coli* or even the much larger genome derived from a eukaryote such as humans. In both these cases, there will likely be hundreds or even thousands of restriction sites where cuts will be made by whatever restriction enzyme is employed, and hence there will be hundreds of **restriction fragments** of varying lengths[*]. How can one identify and, if necessary, isolate a particular fragment from among this large number? If one has available already a sample of the DNA (or the complementary RNA) one wants to

---

[*] Some restriction enzymes recognize shorter and others longer restriction sites within the DNA. The longer the restriction sequence, the less likely that sequence will appear in the sample of DNA being cut and hence the longer and less numerous the will be the fragments that are produced. Generally speaking, it is more desirable to use a "rare cutter" than one that recognizes a shorter and therefore more commonly found sequence. However, even the best rare cutters still recognize a fairly short sequence of eight base-pairs in length. This means that the genome of *E.coli*, which is fairly small as genomes go, will be cut about 72 times, that of yeast about 230 times, and the human genome (which has about three billion base-pairs) a whopping 50,000 times. But there are developments under way that may permit cleavage at only one of the many restriction sites for a particular enzyme. The most promising of these uses the new-found ability to place specific sequences of DNA at virtually any pre-determined location within a DNA sample. If, for example, one is interested in cutting within Gene X and only there, the operator for the *E.coli* lac operon can be positioned next to or within Gene X. (The DNA containing Gene X can be from any source; it does not have to be from *E.coli*.) Next, the lac repressor is added, which complexes with the introduced lac operator and in so doing also overlays some of the surrounding DNA, including the restriction sites within the target Gene X. Then an enzyme (a methyltransferase) is added that alters the methylation pattern of the DNA and therefore destroys all restriction sites—except the ones protected by the lac repressor. When the repressor is got rid of, the DNA can then be cut just at the Gene X site.

identify, the procedure for doing so has become fairly straightforward. It is known as **Southern blotting**, after E. Southern, who devised the technique.

The first step is to separate the restriction fragments by size. This can be done by electrophoresis on gels composed of agarose. Because the current forces the fragments to migrate through a tortuous complex of channels within the agarose block, the smaller ones will travel further along the gel than the larger ones. Differences of only a single nucleotide will influence the distance the fragment will travel in relation to the other fragments. After the fragments have been separated by length in this manner, they are denatured on the gel, usually by the addition of sodium hydroxide (NaOH). The next step involves the transfer of the denatured fragments to nitrocellulose paper in a way that preserves on the paper the same arrangement the fragments had on the gel. This is usually accomplished by putting a substantial layer of ordinary filter paper under the nitrocellulose and the block of agarose gel on top. A weight is then applied to the gel, which forces the liquid and the suspended DNA fragments through the nitrocellulose and into the filter paper. Single-stranded DNA, however, binds to nitrocellulose and does not pass into the filter paper with the rest of the liquid. In fact, every fragment remains trapped in the nitrocellulose directly opposite where it had originally been in the gel. Hence the separation pattern of the fragments is preserved.

The task is now to identify the fragment that matches in whole or in part the reference DNA, which in this context is usually called a **probe**. This can be done by denaturing the probe and attaching to it a radioactive label. (A number of techniques are available to do this readily; the label is usually P^{32}) A solution containing the radioactive probe is applied to the nitrocellulose under conditions that favor the annealing of complementary DNA molecules. Afterwards, the nitrocellulose is washed to remove uncomplexed probe and then autoradiographed on photographic film. The exposed areas of the film identify the fragments (if any) that had annealed with the radioactive probe.[*]

It should be noted that the probe does not have to be complementary to the whole of the restriction fragment in order for annealing to take place. It could be considerably smaller (the usual case) or larger; it is only necessary that there be complementarity between substantial runs of bases in the two molecules.

COMBINING AND AMPLIFYING DNA

We can now go back to bacterial plasmids. Although they often consist of only a few thousand nucleotide pairs, many of them still have, presumably by chance, restriction sites for known

---

[*]A similar technique, jocularly known as **Northern blotting**, attaches RNA molecules to a special paper by covalent bonds, followed by hybridization with DNA (or sometimes RNA) radioactive probes. Carrying the joke a little further, **Western blotting** attaches proteins covalently to paper and uses probes such as DNA sequences capable of binding to particular proteins or else antibodies that are tagged in various ways.

restriction endonucleases. When such plasmids are extracted from their host cells, they can be cut by that particular restriction endonuclease, which will have the effect of opening up the circle and leaving each end of the now linear molecule "sticky". Suppose, now, that one had available fragments of DNA from humans that had also been produced by the action of this same restriction endonuclease. All these fragments would have sticky ends complementary to one or the other of the ends of the plasmid DNA. If they were annealed with the plasmid DNA, they would join up with it and the recombinant DNA molecule could be sealed with ligase. When the plasmid is re-circularized (and itself resealed with ligase), what one has in effect done is to insert a piece of human DNA into the plasmid. This recombinant plasmid can then be reintroduced into a host bacterium. Suppose, furthermore, that the piece of "passenger DNA" derived from human cells actually contained a complete human gene that codes for some product normally made only by humans. If that gene functioned in the bacteria, bacteria would now be able (in theory) to make that product, too. It has been this goal, now largely realized, that has guided much of the activity in the field of recombinant DNA research for several years. However, it must be stated that many of the "ifs" and "supposes" in the foregoing account neatly paper over some very deep conceptual holes, which now must be filled in more satisfactorily.

Among the difficulties glossed over in the above account are the following two in particular: How can we isolate specific genes from other organisms for insertion into a plasmid? and, How can we get these genes to function in an environment so different from the one in which they normally function? One might also ask whether there are ways to introduce foreign genes into organisms other than bacteria, into plants and animals, particularly. We shall deal with these matters in turn.

First the question of how to isolate specific genes from particular organisms for splicing into a bacterial plasmid. It would be nice, of course, if the gene we were interested in—say, by way of example, the human insulin gene—were conveniently flanked by restriction sites for which we had a restriction enzyme handy. Dream on! For most genes we have no idea even where they are within the human genome, and even if we did know with precision where they were, it is unlikely in the extreme that they could be neatly and specifically excised by any restriction enzymes that we might have at hand. We almost always have to rely on other methods, therefore, in order to isolate the gene we wish to splice into a plasmid.

"REVERSE GENETICS"[4]

One common way around this problem is to start at the other end of the story, so to speak. If we want a bacterium to make a human protein, we usually have been first able to isolate that protein in sufficient quantity to characterize it chemically. Specifically, we can often determine by fairly routine laboratory procedures the exact sequence of amino acids that constitutes the protein. Knowing this, we can then deduce, because we know the genetic code, the precise sequence of ribonucleotides in the m-RNA that would have directed the synthesis of this protein *in vivo*. Knowing the sequence of the relevant m-RNA is nowadays only a short step to actually synthesizing this m-RNA in the laboratory. What we want, however, is not the m-RNA that encodes the protein, but the DNA that produces this m-RNA. Recall, however, that there is a enzyme, reverse transcriptase, that can make a DNA copy of an RNA molecule. Using this enzyme with the m-RNA we just finished synthesizing yields finally what we are after—the gene for the protein in question. (DNA synthesized in this manner is often called **c-DNA**.)

Well, not quite. This piece of DNA, the gene, has blunt ends, and what we need is a gene with sticky ends, in particular, ends that are complementary to those of the opened plasmid into which the gene is to be spliced. Fortunately, it is not too difficult at present to synthesize these termini chemically and to attach them covalently to the ends of the synthetic gene, which can now, at last, be spliced into the plasmid. The recombinant plasmid can then be recircularized and introduced into an appropriate bacterial host.

But will this human gene function in its new environment to produce human insulin? No. The activity of genes is precisely controlled by various initiation signals found on the DNA adjacent to the genes being regulated, and it happens that the initiation signals recognized by the bacterial translation apparatus in the cell are different from those found in human cells. In consequence, although the human gene can be replicated whenever the plasmid containing it is itself replicated, it will not make m-RNA and hence cannot make its product, insulin. The way around this dilemma, however, is very straightforward: before the human gene is spliced into the plasmid it is outfitted with flanking sequences of DNA that correspond to the initiating and regulating signals normally found in bacteria. These, of course, the bacteria can recognize, and when they do, the human gene functions to produce its m-RNA, which in turn gets translated into human insulin within the bacterial cell.

All these steps have been taken, not only with the insulin gene, but also with several other human genes, and the result has been the production in bacteria of products normally found only in the human body. As one might imagine, it is usually much cheaper, safer and more practical to harvest these products from bacteria than from humans themselves.

In the foregoing account, the foreign DNA has always been spliced into a plasmid, which in turn replicates in a bacterial host. In replicating, the passenger DNA is also replicated. Since these modified plasmids can later be isolated in large numbers from their new hosts and their descendants, what one has in effect done is to produce many copies of the passenger DNA. The first copy of the synthetic gene requires considerable time and effort to achieve (see above), whereas the next couple of million copies come quickly and easily simply as a result of the normal multiplication of the modified plasmid inside its host. This amplification of the original gene or other piece of DNA inserted into a plasmid is called **cloning**. As a research procedure in itself, it is exceedingly useful, for it permits investigators to obtain a very large number of exact copies of particular DNA segments for sequence determination or other types of analysis. It is in this way that many genes from bacteria, viruses and mammals, including humans, have had their complete base pair sequence revealed.[5]

REFERENCES

1. Grobstein, C., 1977. The recombinant DNA debate. Sci. Amer. (July).

2. Cohen, S., 1975. The manipulation of genes. Sci. Amer. (July).

3. Moffat, A.S., 1991. Science 252: 1374-5.

4. Varmus, H., 1987. Reverse transcription. Sci. Amer. (Sept.).

5. Brown, D., 1973. The isolation of genes. Sci. Amer. (Aug.).

LECTURE 39

RECOMBINANT DNA: II

VECTORS FOR RECOMBINANT DNA

The agent into which the foreign DNA is spliced—in the cases discussed so far, the plasmid—is technically known as a **vector**. Several vectors other than plasmids are also known and often employed in gene splicing experiments. One difficulty frequently encountered with plasmids as vectors is that they are not efficiently reinserted into new host bacteria once they have been extracted and manipulated. Obviously, given the time, effort and expense involved in obtaining specific DNA molecules and splicing them into plasmids, one would like to have some reasonable expectation that the engineered plasmid will successfully re-establish itself in a host cell. In the procedure for re-establishing them outlined above, far fewer than 1% would actually be taken up by a host cell. To increase efficiency, several newly developed tricks can be used. One makes use of the fact that certain bacteriophage, called lambda, can be tricked into encapsulating inside their heads a modified plasmid instead of their own phage DNA. Since phages are very efficient at delivering their DNA into the cell they attack, these peculiar phages, now called **cosmids**, make very useful vectors.

When the host cell is not a bacterium, other types of vectors altogether must be employed. Certain viruses, for example, can have pieces of passenger DNA from diverse sources spliced into them, which then are inserted into whichever cell the virus infects. Obviously, one would choose for a vector a virus that does not do significant harm to the host, either because it has been deliberately crippled or because it is naturally benign. For plants, it is possible to make use of a particular bacterium (crown gall bacterium: *Agrobacterium tumefaciens*) that has been found to introduce into its plant host cell a piece of DNA similar to a plasmid. This piece often integrates into the chromosomes of the cell, where novel genes on the inserted piece can function. (Before using this particular vector, investigators remove the genes for particular plant hormones that normally reside on the plasmid; in this way, the formation of tumors in the host plant is avoided.)

This technique works well only with the class of plants known as **dicots**, as the crown gall bacterium will not normally infect the other large category of plants known as **monocots**. Unfortunately, the kinds of plants that are primarily responsible for feeding the world are monocots such as corn and all the grains, including wheat. To get around this problem, other means of introducing foreign DNA into monocots have been devised. The most successful of these so far is **electroporesis**. If the cell wall of plant cells is digested away enzymatically, one is left with the plasma membrane as the only thing keeping the cell together. If these **proto-plasts** are subjected to an electric current in the presence of DNA, a small percentage (less than 5%) of them will take up that DNA and integrate it into their chromosomes, where it may function. It has, however, proved impossible to get whole monocots to regenerate from single protoplasts, but recently C. Rhodes has succeeded in doing just that with corn protoplasts. If her techniques have general applicability, the goal of obtaining "miracle" plants capable of producing their own fertilizers and of poisoning their natural enemies (except man, of course) seems within reach.

Another novel technique for introducing foreign genes into plant cells retains the cell wall of the host and hence avoids the difficulties involved in regenerating plants from protoplasts. The technique also works with animal cells. It involves shooting DNA into the cell! Small metal fragments of tungsten are bathed in the DNA preparation that is to be introduced, and then they are fired by a "particle gun" at the host cells. The microprojectiles penetrate the cell wall and some lodge in the nucleus, where the adhering DNA can integrate into the host genome. Since the velocity of the projectile can be regulated precisely, it is possible to imbed the particles into tissues that lie several layers below the surface of the target.[1]

Yeast cells are often a desirable host for transplanted genes, since, unlike *E. coli*, they are eukaryotes and might be expected to have protein synthesis and control mechanisms that more closely resemble those in mammals and the like[*]. The incorporation of exogenous DNA into yeasts has been facilitated by the 2mu (2μ) plasmid, the only one known to exist in yeast. (The normal function of this plasmid is unknown.)

TRANSGENIC ANIMALS

When it comes to introducing genes into mammalian cells, the commonest way of doing this is by **transfection**, which is essentially transformation under another name. Although animal cells on occasion will take up DNA from the culture medium, this method is highly inefficient. The usual procedure with mice is to inject the gene or the whole plasmid carrying the gene directly into the nucleus of the oocyte. The oocyte is then implanted into a mouse made artificially pregnant. When carried out in this way, around 15% of the progeny mice carry the transfected gene, often in multiple copies. Moreover, there is a reasonable chance such genes will function in their new environment if care has been taken to match up the controlling sequences appropriately. It is also possible to enhance the likelihood that the transfected gene will become integrated into a particular region of a chromosome if flanking sequences that match those at the desired chromosome location are first added to the transfected gene before injection. Animals that carry one or a few genes from another species are said to be **transgenic**.

Sometimes germinal stem cells are used as the hosts for transfected genes. If these stem cells contribute to the production of germ cells, many of the fertilized eggs will contain the exogenous gene.

---

[*]In fact, yeast genes function in *E. coli* more commonly than anticipated. It has been estimated that about 30% of yeast genes will function in *E. coli* without substantial modification.

Some recent observations by an Italian research team headed by C. Spadafora suggest that spermatozoa from many different species have a heretofore unexpected ability to pick up exogenous DNA from a culture medium and to introduce it to the eggs they fertilize. The foreign DNA functions in the resulting zygote and will often become part of the germinal line. This simple and efficient technique for generating transgenic animals could become the method of choice for most future studies—if it is repeatable. Many scientists are yet to be convinced of the validity of Spadafora's report.

A novel twist in the search for the efficient use of large mammals as "factories" for the production of biochemical products needed by humans has recently proven successful with sheep and goats. In this case, the transfected gene is first nested inside the gene for beta-lactoglobin, a gene that is expressed only in the mammary glands. If the transfected gene is expressed, the gene product is secreted into the milk, where it can be harvested in large amounts without inconvenience to the animal. This has proved successful with the genes for certain human blood clotting factors (sheep) and for a protein (tissue plasminogen activator) that dissolves blood clots that cause heart attacks (goats). It has been suggested that a herd of about 300 transgenic goats would provide the entire world with its supply of TPA.

It has recently proved possible to transfect the gene for human haemoglobin into very early pig embryos to yield adult pigs that manufacture human haemoglobin. About 0.5% of the attempted gene transfers succeed, and of those that do, only about 15% of their haemoglobin is the human variety, although there are hopes of increasing both the efficiency of the gene transfer and the yield of the haemoglobin. Separation of the pig and human haemoglobins is accomplished electrophoretically. The use of raw haemoglobin instead of whole red blood cells for transfusions, especially under emergency conditions, has many advantages. Whole red blood cells can be stored only under refrigeration and then only for about 30 days before they deteriorate. Tissue matching must also be carried out to avoid transfusion reactions. Whole cells can also be contaminated with the disease agents present in the donor, especially viruses. Raw haemoglobin, being a pure chemical substance, can be stored longer and under less fussy conditions. No tissue matching is necessary, and being of non-human origin, there is little likelihood of contamination with disease agents affecting humans. However, raw haemoglobin, in contrast to whole cells, is effective in carrying oxygen for only a few hours when transfused into the bloodstream, but under emergency conditions, that would usually be sufficient to sustain the patient until better care is available.

ANTI-SENSE MESSENGERS[2]

Sometimes the gene that is introduced into a host cell is not a real gene at all, but an "anti-sense gene". In being introduced into a cell, a normal gene can deliberately be flipped over so that the normal non-coding strand of DNA becomes the one transcribed into RNA. Because this novel RNA would be the complement of the normal gene product, it is called **anti-sense m-RNA**. And being complementary to the normal gene product, it can complex with and hence inactivate the normal messenger RNA. This provides a means, therefore, to in effect inactivate particular genes within the organism. This technique has already been exploited to turn off the production of certain undesirable enzymes in tomatoes that contribute to mushiness or pigment-producing enzymes that affect the coloration of flowers. Most strikingly, an anti-sense gene has been inserted into tomatoes that prevents the production of ethylene gas, the substance that is essential for ripening. Tomatoes that do not produce ethylene grow large but remain green and therefore do not spoil, but ripening can be induced at will by exposing the green tomatoes to ethylene. There artificially ripened tomatoes are reputed to retain the full flavor, aroma and texture of naturally ripened ones, but they can be shipped over long distances in their green state.

In some organisms the anti-sense messenger is simply injected into the egg cell instead of being produced internally by an introduced anti-sense gene. In these cases the effects are somewhat short-lived, since the anti-sense messenger will eventually be degraded. Some scientists believe such an approach will prove useful in anti-viral therapies.

GENE THERAPY IN HUMANS[3]

Although many inherited diseases in humans can be effectively treated, it has only been recently that medical researchers have begun to talk of "cures" for such diseases, since a cure would have to involve the functional replacement of a defective gene with a normal one. The first disease where this approach was tried was SCID—severe combined immunodeficiency syndrome. This rare disease is caused by the inability on the part of lymphocytes to produce the enzyme adenosine deaminase (ADA), and this in turn makes it impossible for the patient to mount any kind of immune response. Needless to say, the disease is invariably fatal. In 1990, doctors first began treating patients with SCID by withdrawing some of their defective lymphocytes and infecting them with a modified (and presumably harmless) retrovirus that carried the human gene for adenosine deaminase. The treated lymphocytes were then returned to the patient. Some of them began making ADA, thus effecting a "cure" for the SCID. There is, however, a major drawback to this approach: the treated lymphocytes, like all lymphocytes, live only a few months and are then replaced by the maturation into lymphocytes of mitotically dividing **stem cells**. This means that engineered mature lymphocytes must be continually introduced into the blood stream in order to maintain the "cure". Hence, the quotation marks. Currently (1992) an Italian medical team is pursuing a plan to infect the lymphocyte stem cells themselves with the retrovirus carrying the human ADA gene. If this effort is successful, it should obviate the need to keep replenishing the engineered lymphocytes, since the lymphocytes that naturally derive from the treated stem cells should carry a functional ADA gene. Then the quotation marks around "cure" can be legitimately dropped.

OTHER APPLICATIONS OF THE TECHNIQUE[4]

Among the more unusual products that both bacteria and plants are now being engineered experimentally to produce is—plastic! Although at first blush our natural reaction is to exclaim that there is already too much of that stuff around and we certainly don't need any more, on sober reflection our minds may change. Aesthetics aside, current plastics have two principal disadvantages: the are derived from a non-renewable resource (petroleum), and they are not biodegradable (which is mainly why there is too much of them around). There is, however, a bacterium (*Alcali-*

genes eutrophus) that makes a polyester compound as an energy storage molecule, just as other organisms synthesize starch or glycogen for the same purpose. Under certain environmental circumstances this plastic substance can account for about 80% of the dry weight of the cells. This plastic, or derivatives of it synthesized under different environmental or genetic circumstances, can be extracted and molded into stiff, brittle plastic useful for containers such as soft drink bottles. These products, unlike most plastics now in use, are completely biodegradable into water and carbon dioxide. The genes that govern synthesis of the plastic in *A. eutrophus* have been cloned and transferred to *E. coli*, where they also function, although not quite so efficiently as in the original host. Work is now under way to transfer them to higher plants such as potatoes, turnips, and corn in order to engineer these crops to produce plastics instead of starch as their energy storage molecules.[5] A more difficult task will be to persuade the public that plastics are tasty and nutritious to eat! (Relax. The intention is not to replace the starch-producing crops with plastic-producing ones, but to give farmers an additional cash crop for sale to industries involved in plastics manufacture.)

It is very likely that the use of plants as hosts for a number of other transplanted genes will become more common in the future. Plants offer the advantage of being largely self-sufficient and easy to grow in quantity, whereas bacteria and yeasts need constant attention and even cows and sheep need regular feeding and fussing over. But give a plant some sun, rain and a few other obvious necessities, and it will do very well on its own. Many plants (e.g., tobacco, carrot, petunia, and potato) can be grown from a single somatic cell so that the tricky problem of engineering germinal cells is obviated. Finally, the metabolic processes of a plant cell, especially those involved in protein synthesis, more closely resemble those of a human cell than do those found in bacteria and yeast. This means that human genes will be more accurately translated and the products more accurately processed or assembled in plants than in bacteria or yeast.

Already human antibody (one used in the treatment of lung cancer) has been produced in tobacco plants[*], serum albumin (a blood protein widely used in surgery) in potato plants, and enkephalin (a painkiller produced by the brain) in rape plants. If the relevant gene had been introduced into a somatic cell and an entire plant derived from this one cell, then the whole plant and usually its progeny, as well, will carry the new gene. Sometimes instead of genetically altering the plant, the desired gene is attached to an infectious RNA molecule akin to an RNA virus. In these cases, only the infected plants, but not their progeny, produce the product. So it might be someday the case that the tobacco plant, now responsible for an estimated 300,000 deaths annually in the United States may be able to redeem itself in the eyes of humanity.

---

[*]Two genes are necessary in order to produce a functional antibody, one for the heavy chain and one for the light. Each is introduced into a somatic tobacco cell using a plasmid vector such as that from *A. tumefaciens*. A complete plant is then grown from each cell, and the adult plants crossed to produce F1 plants, some of which contain both genes. To ensure a good yield of antibody, the appropriate signal sequences are added to the antibody protein that will cause it to be transported to the membrane and secreted to the outside of the cell.

Although this account has focused on the use of plants in the large-scale production of products useful to man, it should not escape notice that many of them could also be useful to the plant itself. For instance, a plant that produced an antibody to phytopathogenic bacteria, molds and viruses would have obvious commercial use in agriculture.

SAFETY CONSIDERATIONS

Finally, some mention must be made about the safety of gene splicing experiments, especially in view of the wide publicity given to this aspect of recombinant DNA research in the late 1970s. It was thought, for example, that bacteria could accidentally be engineered to produce a human substance, like growth hormone. The investigator, not aware that the bacteria had this ability, discards them down the drain, where they infect the water supply—with the untoward results that you can conjure up for yourselves. As a result of the widespread dissemination of such horrific scenarios in the public media, strict regulation of recombinant DNA research and the technologies derived therefrom was put in place through the various government bodies that fund scientific research or control what can be released into the environment. We now see, however, that human genes cannot normally function in bacterial cells, and it takes considerable fiddling to get them to do so. In recent years, therefore, the original regulations have been considerably relaxed, although there are still some who oppose the testing and use of genetically engineered organisms. Obviously, one must always be careful to prevent the escape of deliberately engineered bacteria that carry mischievous genes, but that is the case when dealing with any deadly organism in the laboratory, whether of artificial or natural origin.

REFERENCES

1. A good review of progress in crop improvement through genetic engineering is the following: Gasser, C.S. and R.T. Fraley. "Genetically Engineering Plants for Crop Improvement". Science 244:1293. This entire issue of Science is devoted to articles on various aspects of genetic engineering in medicine, agriculture and research organisms.

2. Moffat, A., 1991. Science 253: 510-11.

3. Miller, A., 1992. "Human gene therapy comes of age." Nature 357: 455-460.

4. Gilbert, W. and L. Villa-Komaroff, 1980. Useful proteins from recombinant bacteria. Sci. Amer. (April).

5. R. Pool, 1989. In search of the plastic potato. Science 245:1187-1189.

LECTURE 40

GENETIC SCREENING IN HUMANS

MAPPING HUMAN CHROMOSOMES[1]

The use of restriction fragments to map human chromosomes (and those of other higher eukaryotes) is an obvious and important extension of the work that has been presented in the previous two lectures. In humans, this mapping technique has particular significance as a diagnostic tool when the actual product of a defective gene is as yet unknown and hence when it is impossible to discover the existence of the gene by direct chemical analysis of a tissue sample (as in amniocentesis, for example). In essence, what this method seeks to do is to establish the genetic linkage of a defective gene with a pattern of restriction sites surrounding the gene. An individual carrying this particular pattern would then most likely also be carrying the defective gene in question. A good candidate for this type of analysis is the gene causing Huntington's disease, a progressive and fatal dominant nervous disorder that strikes affected individuals starting around their mid-40s. The gene product of Huntington's disease is unknown, making it impossible to inform carriers early not only of their own genetic status but also of the likelihood of passing the gene on to their offspring. Let's see how this method works in practice.

RESTRICTION FRAGMENT LENGTH POLYMORPHISMS

The technique is predicated on the observation that individuals within a population display considerable variability in the number and distribution of restriction sites that can be attacked by the several restriction enzymes biochemists have at their disposal[*]. In consequence, when the DNA from one person is digested by a particular restriction endonuclease, there will be some differences in the length of the restriction fragments produced when compared to those obtained from someone else. The hope is that some of this variability can be traced to the presence or absence of restriction sites in the neighborhood of the mutant gene one is interested in identifying. For the sake of illustration, let us suppose that the gene we are interested in (x) lies in the neighborhood of several restriction sites (R1 to R6). Let us further suppose that the R3 site, which lies near gene x, is missing in individuals carrying x and present in individuals who have the wild-type allele of x. (This is just another way of saying that x is linked to R3; "near",

---

[*]It has been calculated that individuals will differ on average in 1 of every 500 nucleotides; about 5% of this variability will result in a alteration to a restriction site, either creating it or abolishing it. Since the human haploid genome contains 3×10^9 nucleotides, the potential for extensive heterogeneity in restriction sites is obviously very great.

in this context, can mean several thousands of bases[*] away and suggests only that recombination rarely occurs between x and R3[**].) This situation can be diagrammed in the following way, where the numbers indicate the numbers of bases (given in thousands of bases) that separate the restriction sites.

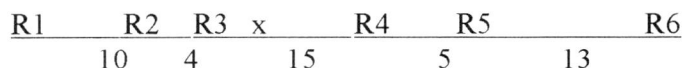

R1		R2	R3	x		R4		R5		R6
	10		4		15		5		13	

If this DNA were cut when R3 is present (and therefore the x[+] allele is present), the five restriction fragments for that region would have the following lengths:

> 10 kilobases
> 4 kilobases
> 15 kilobases
> 5 kilobases
> 13 kilobases.

On the other hand, if R3 is absent (and therefore the x gene is present), there would be only four restriction fragments, and they would be of the following lengths:

> 10 kilobases
> 19 kilobases
> 5 kilobases
> 13 kilobases.

This is obviously a convenient way of identifying whether gene x is present or not—but there is still one important piece information missing from the puzzle before the analysis can be carried out as described. A means must exist to identify just these restriction fragments from among the zillions that would result from digesting the total genome with the restriction enzyme. Here an element of luck must enter the investigation, for identifying these particular fragments rests on

---

[*]Strictly speaking, it should be base-pairs, but since much of the analysis deals with restriction fragments that have been made single-stranded, bases rather than base-pairs are given.

[**]The estimated recombination frequency in humans is 10^{-8} per nucleotide per gamete. If, therefore, the distance between the gene in question and the restriction site is 5 kilobases (5kb), the probability of crossing over occurring between them is 5000×10^{-8}, or 5×10^{-5}. A million base-pairs represents a recombination frequency of about 1%. The statistical error in RFLP analysis is largely that attributable to recombination. Even if the restriction site is fairly distant from the mutant gene so that recombination occurs, say, 5% of the time, the resultant diagnostic error rate of 5% has to be compared to the 50% error rate of the alternative (i.e., guessing or hoping).

finding a DNA probe that binds to a significant part of just these fragments. And since one often does not know where in the genome gene x lies, one usually does not have a known probe handy for the analysis. So "brute force" must be used. By this is meant that often hundreds of different probes must be used in Southern blots of the digests before one is discovered that demonstrates a difference in fragment distribution between definitely affected and definitely unaffected family members. In the case of Huntington's disease, one useful probe has been discovered by J. Gusella and his coworkers that succeeds in differentiating the restriction fragment patterns from the two groups and hence in diagnosing who will get the disease and who will not. The work has also let to the chromosomal location of the mutant gene to chromosome 4. Other genetic diseases whose chromosomal locations have been established using this method are Alzheimer's disease (chromosome 21), cystic fibrosis* (chromosome 7), and a manic depressive disorder in Pennsylvania Amish (chromosome 11).

Hence it is now possible to test adults for the presence of a few genes known to cause serious illness that do not manifest themselves until well after birth, in some cases many decades after birth. In addition to the gene that causes Huntington's chorea, genes for adult polycistic kidney disease (progressive loss of kidney function), polyposis (a precursor of colon cancer), and certain forms of cancer (some leukemias, retinoblastoma—a childhood eye cancer—, and small-cell

---

*Cystic fibrosis is the most common lethal genetic disease in Canada and the United States, affecting about one in 2000 white and one in 20,000 black babies. About one white person in 20 carries this recessive mutant gene, meaning that one couple in 400 have the potential of producing affected offspring. The disease is characterized by the accumulation of thick mucus in the lungs; affected individuals usually do not live beyond their 20s. The identification of the chromosomal location of the cystic fibrosis gene has led quickly to the development of a test that detects about 75% of the carriers of the mutant gene. The identified carriers are those that carry a deletion mutation in the gene; the remainder, those whose mutation arises at a single base-pair site, cannot be identified by current means.

The fact that only 75% of the carriers are detected in this screening test gives rise to an awkward ethical dilemma. If the test is applied to all couples wanting children, cases where only one parent carries the mutation will be about 25 times more frequent than cases where both parents do. (The most common result, of course, is that neither parent carries the mutation.) Since, however, the test detects only the most common source of the mutation, there is still the possibility that the seemingly unaffected parent actually carries one of the point mutations in this gene. In this case, there is a one-in-four chance of their having an affected child. Moreover, this child cannot be detected during prenatal screening, which identifies only those fetuses homozygous for the deletion mutation. This couple, now knowing that they have a much higher than average likelihood of having an affected child, are then left in doubt about how to proceed. (The chances of their having an affected child are between 1 in 300 and 1 in 500.) Because of these limitations, some geneticists argue that only those prospective parents with a family history of the disease be screened, whereas others argue that it is still better to prevent many such births than none at all.

cancer of the lung can now also be detected. Screening adults for the presence of mutant genes that affect them directly is obviously different from the traditional form of genetic screening which had as its object determining the risk of producing defective offspring. However, people seem more willing to test their unborn fetuses or to be tested themselves for the sake of their as-yet unconceived children than for the sake of learning their own predestined fate. Before a method for detecting the gene for Huntington's chorea was devised, about 80% of those at risk thought they would take such a test if it were available; when the test did become available, only 15% to 20% of those at risk decided to take it. Of course, there is nothing medically that can at present prevent the onset of this particular disease; perhaps the result would be different for those conditions that can be averted or at least meliorated by changes in diet, lifestyle, occupation, etc.

When a particular trait—such as the length of restriction fragments in the account just given—exists in a variety of different forms in a population, geneticists call such a trait **polymorphic**. This has led to this technique being called **restriction fragment length polymorphism** analysis. In order to save time for research, this has been abbreviated to **RFLP** (pronounced, somewhat preciously, "rifflip") analysis. The fragments themselves are called **RFLPs**.

TANDEM REPEATS

Besides the variability in the distribution of cutting sites, there exits another reason why samples of human DNA from different individuals would subdivide into unique RFLP patterns when treated with restrictions endonucleases. Scattered throughout the human genome are short nucleo-tide sequences that are repeated many times at each site where they occur. The function of these repeats is unknown, but since unrelated individuals commonly differ in the number of repeats found at each site, it is unlikely that they have any function at all. They do, however, provide useful markers for mapping human genes. Because the number of times the sequence is repeated is variable among individuals, the fragments obtained by digesting their DNA will also vary in length, even when there is no variation in the location of the restriction sites themselves. Hence, RFLPs can be generated both by **VNTR**s ("variations in the number of tandem repeats") as well as by variations in the distribution of restriction sites.

OTHER APPLICATIONS

The fact that individuals differ in the distribution of restriction sites has led to the use of RFLP analysis in crime detection. If a sample of a suspect's DNA at a crime scene is digested by restriction enzymes and tested with a sufficient number of probes, the link with the suspect can be made (or refuted) with an accuracy at least as great as that possible with finger prints. Moreover, RFLP analysis can be combined with the polymerase chain reaction in order first to amplify minute amounts of DNA, such as that present in the follicle of a single hair or in a cell or two from the lips shed onto the butt of a cigarette. Therefore, one shouldn't smoke while committing a crime—or even when not, for that matter.[*]

Because DNA samples can be obtained from almost any remaining tissue in a dead body, this technology is becoming increasingly useful in identifying human remains, especially in those situations where the fingerprints, dental structures, or other conventional identifying markers are mangled, decayed, or absent. For this reason, the United States Army is now taking and storing DNA samples from all military service personnel. It has also created a lively interest in the possible genetic diseases that afflicted long-dead historical figures. Scholars, for example, have often suggested that Abraham Lincoln may have suffered from **Marfan's Syndrome**, a genetic disorder causing, besides a characteristic tall, gangly appearance, weakness in the bones, eyes, heart and blood vessels. Now that the gene for Marfan's Syndrome has been isolated, researchers have requested permission to extract DNA from samples of Lincoln's tissues that have been preserved in the National Museum of Health and Medicine and to test it (or, more precisely, cloned DNA derived from it) for the presence of the mutant Marfan's gene. None of this, of course, will do Lincoln any good and neither would it do him any good if he were still alive, since the disease has no treatment. But apparently it would do scholars some good, for it is probable that permission for the study will in due course be granted.

Post mortem diagnosis has its counterpart at the other end of the lifespan, prenatal diagnosis. Some aspects of this topic were covered in an earlier lecture on chromosomal abnormalities. The increasing sophistication of genetic techniques that permit the identification of particular mutant genes (as opposed to entire chromosomes) in individuals during their lifetime or thereafter also permit such identification to take place prenatally, in the fetus. A striking demonstration of this ability has been the identification of the mutant cystic fibrosis gene in human embryos at the four or eight cell stage of development. To do so requires first that the ovum be fertilized *in vitro*. At the four or eight cell stage of embryological development, one of the cells is removed for analysis; this can be done without any apparent affect on the embryo then or later. Although a single cell contains only 6 picograms (6 trillionth of a gram!) of DNA, a modification of the polymerase chain reaction permits the portion of the DNA containing the cystic fibrosis gene to be specifically amplified to the point where enough exists to be probed for the presence of the mutant gene. If the embryo is found to be free of the defective gene, it is transplanted into the mother to complete development to term. And if it does carry the mutant gene...well, let's just say that it is less likely to be transplanted and leave it at that.

------

[*]These techniques have also spawned some interesting commercial applications. For example, a company in France is marketing a pen that uses a special ink laced with unique DNA sequences. Signatures or manuscripts written with this ink cannot be forged, since it would be almost impossible to duplicate the DNA marker. In theory, it would be possible to use the author's own DNA, making the product both personal and absolutely unreproducable—unless someone had access to your DNA. (I won't speculate here how *that* could come about!)

REFERENCES

1. White, R. and J. Lolouel, 1988. Chromosome mapping with DNA markers. Sci. Amer. (Feb.).

LECTURE 41

DEVELOPMENTAL BIOLOGY

INTRODUCTION

An enduring mystery in Biology is how a fertilized ovum, by mitotic divisions, becomes a multicellular organism of a predetermined shape and composed of cells of many different types. After all, mitosis is supposed to ensure that an exact copy of the genetic material from the mother cell is passed to each of the daughter cells—and, barring accidents like mutation or non-disjunction, it does. So how does it come about that liver cells, for instance, are clearly different from nerve cells and they in turn so different from epithelial cells, and so on? And what causes cells to aggregate in specific patterns during development so that organs and tissues of characteristic shape and cellular composition result? There is no one answer to these questions, since in many cases the answer depends on the organism being investigated or on the organ or tissue under investigation within a particular organism. In this account, we shall look at some of the key events in the embryological development of animals, keeping in mind that the details vary considerably from one class of animals to another. We shall also look at some of the ways we know that cell lines during development come to differ functionally and morphologically from others, even though all cells in the organism contain the same set of genes. Much remains to be learned about these processes; for despite the very rapid pace of research and discovery in developmental biology, this increasing volume of information has also served to highlight the concomitant expanding perimeter of our ignorance.

THE GENERAL LINES OF ANIMAL EMBRYOLOGICAL DEVELOPMENT

Development is normally thought of as beginning with the fertilization of an egg cell by a spermatozoa. This view contains the tacit assumption that these two cells contribute equally to the events that follow. This assumption is not correct, for in many cases the early events in embryology are determined exclusively by the egg cell, either by such considerations as the amount and placement of yolk in the cell or by messenger RNA molecules already present in the egg before fertilization and therefore derived exclusively from the maternal genome. The distribution of yolk in the ovum is important because its concentrated presence makes difficult or impossible the completion of cell division following mitosis of the nucleus. In consequence, cells at one end of the embryo (called the **animal pole**) may divide more rapidly than those at the other (the **vegetal pole**) when the yolk is concentrated at one end of the egg. In the extreme case of a bird's egg, the yolk is not partitioned at all, and development proceeds by division of cells in a small disc on the surface of the yolk.[*] Mammalian cells, by contrast, contain very little yolk, and therefore the full mass of the egg can participate about equally in the ensuing divisions.

Following fertilization, the two **pronuclei** (that of the sperm and of the egg) fuse to form a single diploid nucleus, which then begins to divide mitotically. The concomitant cell divisions, called **cleavage divisions**, soon transform the egg cell into a small ball of cells called a **morula**, the individual cells being known as **blastomeres**. Although there is the necessary synthesis of DNA in order to generate daughter nuclei, there is little or no synthesis of cytoplasmic constituents during these early cleavages. This results in a morula that is about the same size as the egg cell that underwent cleavage in the first place. As noted above, the amount and placement of yolk may cause considerable distortion of these cleavages so that a morula in some organisms will contain cells of very different sizes or in others will not be a ball of cells at all, but a flattened disk (the **blastodisc**) on the surface of the yolk, as, for example, in birds. In some animals, the cytoplasmic constituents of the egg have been unequally apportioned among the blastomeres, and the developmental fate of these blastomeres and their descendants will depend on which cytoplasmic constituents they received during these early cleavages. In other organisms, the blastomeres of the morula are still largely equivalent and hence each is still capable of undergoing a variety of subsequent developmental fates. A cell that is still capable of generating an entire organism if it were separated from the embryo is called **totipotent**. In plants, it is common that cells of the mature organism remain totipotent; that is to say, under laboratory conditions, an entire, normal plant can be regenerated from a single (non-germinal) cell of the mature plant, demonstrating that there has been no irrevocable loss or inactivation of genetic material during plant development. Animal cells, by contrast, generally show a progressive loss of totipotency as development proceeds, some animals losing totipotency much faster than others. This subject will be explored in greater depth later on.

When the morula has amassed several thousand cells, the innermost cells begin to migrate to the periphery so that the morula is transformed into a roughly hollow ball called a **blastula**. As the blastula cells continue dividing, it is common in many organisms for the genes of the embryo finally to be switched on at this point so that an increase in the total mass of the embryo begins to occur. By the late blastula stage, cell movements within the embryo become very pronounced as the next stage in development, **gastrulation**, begins. Gastrulation involves the invagination of the blastula in a manner best visualized as the consequence of bushing in a tennis ball from one side in order to form a crescent-shaped, two-layered structure called, in embryology, a **gastrula**. The analogy is not exact, however, because in development, the invagination is accomplished more by the inward-directed continued growth of cells than by some outside force acting on the surface of the hollow ball. Also, the opening that remains on the side of the

---

[*] The "egg" we are now told for health's sake to forego at breakfast is actually much more than an egg in the reproductive sense. Only the yolk is the ovum, the nucleus of which is at the surface just under the yolk membrane. The rest of the breakfast egg, the "white" and the shell, is nutritive and protective material secreted around the ovum during its passage down the bird's oviduct.

246

invaginated ball of an embryo is a small pore, the **blastopore**, instead of the much larger one it would presumably be in a tennis ball. At this point, therefore, the embryo is a two-layered, hollow ball with a blastopore. The hollow interior is destined to become the gut of the mature organism, and the blastopore its anus. The innermost layer of cells in the gastrula is called the **endoderm**, from which most of the digestive organs are derived; the outer layer is called the **ectoderm**, and it gives rise to the nervous system and the outer body covering.

Cells at the **dorsal lip** of the blastopore appear to be crucial to the invagination process. If at the very beginning of gastrulation of an amphibian embryo cells from the dorsal lip of the blastopore are excised and transplanted to the other side of the embryo, a second invagination will occur at the new site and a double ("Siamese twins") tadpole will eventually develop with two complete sets of body structures. Because the dorsal lip of the blastopore is so important in the reorganization of the blastula, which in turn has far-reaching effects on further development, it is termed the **organizer**. The (presumed) chemical basis for the unique powers of the organizer is not known.

As the gastrula is forming there is also occurring a growth of cells around the blastopore that seem to have been stimulated by the interaction of the endoderm and ectoderm. These new cells also grow inward and invade the space between the two layers of cells already there, thus converting the gastrula into a structure with three **germ layers**: endoderm, ectoderm, and the new **mesoderm**. Mesoderm gives rise predominantly to the musculature and skeletal systems.

Once the embryo has become three-layered, there is a further growth of cells from the dorsal lip of the blastopore that push inward to form a solid tube stretching over the "roof" of the gastrula backwards from the blastopore. This sheath of cells is known as the **notochord**. In very primitive chordates, those without a vertebral column, the notochord is retained throughout life and serves as a substitute for a vertebral column; in higher vertebrates, it serves as the embryonic organizer around which the vertebral column later develops. But regardless of whether the notochord is retained or not, early in development the ectodermal cells lying above the notochord invaginate and fold over on themselves to form the **neural tube**, from which the spinal cord, brain and all peripheral nerves will eventually develop. Along side of the notochord, the mesodermal cells begin to form segments known as **somites**, from which the musculature and skeletal systems develop. The remnants of this mesodermal segmentation can be seen in the repetitive arrangement of vertebrae and ribs in the adult.

It is important to keep in mind, as has already been pointed out, that in describing the events of early embryology there is considerable variation in detail from one group of organisms to another. In mammals, for example, a **blastocyst** develops from a solid morula. This blastocyst consists of a hollow ball in which a group of cells known as the **inner cell mass** accumulates at one end. All the structures of the embryo are derived from the inner cell mass, which proceeds to flatten out into a two-layered **embryonic disc**. The cells around the periphery of the blastocyst, the **trophectoderm**, implant into the uterine wall and become supportive and protective tissues (e.g., placenta and amnion) for the developing embryo. Within the embryonic disc itself, the third germ layer, the mesoderm, then grows between the two already present in the disc to establish a three-layered embryo superficially similar to that of a bird embryo. Which cells of the morula become part of the trophectoderm and which become part of the inner cell mass seems not to be predetermined; those that happen to be around the periphery become trophectoderm and those in more internal locations constitute the inner cell mass.

REGULATIVE AND DETERMINATE DEVELOPMENT

Given this somewhat hasty generalized description of early development, we are now in a position to ask the more interesting and fundamental questions about the nature and causes of the changes that occur in embryonic cells, individually or in groups, as development proceeds. The first of these concerns how early in development can such changes be detected in cells, and the second the closely related one of how late in development can cells (or cell nuclei) remain unchanged. This question can be put more succinctly by reintroducing the term **totipotency**. Are all the cells of the morula, the blastula, the gastrula, or even later stages totipotent, i.e., capable of regenerating a normal embryo and thence an adult if separated and grown under appropriate conditions? Again, as is so often the case in biology, the answer depends on which organism is being investigated. In molluscs and sea squirts (ascidians), for example, the very first division of the fertilized egg establishes a difference in the two daughter cells that will direct them along different developmental paths. It is likely that this difference is the result of the cells having acquired different *cytoplasmic* determinants rather than different *nuclear* (genetic) information. At the other end of the scale are mammals. Identical twins in humans are the result of the independent development of the initial two daughter cells, and it is clear that each cells has led to the formation of a normal embryo and later adult. This totipotency of the early cleavage products continues even later. In the mouse, it can be demonstrated that all cells of the eight-cell morula are totipotent. These cells can be artificially separated and each will become a normal embryo if allowed to develop individually. Moreover, a group of cells can be removed from a morula, and the remaining cells will divide to make up the deficit without affecting further development; or two morulas (morulae?) can be fused, and the cells will sort themselves out into a single morula that develops normally. This latter embryo with four distinct parents is known as a **chimera**. An embryo that can compensate for the removal or addition of cells is said to have **regulative development**; one that cannot, such as the molluscan embryo, is said to have **determinate development**.

Chimeras can be created by injecting early embryonic cells (or just a single cell) into the blastocyst of another embryo in the hope that some of these cells will become incorporated into the inner cell mass and become part of the embryo proper. If the introduced cells are genetically distinct from those of the host, their subsequent fates during development can often be followed precisely. For example, cells from a white mouse can be fused in this way with an embryo from a black mouse, and a piebald black and white mouse will develop. These quadra-parental chimeras have been very useful in studying mammalian development. One question they answered is whether tissues and organs that develop in later stages of embryology trace their origins to single precursor cells in the early embryo or to groups of cells. By using genetically marked cells in a chimera, it can be demonstrated that cells from both sets of parents wind up

248

in all organs and tissues of the adult. Hence, single "founder" cells of the embryo are not the precursors of later-arising tissues and organs.

THE TOTIPOTENCY OF THE NUCLEUS DURING DEVELOPMENT

However, even in cases where embryonic cells remain totipotent during the early cleavages, there comes a time, usually soon after the eight-cell stage, when this ability is lost. The most obvious reason for this loss of totipotency, given what has already been described, is that very early development is largely governed by maternal messengers pre-existing in the fertilized egg; the dilution of these maternal messengers by continued cell division would be expected to lead soon to a situation where a cell would have an insufficient supply and hence would be incapable of beginning successful embryological development *ab initio*. Although this explanation adequately accounts for the phenomenon, there could be a further reason, as well. Could it be that the genetic material of the developing embryonic cells is altered in some way so that later embryonic cells either no longer can express (more likely) or no longer have (less likely) genes that are active at earlier stages. In other words, could a nucleus from a later development stage, put back into the cytoplasm of a fertilized egg, direct the development of a normal embryo? Does a *nucleus* remain totipotent?

The answer to this question was supplied by the widely-known nuclear transplantation experiments of J. Gurdon. Gurdon took fertilized amphibian eggs and destroyed their nuclei by ultraviolet radiation, a treatment that seems to do little harm to the cytoplasm. Using a micropipette, he then transplanted into these enucleate eggs nuclei from cells at later stages of development. Many of these eggs subsequently went through a completely normal development through the tadpole stage and on to fertile adult frogs. Gurdon demonstrated that even fully differentiated tissues such as the epithelium of the tadpole intestine has nuclei that are totipotent. The success rate in such experiment, it must be admitted, is disappointingly low; only about two percent of the transplantations from tadpole gut cells yield normal adult frogs, and most tissues yield none at all. Nonetheless, the point is still made that fully differentiated (and normally non-dividing) cells have nuclei that contain the entire genome in a potentially active form[*].[1] [2]

---

[*] These transplantation experiments of Gurdon's have spawned a lively genre of science fiction in which adult individuals (humans, say) have been cloned in one or multiple copies. Human twins represent, of course, a small clone, but in this case the genetic potential of the two individuals is not known beforehand. If tissues from an adult of established accomplishments (good or bad, but assuming—a big assumption in itself—that they are to a significant degree genetic) could yield nuclei that could be transplanted into any number of fertilized eggs, these "synthetic" embryos could then be implanted into surrogate mothers for further incubation and birth. (See, for example, Southin, J., 1972. "Cloning: One hundred Einsteins or one hundred Hitlers?" in *S.F.: Inventing the Future*. R.D. Appleford (ed.), Bellhaven House, Scarborough, pp. 213-15.) As it happens, mammalian eggs are just too small to manipulate successfully in this way—thank goodness.

REFERENCES

1. Gurdon, J., 1968. Transplanted nuclei and cell differentiation. Sci. Amer. (Dec.)

2. Gehring, W., 1985. The molecular basis of development. Sci. Amer. (Oct.)

250

LECTURE 42

THE ORIGIN OF TISSUES

DETERMINATION, CELL MEMORY, AND DIFFERENTIATION

Determination in development refers to a process whereby a cell and all its descendants becomes irreversibly committed to a particular fate, which may be manifested only at some later stage. Cells may become determined at various stages of development. Some become determined very early (as, for example, cells of the three germ layers which, once becoming endoderm, mesoderm or ectoderm, never switch fates and assume characteristics of a different germ layer) and some become determined much later (neural cells in the rat retina, for example, can still differentiate into a variety of quite different cell types right up to shortly before birth). This latter example also demonstrates that there are different levels of determination: these cells first become determined as ectoderm, then some of these later become determined as neural tissue, and later still some of these become determined as sensory receptors or as other types of neurones in the eye.

It is important to note that determination is an inherited characteristic. Once a cell has become determined, it or its progeny cells does not have to remain under the influence of whatever caused this determination in the first place; all progeny of the original determined cell remain committed to the same developmental fate, even if they are naturally or artificially moved to some other location in the embryo. For example, if determined but otherwise undifferentiated mesodermal cells from a chick wing bud is transplanted to the seemingly indistinguishable cell environment of the leg bud, the transplanted cells will still develop into wing-like structures. At some early stage during wing bud development, the mesoderm cells became determined as presumptive wing components, and that they and their descendants doggedly remain wherever they may later wind up. This phenomenon is known as **cell memory**. What causes determination is largely unknown, and so is what maintains that state down through the cell lineage. Whatever maintains determination, it cannot involve an irreversible change in the DNA, since, as noted in the previous lecture, the nucleus remains totipotent right up to and beyond terminal differentiation.

It is also important to note that determination is a subtle internal alteration of a cell; there is typically no outward change in the appearance of a determined cell or in its metabolism, at least for some time after the event. Whether and when a cell (or, more commonly, a group of cells) has become determined is usually discovered in transplantation experiments. In an early amphibian gastrula, a patch of presumptive neural ectoderm can be relocated to a different part of the embryo and, while remaining ectoderm, will contribute to whatever ectodermal derivatives are characteristic of its new environment. If the graft is done on a later gastrula, the transplanted tissue later becomes misplaced neural tissue regardless of its new environment. Sometime, then, during the gastrula stage this group of cells became determined as neural ectoderm, and that is what they will become no matter what or where.

The eventual manifestation of the determined state is **differentiation**, a term already much used in these lectures. A differentiated cell has a characteristic shape or metabolism that distinguishes it from other cell types. There are about two hundred different cell types in a typical vertebrate adult. Almost all of them share a large number of functions in common (sometimes called **housekeeping functions**), such as the obtaining of energy by glycolysis and the Krebs cycle and the synthesis of amino acids and proteins by the conventional metabolic pathways. But a differentiated cell or tissue will also have unique products and structures not found in other cell types. Red blood cells contains haemoglobin; other cells do not. Neural cells, because of their unique shape, structure, and components, are capable of transmitting an electrical impulse; other cells cannot. This differentiated state is normally stable (and hence the term "terminal differentiation") and in many cases a terminally differentiated cell does not divide further.* Nerve and muscle cells are obvious examples of non-dividing differentiated cells, and an even more extreme case is the red blood cell, which loses its nucleus on becoming differentiated and hence not only cannot divide but cannot live for long, either (about 120 days). Given that a differentiated cell (other than a red blood cell) still contains a totipotent genome, it follows that for most of its life—often a matter of many years in vertebrates—a cell expresses only a tiny fraction of its genes; in fact, most of its genes have never been expressed at any time during its lineage from a fertilized ovum to its terminally differentiated state. The molecular basis for the stability of this differentiated state is not known, but some evidence suggests that the answer has to do with the pattern of methylation that is added to the DNA.

Although it is usually true that the differentiated state is stable and is passed on to all progeny cells (if the differentiated cell is capable of division), there are some interesting exceptions. Taste buds in the tongue epithelium are small rosettes of differentiated cells specialized in responding to sweet, sour, bitter and salt. If their nerve supply is severed, the taste buds disappear entirely, but they are reformed in the tongue epithelium if the nerve supply regenerates. Hence, a

---

* Because a differentiated cell such as a muscle cell does not divide, this does not mean it does not renew itself. Muscle cells continuously synthesize new proteins and other functional and structural components to replace those that wear out or are needed for cell enlargement. A distinction should also be made between cells that cannot divide and cells that cannot be replaced. Muscle and nerve cells cannot divide and cannot be replaced. Shortly after birth, you have as many of these cells as you will ever have. Red blood cells cannot divide, but can be replaced. The replacement is accomplished by **stem cells**, cells that are determined in a specific direction but which remain undifferentiated throughout the life of the organism. When replacement of a red blood cell is called for, the appropriate stem cell in the bone marrow divides. One daughter cell differentiates into the RBC, whereas the other daughter cell remains an undifferentiated stem cell, capable of continuing this process *ad infinitum*. Finally, some differentiated cells such as hepatocytes (liver cells) or pancreatic cells divide in an unexceptional manner to yield two differentiated daughter cells like the parent.

continuing interaction between nerve and taste bud is necessary for the maintenance of this particular differentiation.

INDUCTION

In organism that follow a regulated developmental sequence, the **induction** of a new cell type by a pre-existing tissue is a ubiquitous phenomenon. The most well-known example is the induction of the neural tube from the ectoderm of developing amphibians by the underlying mesodermal notochord. Early in this century, the embryologist H. Spemann and his students demonstrated that transplanting parts of the normally dorsal notochord to a ventral location causes the formation of a misplaced neural column from the ectoderm overlying it in its new location. Moreover, if the transplanted notochord is from an anterior location, a forebrain is induced; if it notochord from further back, a spinal column is induced. Despite almost a century of study, the exact nature of the signals emanating from the mesodermal tissue to the ectoderm is still unknown.

Another interaction extensively studied by Spemann was the development of the optical lens in the salamander. This structure, which develops from the ectoderm of the head, must come to be positioned in a precise way with respect to the retina of the eye in order that light can be properly focused. This retina develops at the posterior of the optic cup, which is an outgrowth of the forebrain. Spemann demonstrated that the lens is induced by the retina. If the developing optic cup is excised, no lens develops; if the developing optic cup is transplanted under the ectoderm of the belly, it induces a lens above it, and this lens fits perfectly into the optic cup in its new location.

The interplay between mesoderm and ectoderm has also been extensively studied in the developing chick embryo, particularly with respect to the development of feathers and scales on the skin. Skin, which we are accustomed to thinking of as a single tissue, is actually a bipartite structure with two distinct origins. The outer layer, the **epidermis**, derives from the ectoderm, but the deeper layer, the **dermis**, comes from mesoderm. Accessory structures such as feathers, scales, hair, and many types of glands arise in the ectoderm, but, as we shall see, are very much influenced by the nature of the underlying mesoderm. This can be shown by grafting epidermis from the leg of chicks, where it normally produces scales, to the back, where epidermis normally produces feathers. In its new location, however, in contact with back dermis the transplanted epidermis produces feathers instead of scales. To show that it is the dermis that influenced this change, back dermis can be transplanted to underlie leg epidermis, and feather buds will now appear on the leg. Even the pattern of the feathers will be characteristic of the feather pattern of the back instead of that of the leg scales. Although, like so much else in development, the nature of the signals from the dermis is unknown, it seems clear that they have been preserved over long periods of evolution. It is possible to take, for example, dermis from the snout of a mouse embryo and graft it over the epidermis of a developing chick. The epidermis will still produce feather buds, but they will be arranged in the pattern characteristic of a mouse's whiskers instead of a chick's feathers.

The interaction between dermis and epidermis is not limited to the embryonic state but continues to some extent in the adult, as well. If epidermal cells from the adult ear (we are obviously no longer talking about chickens, but mammals like rats) are grafted above the dermis at the sole of the foot, the ear epidermal cells will take on the specialized characteristics of the thickened foot pads. As in the case of the taste buds mentioned above, we note that the differentiated state is occasionally capable of being modified.

MORPHOGENESIS AND MORPHOGENS

Besides the development of tissues, embryology must also ensure the development of a prescribed shape, and the manner in which this is accomplished has proved at least equally difficult to unravel. Part of the answer lies with progressive changes to the cell membrane which enables certain aggregations of cells and prevents others. Embryonic tissues can be dissociated into free cells by various means, including calcium depletion and proteolytic (protein-degrading) enzymes. If the cells that had been separated by calcium depletion are returned to normal levels of that ion, they reaggregate spontaneously. The same thing occurs with cells separated by proteolytic enzymes, except that in this case an interval for protein synthesis must precede the reaggregation. This suggests that surface proteins are important in the recognition process. If dissociated cells from two different embryonic tissues, say kidney and liver cells, are intermixed under conditions that would permit the reaggregation of the homogeneous cells, the cells will form at first a single mass, but will then quickly sort themselves out into two distinct tissue masses. This sorting out also is dependent upon maintaining the integrity of the surface membrane proteins. Various glycosylated proteins (glycoproteins) have been implicated in these cell-to-cell recognition processes.

However, the aggregation of cells into organs and tissues of distinct architecture and shape is more complex than can be explained solely on the basis of surface glycoproteins. Take, for example, the development of digits of several distinct shapes on the hands or feet. Why are thumbs always on the "inner" edge of the hand and the smallest fingers on the "outer" edge? Some interesting answers have come from the study of developing chick wings. Normal chicken wings terminate in three modified digits that correspond to the middle three digits of our own familiar hand. Chicken experts can readily distinguish any one of these three from the others. In the developing early wing bud, there exists a small group of cells that is instrumental in determining in a fairly simple manner the over-all pattern in which the three digits will appear. These cells, known as the **zone of polarizing activity (ZPA)**, do not themselves contribute to the structures of the wing or its digits, but instead secrete an unknown substance that diffuses across the developing wing bud. The different digits develop as a consequence of the concentration of this substance to which they are exposed. Prospective digital mesoderm that receives the highest concentration (i.e., that lies closest to the ZPA) becomes the fourth wing digit; mesoderm receiving the least concentration (because of its being furthest away from the ZPA) becomes the second digit; and that receiving an intermediate concentration differentiates into the third digit. If a second ZPA is grafted into the early wing bud on the opposite side (in the anterior-posterior sense), two complete sets of digits develop in a pattern that is a mirror image: digits 4,3,2,2,3,4. Whatever the chemical basis of the **morphogen** secreted by the ZPA—and there is mounting

254

evidence that it is a substance related to Vitamin A, retinoic acid[*]—it appears to have conserved over long periods of evolution, since the similar centers from mammalian and reptilian limb buds transplanted into chick embryo behave the same as the normal chick ZPA.[**]

More recent work has revealed that the receptors (there are several different ones) for the morphogen appear to be located not on the cell surface but within the nucleus. Current speculation holds that the morphogen, retinoic acid or a related substance, diffuses into the cell and then into the nucleus, where it binds to one or more protein receptors. These receptors then display an altered propensity for binding to DNA at specific locations and in this way the nearby structural genes are regulated. This model obviously owes much to the operon model for the regulation of prokaryotic genes.

Other positional effects have a different explanation. For example, a piece of undifferentiated tissue from the base of the chick leg bud would normally develop into the thigh. If this tissue is transplanted to the tip of the wing bud underlying the ectoderm, it later differentiates into—toes! Apparently this mesoderm had already been determined as leg, but not as any part of the leg in particular. Which part it does become seems to correlate with its proximity to ectoderm. This conclusion is reinforced by the fact that the normal wing tip, now moved further away from the ectoderm by the graft, no longer differentiates into digits, but becomes forearm (forewing?) instead.

GENETIC CONTROL OF DIFFERENTIATION

It has already been noted that the genome remains intact in almost all cells as they pass through the embryological to the fully differentiated state. This naturally leads to speculations on the nature of the genetic control governing these processes. In theory, such control could occur when m-RNA is transcribed, processed, or translated. Although the latter two controlling points are undoubtedly important in some cases, it is clear that control of transcription is of over-riding significance. When radioactively labelled m-RNA from one differentiated tissue of a rat is hybridized to rat DNA in the presence of unlabelled m-RNA from a different rat tissue, the

---

[*] Retinoic acid injected into an appropriate embryonic tissue produces a series of digits showing the same characteristics as gradients produced by a transplanted ZPA. This could mean that retinoic acid is the substance diffusing from the ZPA—or, of course, that retinoic acid is inducing another ZPA in the tissue. Both interpretations of the data have ardent supporters at present.

[**] The notion that developmental features could be controlled by varying concentrations of diffusing substances was first proposed several decades ago by A. Turing, who is better known to-day as the father of the modern computer (or "Turing machine", as it was once called). He demonstrated mathematically that interacting substances diffusing at different rates could account for the spots and stripes on the coats of tigers, zebras, and the like. It has only been very recently that his speculations have proven amenable to experimental proof.

unlabelled m-RNA only partially blocks the labelled RNA from binding. This suggests that some of the labelled RNA is transcribed from DNA sites that are different from those that transcribed the unlabelled RNA, and *vice versa*. This is consistent with the interpretation that whereas there is a set of m-RNA molecules common to a variety of cells and serving to maintain the "housekeeping functions" mentioned earlier (protein synthesis, intermediary metabolism, and the like) there are also unique m-RNA molecules in each differentiated cell type. Indeed, direct quantitative measurements of specific m-RNA species in terminally differentiated cells demonstrates that they can be as common as ten to twenty percent of all messengers in one cell type and as rare as one-tenth of a copy per cell in another.

The manner whereby this tight control of transcription is exercised is not entirely understood for eukaryotic organisms. The mechanism of transcriptional control in bacteria has served as a take-off point for these speculations (see Lecture 37).

An organism from which a great deal of information on the genetic control of differentiation has already been gleaned and which will become even more important for such studies in the future is the lowly nematode, *Caenorhabditis elegans*, a transparent worm about one millimeter in length. The complete lineage of every one of the 959 cells of the adult is already known right back to the fertilized egg. Similarly all of the worm's 302 nerve cells are identified, as are all of their connections. Studies are well under way to sequence its entire genome of 100 million bases. It is expected that before too many years have past the identity and function of every one of the worm's genes will be known.

LECTURE 43

THE IMMUNE SYSTEM: 1

HUMORAL IMMUNITY[1]

When laymen think of the human immune system, most commonly it is the production of antibodies that they have in mind. Technically, this aspect of immunity is known as **humoral immunity**. Although antibodies are extremely important for maintenance of health, they are only one aspect of the complex system that safeguards us against infections. In the case of infection with HIV, the other parts of the immune system, collectively known as **cellular immunity**, are of equal if not greater importance. But because almost everyone has at least some knowledge of antibodies, this is a good place to start our discussion.

Antibodies are protein molecules that complex (bind) in a specific manner with foreign matter encountered in the body. This "foreign matter" is called **antigen**; it is commonly a protein or polysaccharide (a complex sugar) that is not a normal constituent of the body, but other types of chemicals (e.g., nucleic acids) can also be antigenic, and sometimes antibodies will form against molecules that do belong normally within the body. In this latter case, a pathological condition known as an "autoimmune syndrome" results, the diseases known as **myasthenia gravis, lupus, rheumatoid arthritis** (and, as now appears likely, **multiple sclerosis**) being some examples, several of which are often fatal.

It is necessary to emphasize the very specific nature of the binding between an antibody and its antigen. Often even minute changes in an antigen—changing just one amino acid in an antigenic protein, for example—will render it incapable of being recognized by the antibody that binds to the unaltered molecule. Yet it is also possible to introduce changes in the antigen that will weaken, but not wholly eliminate, its binding to an antibody. This altered molecule is said to **cross-react** to some degree with the antibody to the original molecule.

An antibody does not usually recognize the whole of a given antigen; instead it recognizes particular parts called the **antigenic determinants**. A complex molecule will have many different antigenic determinants. In consequence, a complex antigenically active molecule will often elicit many different antibodies, each of which is specific for a different antigenic determinant on the molecule and some of which are specific for the same determinant, but bind to that determinant in different ways. Such a diverse antibody response to a single antigen is said to be **polyclonal**; by contrast, if only a single type of antibody is elicited by the antigen (a rare event in nature, but an exceedingly useful one when engineered artificially in the laboratory), the response is said to be **monoclonal**.

Antibodies are produced by the **B-cells** of the immune system. These cells are a subclass of a group of cells called **lymphocytes**, which circulate freely through the blood and lymphatic system of the body. The B-cells are produced in the bone marrow and (in mammals) also complete their development there. In birds, however, they complete their development in the Bursa of Fabricius, from which the designation "B-cell" is derived. When these cells are primed to produce and secrete antibody into the blood stream, they do so in prodigious quantity: about 2000 antibody molecules per cell per second. This effort is not without cost, however, for these antibody-secreting cells, now terminally differentiated and properly called **plasma cells**, die from the effort within a few days.

For many years scientists labored to discover how B-cells could be stimulated seemingly on command to differentiate into plasma cells that in turn produced specific antibodies. Since the shape of an antibody must complement that of the antigenic determinant in order for binding to take place, the particular shape of the antigenic determinants would presumably elicit only those antibodies with complementary shapes. But shapes in a world of molecular proportions are about as variable as shapes we see in our own world. Moreover, because we can synthesize in the laboratory molecules that are fully capable of eliciting a strong antibody response and yet have never appeared on earth before, it is clear that the immune system must be able to produce as needed an almost infinite variety of antibody molecules[*]. How is such a diversity possible?

For many years it was believed that antigens "instructed" the B-cells to produce the appropriate antibody. In this view, any B-cell (we'll usually refer to B-cells producing antibodies, although correctly it is their progeny, the plasma cells, that actually secrete antibodies) could be "taught" to produce whatever shaped antibody was needed. Those that successfully "learned" would be the ones that differentiated into plasma cells and secreted, for a time, the necessary antibody. Given that antibodies are proteins and that proteins are encoded within the DNA, this hypothesis would require that proteins (or whatever material the antigen was made of) could modify the genetic material (DNA) in a specific manner. This was a formidable conceptual requirement which, if true, would have made necessary considerable revision of our understanding of how genes function. The immunologists would have solved their problem, perhaps, but in so doing would have vexed the geneticists unconscionably.

An alternative view of antibody generation, the one we now firmly believe to be correct, is termed **clonal selection**.[2] This view holds that the large population of unstimulated B-cells collectively are potentially capable of producing antibody to any conceivable shape, but that only one (or a few) are capable of producing the one of a *given* shape. That is, a particular B-cell produces only one kind of antibody. When that cell comes in contact with the appropriate antigen, that B-cell is stimulated to divide, increase in number, and eventually (as a plasma cell) secrete the antibody that complexes with the antigen that had stimulated it into activity. This theory, then, does not require that the B-cell acquire new genetic information derived from an antigen, but it does engender the almost equally formidable notion that B-cells collectively

---

[*]It is estimated that a mouse can produce more than 10^7 different antibodies.

already have the genetic information needed to make an antibody of any required shape. It becomes simply a matter of selecting the appropriate B-cell to expand into a clone (a large number of genetically identical cells) and produce the needed antibody. In recent years advances in molecular genetics have succeeded in explaining how the required genetic diversity for antibody coding is accomplished, but understanding the mechanism is beyond the requirements of these lectures.[3]

The exact manner by which an antigen can stimulate a B-cell is not entirely understood, but still many features of the event have been learned in recent years. Key to an understanding of these events is the observation that resting B-cells already carry on their cell surfaces copies of the particular antibody they will be able to make in quantity after being stimulated to become plasma cells. These bound antibodies are the actual receptors for antigens, and it is the complexing of the bound antibody with its complementary antigen that provides the first step in the clonal expansion of the particular B-cell. The acquisition of receptors, which is an antigen-independent process, occurs first in the liver of the fetus but is maintained continuously afterward in the bone marrow. When the B-cell leaves the bone marrow to circulate in the blood and lymph systems, it has not yet encountered specific antigens for which it carries the receptors, and is usually called a **virgin B-cell**.

After leaving the bone marrow, the virgin B-cell must encounter its appropriate antigen within a few days or else it will die. This encounter, which usually occurs in the spleen or lymph nodes, then initiates a series of internal events that may lead to the clonal expansion of the virgin B-cell into a large population of plasma cell secreting antibody. Only a very few antigens, however, have the capability of initiating this clonal expansion; only antigens that have many copies of a single determinant ("multivalent antigens") are effective. The vast majority of antigens, those with one or just a few copies of a determinant ("monovalent antigens"), need help in stimulating the virgin cells, and this is provided by another type of cell altogether, the **antigen presenting cells**, of which the **macrophage** are the best understood.

Macrophage are cells in the blood and lymph systems that have the capacity to engulf and destroy foreign antigens that they encounter in their peregrinations through the blood and lymph vessels. They do so non-specifically; that is to say, any foreign antigen (including molecules, and whole bacteria and viruses) can be engulfed. The macrophage do not, however, destroy the entire antigen; rather they usually "process" it and display the determinants of the antigen on the cell surface of the macrophage, like souvenirs of the kill. When an antigen is monovalent, the virgin B-cell must re-encounter that determinant displayed on the surface of a macrophage in order to develop further. At this point the virgin B-cell is said to mature, but it cannot yet expand into a clone without at least two additional stimuli.

The first of these is given by macrophage (regardless of whether the initiating antigen was mono- or multivalent). The B-cell will not expand antibody into a clone unless a particular hormone-like molecule (properly, a **lymphokine**), called **interleukin-1** (IL-1), is secreted by macrophage, which do so when they encounter mature B-cells. The second stimulus required by the mature B-cells is given by another cell type, the **helper T-cells**, which will be described later on in this section. Both stimuli are absolute requirements for the expansion of B-cells into a large number of plasma cells capable of secreting vast quantities of specific antibody.

But not all fully stimulated B-cells go on to produce antibody. Some become instead **memory B-cells**. These memory cells are very long-lived and remain in the circulatory system for many years. Should at some future time they encounter the antigen that primed their initial development, they can very rapidly differentiate into antibody secreting plasma cells. It is for this reason that one often "acquires" immunity for a lifetime after surmounting an initial infection. Hence one usually gets diseases like mumps, chicken-pox, measles, etc. only once and is immune to reinfection thereafter. This aspect of immunity was known even to Thucydides, who noted in his journal on the Peleponnesian Wars that those who had managed to recover from the plague could safely attend the sick thereafter without fear of contracting the disease a second time.

ANTIBODIES

Before leaving a discussion of antibodies, we should know a bit more about their structure and what they actually do. There are actually five different classes of antibodies, depending upon certain details of their structure and their precise role in the immune response. The are named IgA, IgD, IgE, IgG, and IgM. ("Ig" means "immunoglobin".) In general, they are Y-shaped proteins. For a particular class of antibody, the parts that bind the antigenic determinant, and therefore the parts that have the variable structure, are at the ends of the "arms" of the Y; the 'stem' is largely invariant for a particular class and determines such properties of the antibody as, for instance, whether it can pass through particular membranes, attach permanently to particular cell surfaces, or circulate in the blood stream. Each arm of the basic antibody can bind a single antigenic determinant; if each has snared a determinant, then two molecules of antigen have become cross-linked (unless it is two determinants of a multivalent antigen that have become attached). Some classes of antibodies consist of two or five copies of the basic Y-shaped antibody; these can obviously bind many copies of the antigen simultaneously.

The principal antibody type in the body is IgG, accounting for about 80% of all antibody in the body. It is the type we usually have in mind when thinking about "antibodies". IgA occurs in certain secretions such as saliva, tears, mucus of the respiratory tract, and milk. (Nursing infants acquire an array of maternal antibodies in this manner, which protects them until they develop their own.) The particular cells that produce these secretions take the antibody up by endocytosis from the blood and then release it together with its particular secretion. IgE is responsible for many allergic reactions. The antibody attaches to the surface of **mast cells** that occur in the epithelial layers of the respiratory tract and other places. When the corresponding antigen is encountered, the mast cells release large quantities of **histamines**, which dilate the nearby blood vessels and increase the blood supply to the area. Oftentimes this is more of a nuisance than a benefit, as sufferers of hay fever, hives, and asthma can attest. IgM antibodies are produced early in the immune response and are gradually superseded by the IgG type. The final type, IgD, are not well understood at present.

For some antigens (for example, toxic molecules), being bound to antibody and cross-linked into an unwieldy mess is sufficient to inactivate them; here, the binding of antibody itself explains how the organism is saved. However, when the antigen is something large (relatively) and living, such as a bacterium (which, of course, is not a single antigen, but a whole slew of different antigens done up in one package), the binding of antibody to its surface, while unattractive and possibly even inconvenient, is not likely fatal to the cell. In these cases, the antibody is really just serving as a marker to direct the attention of other processes to the cell. Macrophage, for example, are much more strongly attracted to foreign objects coated in antibody than to the naked object itself. They prefer, if possible, antigens to be frosted in antibody before they will engulf them, but when it comes right down to it, they will settle for it plain. There also exists another immune cell type, called **killer cells**, that will only attack cells tagged with antibody. They seem to have the ability to drill holes in the membrane of such tagged cells, effectively killing them, although the exact mechanism of their action needs further study. And finally, there exists in the blood and lymph systems a collection of about a dozen proteins collectively known as **complement**. These proteins are attracted to antibody-coated cells (this attraction is a function of the stems of the attached antibodies), where they cooperate in a series of reactions that also result in holes appearing in the membrane of the foreign cell, and its death in consequence thereof.

MONOCLONAL ANTIBODIES

As mentioned above, most antigens comprise several antigenic determinants and hence will elicit a polyclonal antibody response. For scientific or medical purposes, it is often necessary to prepare large amounts of antibody to a single determinant, a so-called **monoclonal antibody**. Preparing a monoclonal antibody makes use of the fact that cancerous B-cells (called **myelomas**) can be readily induced in mice, and these cells, unlike normal B-cells, will grow indefinitely in cell culture. The antigen to which one wishes to prepare a monoclonal antibody is injected into the mouse in order to initiate the expansion of the appropriate B-cells. The mouse B-cells are then removed and fused *in vitro* with myeloma cells to produce **hybridomas**—cells that are immortal in culture and which secrete large quantities of specific antibody. A clone can be derived from each hybridoma cell, and the one that produces the required antibody identified. This clone will continue producing large quantities of that one antibody indefinitely.[4] More recently a simpler method for producing monoclonal antibodies has been developed. In this process, the actual gene for the production of the human antibody is transferred to a bacterium, where it can be made to produce its normal product, the antibody, in large quantity. Because the antibody in this case is a real human antibody (as opposed to a mouse antibody, when hybridomas are used), this will likely be the method of choice for producing monoclonal antibodies in the future.

Because rodent-derived antibodies are themselves antigenic when injected into humans, their use in medical practice has so far been somewhat limited. Although the new technique mentioned above ought to eliminate this problem entirely, there have been developed so-called "designer antibodies" that seek to reduce the antigenicity of monoclonal antibodies in humans to a minimum. The most effective of these, known as **humanized antibodies**, use sophisticated recombinant DNA techniques to create an antibody that has almost all of the molecule derived from a human antibody and only the most variable portion, the part that actually complexes with the target antigen, derived from a rodent.[5]

REFERENCES

1. Tonegawa, S., 1985. The molecules of the immune system. Sci. Amer. (Oct.).

2. Ada, G. and G. Nossal, 1987. The clonal selection theory. Sci. Amer. (August.)

3. Leder, P., 1982. The genetics of antibody diversity. Sci. Amer. (May).

4. Milstein, C., 1980. Monoclonal antibodies. Sci. Amer. 243: 66-74.

5. For a discussion of various designer antibodies and the medical uses to which they can be put, see: Mayforth, R. and J. Quintans, 1990. "Designer and catalytic antibodies". New Engl. J. Medicine 323:173-178.

LECTURE 44

THE IMMUNE SYSTEM: II[1]

CELLULAR IMMUNITY: THE T-CELL[2]

The humoral immunity system just described is one major arm of the immune system; the second is the cellular immunity system based on the activity of circulating cells known as **T-cells**. The name derives from the fact that T-cells, although like B-cells in that they are formed in bone marrow, must complete their development in the thymus gland. Morphologically, they are indistinguishable from B-cells, but can be discriminated biochemically by certain specific proteins they carry on their surfaces. T-cells themselves can be further subdivided into three principal classes: cytotoxic cells, helper cells, and suppressor cells. Each of these classes also has a corresponding memory cell that continues circulating long after the cytotoxic, helper and suppressor cells have completed their activities. The chief role of the cellular immunity system is to guard against the spread of virus infections by killing body cells that have become hosts to viruses. This prevents the virus from further replication and brings the infection to a halt. The T-system also interacts in important ways with the B-cells, which could not differentiate into plasma cells without this cooperation.

THE MAJOR HISTOCOMPATIBILITY COMPLEX

Before the function of T-cells can be discussed, we need to know about **histocompatibility proteins**, which were discovered in the course of research on tissue transplantation. It is now part of everyone's canon of information that it is not possible without artificially suppressing the immune system to transplant tissues between individuals who are not identical twins. When such transplants are attempted, the usual result is that the transplanted tissue is 'rejected' by the host. This rejection, in large part a consequence of the activity of T-cells, pointed out that, whereas one person's liver may seemingly be the same as the next person's, in fact, at the cellular level they are distinct. What makes them cellularly distinct is the existence on the surface of cells unique proteins once called **transplantation antigens** but now more commonly known as **histocompatibility proteins**. These proteins exist in a very large assortment, so large that virtually everyone (who does not have an identical twin) has a different set. They exist on the surface of all body cells that have nuclei.

Actually, there are really two major classes of **major histocompatibility complex (MHC)** proteins: those that all nucleated cells carry—MHC-I proteins—and those that certain cells of the immune system carry in addition—the MHC-II proteins. Of the cells that we already know

263

about, the macrophage T-cells and B-cells carry MHC-II (plus MHC-1) proteins.* Macrophage, as has already been discussed, have the ability to process antigens and to display the antigenic determinants on their cell surface; B-cells can also process proteins in the same manner, but a B-cell can only process the antigen for which it carries the complementary antibody, whereas macrophage can process any antigen and display its determinants. The ability to process and display antigen *originating outside the cell* is therefore related to having MHC-II proteins. The displayed determinant is actually complexed with the MHC-I and MHC-II proteins on the cell surface in some manner as yet poorly understood.** It is generally the case that antigenic determinants of proteins being synthesized inside the cell (e.g., viral proteins during intracellular viral replication) are complexed with MHC-I proteins and determinants of engulfed foreign proteins are complexed with MHC-II proteins. The determinants complexed with MHC-I proteins are surprisingly small—only eight or nine amino acids in length. Those that are complexed with MHC-II proteins are between 13 and 17 amino acids long. The processing of antigen for display with either MHC-I or MHC-II proteins occurs in different locations of the cell, depending upon which MHC protein the antigen will eventually be complexed with. In both cases, however, the complexing with the MHC protein occurs inside the cell and it is consequently the combined protein that is transported to the cell surface for display.

For many years the purpose of the MHC proteins was a puzzle to immunologists; it was particularly puzzling that a complex and efficient system for safeguarding against tissue transplantation would have evolved at all, given the unlikelihood of such an event happening in the natural world. The matter is now making more sense as we begin to see the role these MHC proteins play in the immune response in general.

Like the B-cells, the T-cells are highly specific in their response to antigenic stimulation and are also subject to clonal selection. Immature T-cells, therefore, exist with the capacity to respond, as a whole, to virtually any antigen, but individual cells can respond only to one particular antigen. The three classes of T-cells, cytotoxic, helper and suppressor, apparently require independent stimulation by antigen. When the basis for the phenomenon of graft rejection became known as principally an activity of T-cells, it was then noticed that a very large fraction (around 10%) of the immature T-cells were capable of being stimulated into fighting the transplanted tissue. This observation was seemingly contradictory to the notion that only a small fraction (much less than 1%) of immature T-cells ought to be stimulated by a particular antigen, in this case the foreign MHC-1 glycoproteins. This, together with the almost obsessive concern T-cells

---

* In humans, the MHC antigens are called **HLA**, meaning **human leucocyte antigens**.

** Since the number of different possible antigenic determinants is much larger than the number of different MHC-I molecules on a cell—thought to be about six—it was very puzzling how so many different small peptides could be tightly bound to a single MHC molecule. The trick, apparently, is the ability of the MHC molecule to form hydrogen bonds with the *backbone* of the determinant, thus making the actual side-groups that distinguish one amino acid residue from another largely irrelevant to the binding process.

display towards foreign MHC antigen, underscores the key role MHC glycoproteins play in the immune response—but it took a great deal of scientific effort to understand exactly what that role was.

The important thing to understand about T-cells is that they can only respond to antigen found on (self) cell surfaces. They cannot, therefore, respond to free, circulating antigen (for example, to free viruses) nor to antigens on the surface of bacterial cells (which, of course, are not "self"). The initial stimulus for a cytotoxic T-cell to begin the process of clonal expansion is its encounter with an antigen-presenting cell (such as a macrophage). In order for this to be an effective stimulus for the cytotoxic cell, the displayed processed antigen must be the one for which the T-cell is specific. Equally important, however, is the requirement that the displayed processed antigen be complexed with self MHC-1 proteins on the surface of the presenting cell. The cytotoxic T-cell recognizes a complex of antigen and self MHC-1 proteins, not each separately*. The macrophage then releases interleukin-1, which also is a required stimulus to the T-cell. (The strong and multicellular response of cytotoxic T-cell to foreign MHC-1 proteins by themselves—as happens in transplantations, for example—suggests that T-cells are fooled into thinking that the uncomplexed foreign MHC-1 proteins, being different in structure from self MHC-1 proteins, are really self proteins complexed with foreign antigen.) Once the cytotoxic T-cell clone has expanded (and this involves several additional interactions we have yet to examine), the cytotoxic T-cell descendants will kill any self cell having that antigen on its cell surface. The exact mechanism of this killing is not yet fully understood, but it is rapid and efficient and requires direct contact between the T-cell and its victim.

The cell that is killed is often one that is infected with a virus. But how does the T-cell know this? Well, normally when a virus replicates inside a cell, pieces of its various protein components are shipped to the cell surface as processed antigens complexed with MHC-I proteins. (Cells normally do this for *all* the proteins they manufacture, whether of viral origin or not.) There these viral antigens function to the immune system like little signs or tags indicative of the fact that a virus is replicating inside that cell. Another representative of that virus, however, may have already been discovered circulating in the blood stream by a macrophage,

---

*Since an individual expresses only about six different MHC-1 molecules and about 20 different MHC-II molecules, it might seem unlikely that they would be able to bind the great diversity of antigens that exist. In fact, they couldn't. However, it is not intact, but processed, antigen to which they bind. Processed antigens are short peptides (sequences of amino acids) which generally would be capable of binding to one of the MHC-I and MHC-II proteins. Direct binding of processed antigen to MHC protein has been demonstrated. Moreover, free MHC protein bound to processed antigen (but not intact antigen) can directly stimulate the appropriate T-cell in the complete absence of antigen-presenting cells. It appears that antigens are processed in the cytosol, after which the antigenically active fragments are transported to the endoplasmic reticulum where they complex with the MHC-1 proteins during their assembly. The MHC/peptide complex then is carried to the cell surface for presentation to T-cells.

which would engulf ("phagocytize") it, processing its antigenic determinants and displaying them on its own cell surface complexed with its MHC-1 and MHC-II proteins. The immature cytotoxic T-cell that can react to this antigen is then selected for clonal expansion, and its descendants search out and kill self cells with that same foreign antigen on their surfaces[*]. Some of these descendants become memory cells, just as some B-cells do.

We have previously noted that the surface receptor that B-cells employ is actually a membrane-bound representative of the antibodies its descendants will later produce and release into the circulatory system. In the case of the T-cell, the receptor is not an antibody, although it shares some of its characteristics. The exact nature of this receptor is only recently becoming clear. The same receptor functions both in the recognition of the antigen presented by the macrophage and the later recognition of that antigen on the surface of self cells. In both cases, the antigen must be recognized jointly with self MHC-1 glycoproteins.

HELPER CELLS

The real orchestrator of the immune response, however, is the **helper/inducer cell**, which is necessary for both cytotoxic T-cell and B-cell responses to foreign antigen. (Some accounts make a finer distinction between helper and inducer cells; most, like this one, treat them as a single class usually just called **helper cells**.) They carry on their surfaces a particular protein called the **T4 marker**, which enables them to be distinguished biochemically from other types of T-cells to which they are morphologically identical. Helper cells are often called **T4 cells** and the cytotoxic and suppressor cells **T8**, after one of their own unique surface proteins. (The marker proteins are also called "CD4" and "CD8" proteins. The exact role of there marker proteins on the cell surface is unknown; although they are not themselves the T-cell receptors, they clearly play an "accessory role" in the recognition of antigens by the actual receptors.))

Before an effective immune response is mounted against foreign antigen by either T or B cells, the helper cells must recognize and be stimulated by this antigen. This stimulation eventually leads to clonal expansion of that particular cell that carried the appropriate receptor for the antigen in question. Helper cells usually recognize it on the surface of macrophage (or one of the few other types of antigen-presenting cells). Like the cytotoxic T-cells, the helper cells must recognize it in the context of self MHC proteins, but in the case of helper cells, the required proteins are the *MHC-II* type. This means that helper cells are restricted to interactions with

---

[*]Recently (1991) it has been discovered that neuronal (nerve) cells appear to lack MHC-I antigens on their surface. This solves a long-standing puzzle over how many viruses seem to be able to "hide" for long periods in nerve cells to cause trouble years after the initial infection. Herpes viruses, for example the one that causes chicken pox, often persist harmlessly in nerve endings for many years, only to cause problems much later in life as the disease "shingles". It is advantageous to the organism not to have its neurons destroyed by the immune system, since neurons, unlike most other cell types, cannot be replaced. It is presumably better to tolerate infected (and possibly somewhat impaired) neurons than to destroy what cannot be replaced.

other immune-related cells, for only they carry the MHC-II glycoproteins. The macrophage will release interleukin-1, which serves to further stimulate the T4 cell. In turn, the helper cell then produces another hormone, **interleukin-2**, and also acquires the appropriate receptors to enable it itself to respond to this hormone—such hormones that cause self-stimulation are called **endo-hormones**—and begins expanding into a clone of active, mature helper cells.

The activities of mature helper cells are numerous and varied. As we have already noted, the helper cell must interact with the activated B-cell of helper in order for the B-cell to expand and differentiate into antibody-secreting plasma cells. Hence, there is a firm and necessary connection between the humoral and the cell-mediated immune systems. The helper cell must also interact, probably by direct cell contact, with the activated cytotoxic T-cell in order for them to expand into a clone. The T-cells must have been primed for this interaction by having already encountered on macrophage the same antigen. This interaction involves the release by the helper cells of interleukin-2, which stimulates the T8 cells. It is also likely that the activated T8 cell in some way further stimulates the T4 cell, too. In any case, this T4-T8 interaction is absolutely necessary for the expansion and activity of cytotoxic cells.

Helper cells also have an important influence on macrophage. We have previously noted that macrophage weave their way through the blood vessels and capillaries engulfing, processing, and displaying whatever foreign antigen they encounter. Some of these antigens, however, can be pretty prepossessing, and the tendency of many macrophage is, naturally enough, to run away from the site of infection. Helper cells prevent this by secreting **Macrophage Migration Inhibition Factor**, which prevents them from moving away and may additionally, according to some, attract them to where they are needed. Whatever the exact details, it is clear that helper cells do function to remind timid macrophage of their duty in times of danger.

Some helper cells become memory cells to await future encounters with the same antigen.

SUPPRESSOR T-CELLS

The final type of T-cell is not as well understood as the other two; it is the **suppressor T-cell**. It is more slowly aroused than the other two types of T-cells. It recognizes antigen on macrophage in connection with MHC-1 proteins and in consequence of this interaction expands slowly into a clone. In due course it interacts with helper cells, which activate it into its suppressor function, which is to dampen down the immune response by interacting with cytotoxic T-cells (and probably with B-cells, too), turning them off. In this way, an immune response runs its course, but the memory B- and T-cells remain ever vigilant to a return of the same antigen.

OTHER IMMUNITY CELLS

The immune system encompasses several other types of cells besides the three (B-cell, T-cell, and macrophage) discussed already. Perhaps the most important of these others are the **natural killer cells (NK cells)**, which constitute 5 to 8 percent of all circulating leucocytes and whose role is only now becoming clear. Natural killer cells (not to be confused with cytotoxic T-cells, which are sometimes called simply "killer cells") are able to destroy virus-infected cells on their own, so to speak, without having to have foreign antigens "processed" or"presented" and without the complications of having to recognize them in the context of particular restriction proteins. Although it is not necessary for their activities, the killing function of NK cells is enhanced when the target cell is coated with antibody. Because of the seemingly uncomplicated way in which NK cells become activated, they are able to mobilize much more quickly against virus attack than are T-cells, whose response takes about two weeks to become effective. Natural killer cells seem particularly effective against **herpesviruses**, a large group of viruses that cause many more human illnesses than the well-known one the name suggests. There is on record, for example, an individual born without the ability to make any NK cells at all. Although the patient's antibody and T-cell response eventually cleared a herpes infection, without early massive medical intervention, the herpes infection would have been fatal well before the B- and T-cell responses could become effective.[3]

Another exceedingly important role for NK cells is in **immune surveillance** against cells that have become cancerous. It is thought that NK cells are constantly on the lookout for newly-arising cancer cells and can destroy them when found. As such, NK cells would be the first line of defence against cancer. Several therapies against cancer currently under development have as their aim the bolstering of the NK cell response to cancerous cells in the body.

LEARNING TOLERANCE

There are many mysteries surrounding the functioning of the immune system left unprobed by this account. Some are too complicated for inclusion in this introductory narrative (e.g. the genetic origin of antibody and T-cell receptor diversity), and some are simply not understood. Most are both! One such mystery—why our immune system doesn't attack our own cells (or, as an immunologist would say, the origin of "immune tolerance" to self) is just too important to ignore altogether, despite our sketchy knowledge of the processes involved. The best supported view of this phenomenon is that many T-cells as they are produced in the bone potentially do have the capacity to react with self proteins and thereby cause the body great harm. Before they become active, however, T-cells must mature in the thymus gland. It is thought that cells that develop the capacity to respond to self proteins are simply killed off and hence for them the trip to the thymus is a one-way journey to oblivion. What triggers this destruction is unknown, although there is evidence that certain macrophage-like cells migrate to the thymus from the bone, where they tempt the maturing T-cells to bind to them. Those that do are killed ("deleted from the repertoire", in scientific newspeak). The process is sometimes called "thymic education". "Madam, we guarantee an education or we return the boy," once wrote a president of Princeton University to an inquiring parent. At Thymus U. they are returned in a box.

Besides this process of **negative selection** of self-reactive T-cells, there must also be a **positive selection** of T-cells that *do* react with self proteins, in particular with the self MHC antigens. This is necessary because presumably many (most?) immature T-cells would carry receptors that would not recognize the specific MHC antigens of the host, yet when they become functional they must respond to foreign antigen only in the context of self MHC proteins. It is now thought that unless immature T-cells bind to self MHC proteins in the thymus they die a "programmed death". Those that are stimulated by this encounter with self MHC proteins are then further selected, this time negatively, on the basis of their response to other types of self antigens. One cannot help noticing a mind-boggling paradox in all this, for we are to now believe that the engagement of a single T-cell receptor will under one set of circumstances rescue the cell from programmed death (positive selection), under another cause its death (negative selection), and under a third—after it matures as a helper or cytotoxic T-cell—cause it to expand into a clone and become functional (activation). However, it is often the lot of scientists to be called upon, like Alice, to believe a quota of impossible things each day.[4]

Because the activation of B-cells requires commonly the participation of activated T-cells, the self-tolerance of B-cells has usually been viewed as a secondary consequence of the learned tolerance of the T-cells. It is probably more complicated than that, for there is considerable evidence that other mechanisms also work to keep B-cells that could potentially produce self-reacting antibodies from actually doing so. What is clear is that there exist in the body many small clones of B-cells that still have but do not express the capacity to produce anti-self anti-bodies.[5]

The fact that T-cells learn tolerance in the thymus gland has recently led one investigator to the idea that cells from immunologically incompatible individuals could be injected into the thymus gland and thereby become "invisible" to T-cells that mature in the presence of these transplanted cells. Tissues from the donor individual could then later be transplanted to any location of the host animal and escape rejection. To minimize rejection in the thymus gland itself, the gland was first injected with a substance that killed off most of the T-cells that were initially there. The experiment seems to have been successful in rats. If these finding are confirmed and extended to humans, the way may be open for widespread transplant surgery that does not require continued follow-up treatment with damaging immune-suppressive drugs.[6]

REFERENCES

1. Laurence, J., 1985. The Immune System in AIDS. Sci. Amer. 253:84-93 (Dec.).

2. Marrack, P. and J. Kappler, 1986. The T Cell and its Receptor. Sci. Amer. 254:36-45 (Feb.)

3. Biron, C., et al., 1989. Severe herpesvirus infections in an adolescent without natural killer cells. N Engl J Medicine 230: 1731-5.

4. The 15 June, 1990, issue of Science (vol. 248: pp 1335-1393) has a series of interesting articles on immunity and specifically on the mechanism of tolerance of self.

5. Nossal GJV, 1989. Immunologic tolerance: Collaboration between antigen and lymphokines. Science 245: 147-153.

6. Posselt, A. et al., 1990. Science 249:1293-1295.

LECTURE 45

DARWIN AND EVOLUTION

Another curious aspect of the theory of evolution is that everybody thinks he understands it!

J. Monod

You can't reason someone out of something he wasn't reasoned into.

M. Twain

INTRODUCTION

Without a knowledge of biological evolution and the mechanisms that underlie it, it would be impossible to make sense of the living world. Without the unifying power of this discovery, describing living matter would be like a series of anecdotes seemingly without connection one to another. Fortunately, since 1859, the year Darwin published his pathbreaking book, *The Origin of Species*, biologists have had at hand the schema that untangles the myriad of facts that constitutes their discipline—although this is not to suggest that this untangling has been an easy or an uncontentious process. As we shall note, many tenets of the schema itself have been and still are sources of controversy among biologists themselves, not to mention among the know-nothings who even to-day like to imagine that biological evolution does not occur.

DARWIN, EVOLUTION, AND NATURAL SELECTION[1]

At the outset of any discussion of biological evolution, it is well to distinguish firmly between two related aspects of evolution that are often confused in the popular mind. One is the fact of evolution itself, and the other is the mechanism (or mechanisms) that promotes it. Both ideas are often thought of as the unique contribution of Charles Darwin, whereas in fact the suggestion that living things are capable of evolution over time has a long history that considerably predates Darwin. Many earlier thinkers had proposed that living things evolve, most notably Jean Baptiste de Lamarck, who was writing near the beginning of the nineteenth century. His views on this question were strongly held, well supported by his own researches, and clearly enunciated, and there were many other naturalists both before and after Lamarck who held similar opinions. Lamarck, however, deserves to be singled out because he also proposed a plausible mechanism—plausible at the time, that is—to account for evolution. His notion was that evolution occurs through, first, the soliciting of biological change by the changing environment, and second, the inheritance of these acquired characteristics. He also believed in the "use and disuse" theory that parts of the body subjected to constant use increase in size and those left unused diminish or even disappear. Use and disuse, together with the mutagenic effect of the

271

environment, become a means to promote evolution when coupled with the inheritance of acquired characteristics.

These ideas are commonly illustrated by Lamarck's own famous example of the giraffe, whose long neck Lamarck supposed evolved as a consequence of giraffes' continually stretching their necks in order to reach progressively higher leaves in the trees on which they feed. The longer neck that resulted from this stretching would be inherited by the offspring, who in turn would add incrementally to their own neck size and hence pass an even longer neck to the next generation. And so on. A change an the environment (taller trees) summoned up a compensating response (longer necks) which was passed on to future generations (inheritance of acquired characteristics). None of these suppositions, as we shall see, is valid, but there is an undeniable folksyness about them that even now bedevils discussions about evolutionary mechanisms.

In the lectures dealing with the origin of spontaneous mutations, we show that the environment does not induce compensating or beneficial mutations. The role of the environment is solely to *select* any randomly-arising mutations that by chance are beneficial under the circumstances that prevail. Hence the increasing height of the leaves did not *induce* a beneficial response in the giraffe. As with people, progenitors of the modern giraffe would normally differ somewhat in the length of their necks. As the lower vegetation became scarce, giraffes which by chance had already a genetically determined longer neck became more viable and hence more capable of leaving offspring. As this **selection pressure** continued over time, giraffes with increasingly longer necks became the norm.

Characteristics acquired by an organism during its lifetime are never inherited. This is amply illustrated in human populations where, for example, certain ethnic groups have practiced mutilations such as male circumcision or female foot binding over untold generations without there having occurred any perceptible need to abandon the procedure because men's foreskins or women's feet grew smaller. It is only by modifying the genes governing a characteristic, not by modifying the characteristic itself, that an inherited change will occur. We shall return to this notion later in the discussion.

The real contribution of Darwin, then, is not the proclamation that biological evolution occurs, but the enunciation of a plausible mechanism to explain how and why it occurs. Because the proposed mechanism was so plausible and because Darwin had amassed so much evidence in favor of it, the idea that all current life forms have evolved from earlier different ones gained a respectability that had previously eluded it. Those whose prejudices conditioned them to believe otherwise were for the first time put strongly on the defensive. The ensuing controversy has been both acrimonious and protracted, both within and without the profession of serious biologists. Within the profession, the arguments have been largely over details of the evolutionary mechanism, and these arguments continue to this day. Outside the profession, the controversy has been over whether evolution occurs at all; it is embarrassing to note that 130 years after Darwin's publication there is still a sizable fraction of the population who believe in the fixity of species. Evolution is, paradoxically, a victim of its simplicity, a simplicity that permits anyone to imagine he has a democratic right to an opinion on an equal footing with anyone else's.

Physicists, by contrast, are not obliged to be patient with untrained kooks and cranks who want to waste their time arguing against relativity or the second law of thermodynamics. But anyone who possesses a body, if not a mind, feels himself an expert on biology, alas.

THE VOYAGE OF THE BEAGLE

The origins of Darwin's revelation lie in his famous five-year-long voyage on *H.M.S. Beagle*, a British naval ship that sailed in 1831 on a mapping and collecting expedition around the world. Darwin, who had formal theological training but had become seriously interested in geology and natural history before sailing, was employed as a naturalist on this expedition. On returning to England in 1836, he settled down to a life of experimentation, contemplation and writing (not to mention severe and almost continuous ill health) that was to last until his death in 1882. The idea that **natural selection** powered evolution came to him about two years after returning from his travels, but he did not publish his theory until about 20 years later, in 1859. In the meantime he patiently amassed additional evidence to support evolution and his theory of natural selection.[2]

While in South America on the *Beagle*, Darwin was powerfully influenced by a number of observations that compelled him to entertain the notion that species were not fixed entities, but were capable of evolution over time.[3] These conditioning influences can best be stated in his own words.

> During the voyage of the *Beagle* I had been deeply impressed by discovering in the Pampean formation great fossil animals covered with amour like that on exist- ing armadillos; secondly, by the manner in which closely allied animals replace one another in proceeding southward over the Continent; and thirdly, by the South American character of most of the productions of the Galapagos Archipelago, and more especially by the manner in which they differ slightly on each island of the group; none of the islands appearing to be very ancient in the geological sense.
>
> It was evident that such facts as these, as well as many others, could only be explained on the supposition that species gradually became modified; and the subject haunted me. But it was equally evident that neither the action of the surrounding conditions, nor the will of the organisms (especially in the case of plants) could account for the innumerable cases in which organisms of every kind are beautifully adapted to their habits of life....[4]

He returned to England, therefore, a convinced evolutionist, but what eluded him was a mechanism to explain it all. This remaining piece of the puzzle dropped into place unexpectedly in 1838 as he was reading "for amusement" the *Essay on the Principle of Population* by T. Malthus. In this scarcely amusing essay, Malthus pointed out that humans tend to produce many more offspring that can be supported by the food supply and that starvation, disease, war, etc. act to keep the population size in check. Darwin was immediately struck by the applicability of this principle to all living things. Every organism potentially produces far more offspring than can be supported under the prevailing circumstances, yet for the most part the over-all numbers in every population remain fairly constant from one generation to the next. The few offspring that do survive and reproduce will be those that are best adapted to their environment; hence these beneficial adaptations will increase in the population and become the norm. This "struggle for existence" is never-ending, however, for these better-adapted organisms will themselves produce more offspring than can be supported and only those among them with still better adaptations will survive and reproduce. The environment is constantly selecting only a few organisms to contribute their heritable beneficial adaptations to the next generation, a phenomenon Darwin called the principle of **natural selection**.[5]

THE ORIGIN OF SPECIES

As Darwin assiduously went about over a 20 year period amassing more and more evidence for his theory of evolution through natural selection, his intention was ultimately to publish his conclusions in a multi-volume work wherein the sheer mass of the evidence would overwhelm the critics. He was not given this opportunity, for in 1858 he received in the mail a manuscript from a naturalist previously unknown to him, A. Wallace, which succinctly outlined his own theory of evolution through natural selection.[6] Moreover, Wallace had been influenced by reading the very same essay of Malthus's that had so stimulated Darwin 20 years earlier. Wallace petitioned Darwin to submit the manuscript for publication, if he thought it had merit. This obviously put Darwin in an ethical dilemma, and his initial intention was to forward the Wallace manuscript for publication without mentioning his own monumental work on the same subject. Fortunately, Darwin had through those many years of work kept several others apprised of his theory in both conversations and letters, and the more influential among them, principally his close friend the geologist C. Lyell, persuaded Darwin to submit a joint paper with Wallace for publication that same year.[*] Wallace readily agreed that Darwin's evidence for their common idea far surpassed what he had been able to gather himself and that Darwin had the idea first; he graciously consented to a joint submission. Darwin then set about to distil his evidence for natural selection into the abbreviated single-volume work published the following year as *On the Origin of Species by Means of Natural Selection* (usually entitled *The Origin of Species*). In many ways it is likely for the best that Darwin's hand was forced in this manner, since a multi-volume work, instead of overwhelming potential critics, might well have smothered his essential conclusions in the huge mass of supporting data.

The ideas of Darwin and Wallace can be expressed very simply by the following five principles.

> 1. All organisms tend to produce many more offspring than can survive under the prevailing circumstances.

------

[*]In a letter to Lyell, Darwin wrote of his astonishment on receiving the Wallace manuscript:

> I never saw a more striking coincidence. If Wallace had my M.S. sketch written out in 1842, he could not have made a better short abstract! Even his terms now stand as Heads of my Chapters.

2. Individuals within any species differ from each other in many slight ways. Some of these differences may be advantageous under the environmental circumstances that prevail; many of them will offer no special advantage or even be detrimental.

2. Some of these differences will have a genetic basis; that is to say, these traits can be inherited.

3. Those offspring that by chance are endowed with characteristics that render them better able to survive and reproduce under the environmental circumstances will be making a greater contribution to the next generation than do those organisms that are less well adapted.

4. If these favorable characteristics have a genetic basis, then over time they will become the norm for the population. The population will have evolved.

There are a number of things to note about these principles, for the moment the most important being the emphasis throughout on survival with *reproduction*. It is only because it is commonly the case that those who survive longest reproduce more that there is any emphasis on individual survival at all. The key to evolution is not individual survival but successful reproduction, or more particularly, making a proportionally greater genetic contribution to the next generation. If this is accomplished by greater individual survivability, so be it. But it might be equally well or better accomplished, as we know it is in many organisms (shad flies, living as adults only one day, come immediately to mind), by a very short lifetime devoted to a great burst of reproductive activity. *Natural selection is differential reproduction.*

It is equally important to note that it is *populations*, not individuals, that evolve, and that evolution is nothing more than a shift in the frequency of genes within that population over time. An individual cannot evolve, since its genetic endowment is permanently established (minor exceptions notwithstanding) at the moment of conception. An individual who, as a consequence of new mutation, recombination, or any other process that produces a novel genotype, gains a genetically determined advantage, however slight, over others of the species is likely to leave more or more successful offspring than will those who are less fortunately endowed. These offspring will themselves be more successful reproducers, and so on. Over time, that novel genotype will have been **selected** and will become the norm. A shift in gene frequency within the population has been achieved; evolution has occurred.

Relying only on intuition, one cannot be blamed for doubting that genes conferring only minor advantages on their possessors could increase significantly in frequency if this were the only mechanism operating within the population to promote this spread. In this case, however, intuition is wrong. The well-known population and mathematical geneticist, J. Haldane, has produced impressive calculations that show otherwise. He has shown that a theoretical new dominant mutation (call it $\underline{X}$) that conferred a reproductive advantage on its possessor as small as 0.001 (meaning that 1000 homozygous or heterozygous $\underline{X}$ individuals reproduced for every

999 homozygous recessive x individuals) would increase in frequency within the population from an initial frequency of 0.00001 to 0.99 in just 23,400 generations. Given sufficient time, even minor genetic advantages are sufficient to transform the genetic basis of the population and hence to power evolutionary change.

Well, how much time has been available? Lots. The earth is approximately 4.6 billion years old, and life seems to have arisen on earth about 3.5 billion years ago. These are impressive time intervals, but unfortunately in these days of multi-billion-dollar defense budgets and deficits, many of us have lost our awe of how much a billion (in the American sense of 1,000 million) really is. Perhaps some perspective can be regained by trying to guess how long a time interval would be encompassed by a billion seconds. The answer is (approximately) 32 years!

Before we consider in detail some of the evidence favoring evolution through natural selection, several general comments are in order about the process and the terms used in describing it. The first has to do with the phrase **survival of the fittest** that is often quoted from Darwin and equally often misinterpreted. We have already noted that survival cannot be equated necessarily with longevity; the term "fittest" in this context also has a special meaning at variance with its common one. Its meaning to an evolutionist is "leaving the greatest number of surviving offspring". It has nothing whatever to do with fitness in the jocky, Soloflex, Schwartzenegger sense. An individual who makes no genetic contribution to subsequent generations has a Darwinian "fitness" of precisely zero, whatever admirable qualities of brain, brawn or pelf he or she may possess.

In the foregoing account of evolution the discussion has been couched in genetic terms. This represents the modern interpretation of Darwinism, an interpretation often called **Neo-Darwinism**. Darwin himself was ignorant of genes as the basis of inheritance, as was everyone else until the rediscovery of Mendel's principles in 1900. Natural selection acts directly only on *phenotypes*, but in promoting or discouraging particular phenotypes, the underlying *genotypes* that are responsible for these phenotypes are indirectly selected for or against. The lack of knowledge about the mechanism of inheritance Darwin saw as a great impediment to the full understanding of his theory. Just as Mendel (who published his work about 20 years before Darwin's death) had to work out the movements of what we now call genes through successive generations without knowing about chromosomes, so Darwin had to analyze the evolution of phenotypes through generations without knowing about the genes that specified them.[7]

REFERENCES

1. Mayr, E., 1978. Evolution. Sci. Amer. (Sept.).

2. Eiseley, L., 1956. Charles Darwin. Sci. Amer. (Feb.)

3. Lack, D., 1953. Darwin's finches. Sci. Amer. (April).

4. C. Darwin, 1876. "Recollections of the Development of my Mind and Character". quoted in *The Darwin Reader*. M. Bates and P. Humphrey, eds. Charles Scribner's Sons, N.Y., 1956.

5. Darlington, C., 1959. The origin of Darwinism. Sci. Amer. (May).

6. Eiseley, L., 1959. Alfred Russel Wallace. Sci. Amer. (Feb.).

7. Dobzhansky, T., 1950. The genetic basis of evolution. Sci. Amer. (Nov.).

LECTURE 46

EVOLUTION: CONTROVERSIES AND DIFFICULTIES

It shall be unlawful for any teacher in any of the universities, normals, and all other public schools which are supported in whole or in part by the public school funds of the state, to teach any theory that denies the story of the divine creation of man as taught in the Bible, and to teach instead that man has descended from a lower order of animals.

> State of Tennessee, 1925
> (repealed, 1967)

Well, it is a theory, it is a scientific theory only, and it has in recent years been challenged in the world of science and is not yet believed in the scientific community to be as infallible as it was once believed. But if it is going to be taught in the schools, then I think that also the biblical theory of creation, which is not a theory but the biblical theory of creation, should also be taught.

> R. Reagan

OBJECTIONS TO DARWIN'S THEORY

Among the many objections one hears raised against Darwin's theory of natural selection, none is more common than the view that exceedingly complex organs such as the human eye—an organ that requires the coordinated function of many different parts (lens, optic nerve, brain, cornea, ocular muscles, etc), each one itself highly complex, in order to register an image—is incapable of having evolved to its present perfection, since the earlier stages, lacking all the essential parts, would have been useless or at least highly impaired. You can't have one-third of an eye, the argument goes; you have to have the whole pie or none at all. And since everyone agrees that something as complex as an eye cannot suddenly arise *holus bolus* by a one-step random mutation, it must have been consciously designed and created.* Let's see.

---

*This argument even gave Darwin pause. "To suppose that the eye," he wrote in *The Origin of Species*, "with all its inimitable contrivances for adjusting the focus to different distances, for admitting different amounts of light, and for the correction of spherical and chromatic aberration, could have been formed by natural selection, seems, I freely confess, absurd in the highest degree." One aspect of this perennial theological objection to Darwinism has been termed the
(continued...)

First of all, it should be noted that the human eye probably is not the last word in eye perfectibility, anyway. Hawks, for example, have much greater visual acuity than do humans, being able to spot the movement of very small animals at great distance. A hawk would think that humans have only one-third an eye to begin with. Even among humans, there are many who have only one-third as much vision as the majority, and many who have even less than that. Given the tenacity which people fight to preserve even 5% vision, or 1%, or whatever, it is clear that any degree of vision is considered preferable to no vision at all. So vision is *not* an all-or-nothing phenomenon. Nor is hearing (bats probably imagine we are stone deaf), smell (dogs might well wonder what we do use our nose for), or any other faculty one can point to.

Since we are on the subject of eyes, I might as well add that the vertebrate eye is not really all that well designed anyway. In fact, it is a downright bungled design; an engineer putting such a contrivance on the market now would soon be finding himself the hapless defendant in a product liability suit! The retina, for example, is installed backwards, with the light-sensitive side to the rear and all the "wiring", the zillions of neural connections to the brain, lies on top of the retina, significantly impeding the light from reaching the lower surface. These "wires" then must be gathered together on the surface and brought through the retina at some point. This creates a "blind spot" in each eye, which the brain does as best it can to fudge over for our convenience. The only way such a flawed design can be understood is as a consequence of having to make-do in evolution with the material at hand. "Evolution is a tinkerer", F. Jacob was fond of saying, and this and innumerable other aspects of organisms illustrate their origins not as premeditated designs starting from a clean slate but as add-ons, rearrangements, or similar modifications of existing structures jerry-built from whatever was (embryologically) at hand to best meet the exigencies of the moment.

Finally, vision need not be equated necessarily with forming images. Indeed, in the majority of animals, image formation is not the purpose of eyes, which record only degrees of brightness. In this way, they can be attracted to relative light or darkness. *Paramecia* (one-celled ciliates) have just a light-sensitive spot to clue them in, other organisms have light-sensitive cells inside an invagination or cup for protection. In some cases this invagination partially closes over so the entering light is focused as in a lensless pinhole camera. In fact, within the extant animal world one can discover an unbroken series of mechanisms for dealing with light from the most primitive light spots through and beyond the image-forming human eye. Each of these devises confers some advantage on its possessor and is, therefore, preferable to total blindness. By slow modification over great amounts of time, the coordinated components of the mammalian eye evolved in concert, each modification bringing some improvement over its antecedents. In order

---

*(...continued)
"argument from design": complicated pieces of machinery, displaying an organized complexity, cannot arise spontaneously, but require a maker or designer, if mechanical, or a Maker or Designer, if biological. Another aspect of the same objection has been termed the "argument from personal incredulity": I myself cannot understand how...; therefore, no one else can understand it either.

for these modification to be selected for in evolution, all they need to be is better than their earlier versions. Not perfect, just somewhat better. "We are the products of editing, rather than of authorship," as G. Wald has put it.

Another argument one hears has to do with the randomness of mutation. If, as geneticists claim, mutation is totally random, how can something as obviously non-random as an eye develop? The improbability of a monkey's randomly typing out the works of Shakespeare is often trotted out (curiously, often by those who don't appear to have read much Shakespeare, but that is a different story). What such arguments fail to recognize is that whereas mutation is indeed random, natural selection is not. Natural selection is unpredictable because the complex interplay of things and forces that constitute "nature" is unpredictable, but this is not the same thing as saying that natural selection is random. The term "selection" itself implies non-randomness. At each generation nature selects for propagation those phenotypes that are best adapted for the environment as it then exits. As the environment unpredictably evolves, so in concert do the organisms in it. Those that have not, by random chance, thrown up any mutations that improve their chances for survival in changing circumstances lose out in comparison with those that have.

The "typing monkey argument" is not a valid analogy for another reason, too. As pointed out earlier, major improvements in evolution, such as the mammalian eye—as complex and non-random as Shakespeare's works—do not arise at a single step, but are the result of slow, progressive improvement over vast amounts of time. An eye cannot suddenly spring into existence in all its perfection any more than can a monkey just type out the works of Shakespeare. To credit a (long-lived) monkey with this feat would not strain credulity, however, if it could do at its typewriter what nature does in the field: keep the "best" copy, use only it as the basis for the next attempt, and discard all the rest. If every time the monkey hit a key, the result was only recorded if it brought the work in some measure closer to the goal, Shakespeare's works, then over sufficiently long period of time the manuscript that resulted from the accumulation of these randomly generated lucky strikes would, at the very least, raise our estimation of a monkey's creative intelligence considerably. And this is why it is so easy to imagine a creative intelligence at work in nature. We see around us only the successes, not the vastly greater, incomparably greater, number of failures.[*]

WHAT IS A SPECIES?

Most of us have a folksy notion of what a species is based primarily on the degree of apparent similarity or dissimilarity of organisms. We do not confuse domestic dogs with cats and are also at least vaguely aware that coyotes and tigers, although similar respectively to dogs and cats, are

---

[*]This analogy is not altogether parallel, however, for it should be borne in mind that there is no "goal" in the case of evolution, no predictable end-point, whereas the monkeys in this example save the random hits that best accord with a pre-selected goal, Shakespeare's works. But although evolution has no goal, natural selection does, the goal being the best adaptation to prevailing natural conditions.

really something else again. This level of understanding is sufficient to get us through the day, but is not good enough for a concept so basic to the scientific study of evolution. To a biologist, the definition of a **species** is a population within which the individuals exchange genes but which is genetically isolated in nature from other populations. Without too much loss of precision, this definition might be colloquially rephrased to state that a species is an interbreeding population that does not breed with others in nature. Note that the phrase "in nature" was retained in both definitions, for this points out a very important aspect of the modern species concept. Many species potentially could interbreed, as lions and tigers sometimes do in zoos, but if, again like lions and tigers, for reasons of behavior, geography or whatever they do not do so in nature, they can be considered as separate species.

This definition has several weaknesses, however. Since the definition emphasizes behavior in nature, populations that do not occur together in nature cannot be evaluated in the same way as those that do. Here all we can do is assess as best we can whether the dissimilarities between the populations appear as great as those between populations that do live together (**sympatric populations**) without interbreeding. If so, they are considered to be different species; if not, then the same species. Of course, if they cannot be made to breed together artificially, this strengthens the notion they are separate species, but one must be cautious, since many members of a single species that obviously must interbreed in the wild fail to do so in captivity. At its core, there is actually somewhat of a paradox to the modern definition of species. To be classifiable as a species, organisms must have undergone sufficient divergence that effective interbreeding cannot again occur between them, even if the opportunity presents itself again. But then, if they *are* isolated in nature, how do we know that sufficient divergence has already occurred so that we can now classify them as separate species?

Emphasizing interbreeding or the lack of it in nature clearly presents problems in classifying which fossils belong to a single species and which to a different related one. (This is one reason paleobiology is such a contentious discipline.) It also presents problems in classifying organisms that do not interbreed with anything, including themselves. A good example is the totally asexually reproducing common dandelion. In these and similar cases, biologists muddle through as best they can.

Darwin entitled his famous book *The Origin of Species*. In one sense, this was a misleading title, for the major thesis of the book was that plants and animals have sufficient genetic plasticity to permit, under the influence of natural selection, the evolution of one species into another over time. He did not spend much time directly addressing the problem inherent in the title: exactly what conditions promote the origin of two or more species from what was originally one. In other words, what factors lead to reproductive isolation and thence to **speciation**? Geographical barriers are an obvious means of separating two populations of what originally was a single species so that divergent evolution leading to speciation can occur. This was what was primarily at work on the Galapagos islands. A few founding birds would reach an island where previously there were no finches and there, free to interbreed among themselves but isolated from the parental population from which they came, diverged into a separate species that could no longer interbreed with the parental population even if they were reunited. Other instances of **geo-**

graphical isolation might occur when rivers change course suddenly, when new water systems occur as a consequence of changes on the earth's surface, or when the movements of the earth's crust throw up mountains or otherwise separate previously contiguous land masses.*

Oftentimes speciation can occur sympatrically, that is, without geographical separation of parts of the original population. There is a variety of isolating mechanisms that can account for increasing reproductive isolation in the absence of physical barriers. For example, a subgroup of the population may develop by mutation changes in the reproductive apparatus so that they can no longer mechanically interbreed with the parental population. Or as a consequence of non-disjunction they may become **polyploid** (i.e., they acquire multiple sets of chromosomes, a commoner event in plants than in animals) which often prevents fertile hybrids forming when they try to mate back to organisms with the standard chromosome number. Or they may develop behavioral peculiarities connected with the mating process the isolate them from their more conventional parental population. Or changes in breeding times...there are many ways in which subpopulation can change, at first usually very subtly, which will effectively isolate then from the main group and then permit the further evolution of even more formidable barriers to inter-breeding. And once, for whatever reason, interbreeding in the wild with the original population is no longer possible, a new species has developed.[1]

CONTINUING CONTROVERSIES

Mention was made earlier in this account of continuing controversies in evolutionary theory. In recent years a particularly lively one concerns the rate of evolution. This account has undoubtedly left the impression with the reader that evolution is a slow, continuing, almost seamless process of increasing diversity over time. Many biologists still believe this to be—in very general terms and with a variety of caveats to cover special situations—the case. Others, however, believe that evolution is better characterized as undergoing long periods of little or no change followed by short bursts of rapid evolution and then by another long period without significant change. Because these longer periods of evolutionary equilibrium are punctuated by shorter periods of rapid change, this theory goes by the name of **punctuated equilibrium**. These

---

*That geological features such as new barriers in a previously continuous land mass lead predictably in the fullness of time to speciation on either side of the divide is now a common-place observation. Much less so is the prediction of previously unknown geological features on the basis of the observed speciation that has occurred. A striking example of this began with a study of the iguanas of the Galapagos Islands. In the course of studying these creatures, scientists reached the confusing conclusion that the evolution of these iguanas would have had to take place over a longer time interval than the islands on which they evolved were in existence. The Galapagos iguanas, in other words, seemed older than the Galapagos islands on which they must have evolved. This led Drs. V. Sarich and J. Wiles to predict, in 1983, that there must have been other, much older, Galapagos Islands that are no longer apparent above the surface of the ocean. This prediction was confirmed in 1991 with the discovery of several such former islands in the Galapagos region that are now several thousand feet below sea level.

and other controversies within evolutionary theory are often seized upon by critics of evolution either as examples of scientists denying that evolution occurs or as examples of scientists changing their minds and therefore by implication not really knowing what they had been talking about. Neither of these interpretations is tenable. The arguments are over details of the process of evolution, not about the fact of evolution itself. As stated earlier, no legitimate biologist denies the reality of evolution nor the importance of natural selection in powering it. Changing one's mind under the influence of changing evidence is not an example of faulty science, but is in fact fundamental to the whole scientific enterprise. The openness to new ideas as new data accumulate is precisely what distinguishes science from whatever doctrines compel unfaltering adherence to a notion regardless of negating arguments and facts. Yes, evolutionists do change their minds, and proudly so; creationists are the ones who hold rigidly to the same unsubstantiated opinion come what may[*].

REFERENCES

1. Ayala, F., 1978. The mechanisms of evolution. Sci. Amer. (Sept.).

---

[*]There was no intention in this paragraph to suggest that the evidence adduced by the punctuationists compels a reasonable person to abandon the gradualist interpretation. These two views were mentioned only as examples of a lively current controversy.

LECTURE 47

EVOLUTION: THE EVIDENCE

THE THEORY VS THE FACT OF EVOLUTION

Biological evolution is not a theory, it is a fact. It is as well-established in the canon of scientific facts as the valence of carbon or the Second Law of Thermodynamics. For some reason (not hard to guess) the word "theory" has clung to the concept much longer than with many other scientific facts. One still hears the phrase "theory of evolution", but never the "theory of a spherical earth", although the two facts rest on about equal evidential footing. There is some, but not much, justification for still talking about the "theory of natural selection", for as we have noted in the previous lecture, there are still a number of lively controversies raging within the profession of biologists over some of its details. But the controversies, it should be firmly noted, are over relatively minor *details* of the mechanism of natural selection; it would be impossible to find a serious biologist who rejected natural selection as the principal mechanism underlying biological evolution.

The evidence that living things can and do evolve over time comes from many diverse observations, including comparative anatomy, embryology, biochemistry, geology, and biogeography. In this brief account only a superficial account of the types of evidence will be given; fuller accounts can be found in almost any general biology textbook.

THE FOSSIL RECORD

The evidence from the fossil record is very convincing all by itself. Fossils are preserved parts of prehistoric plants, animals or even bacteria that have lain embedded in rocks, sediments or other preservatives (such as amber or ice) from shortly after their death to the present time. Normally we think of fossils as being bones, shells or the impressions of plants, but almost any physical evidence of prehistoric life, such as preserved excrement or footprints also count as fossils. The very earliest fossils are of bacteria believed to have lived 3.5 billion years ago, but the fossils of large numbers of multi-cellular organisms first make their appearance in the Cambrian period, beginning about 570 million years ago. Over 300,000 different species have already been identified in the fossil record, but this is considered only a minute fraction of the species that have become extinct since life originated on the planet. The reason so few traces remain of the vast chain of life that predates our own era has to do both with the nature of this once-living matter and with the fossilization process itself.

In order to be preserved as a fossil, the once-living matter has to resist decay and dissolution long enough to ossify, that is, to have mineral salts or other long-lived material replace the organic material that is subject to deterioration over time. Since fleshy parts of the body rapidly decay after death, it is generally only the mineralized parts of animals' bodies (bones, shells, etc) that are found as fossils. Of course, there are rare exceptions, as when mammoths are preserved in permafrost, insects in amber, or humans in peat bogs where the process of decay is halted by lack of oxygen and great acidity. Animals that do not have lots of bone or shell usually don't get fossilized. Insects, as just one example, although they are our most abundant living class of animals and have probably dominated the earth for the last 100 million years or so, are very poorly represented in the fossil record for this reason. If it were not for the fact that insects are currently so abundant and are so little changed from the fossil insects we do have data on, we might well have concluded that insects were never very prevalent among living forms.* The vast majority of currently living organisms would have been very poor candidates for fossilization; there is no reason for believing that earlier forms of life were better bets.

Besides having structures that are capable of fossilization, an organism has actually to die under circumstances favorable to fossilization, and this is an exceedingly rare event. It is not germane to our purposes here to go into detail on the processes that promote fossilization, but the fact is that most organisms eventually decay *in toto* shortly (geologically speaking) after their death, and that's that; no record of their ever having existed remains. Even when fossilization does occur, many are later destroyed by movements of the rock, erosion, or chemical dissolution. Finally, of course, if the fossil does survive to the present time, it has to be found and by the right person, and both these events are themselves very chancy. What I'm saying is that the fossil record is far from perfect; one would not expect it to be otherwise.

Even with all these drawbacks the fossil record nonetheless documents an impressive history of evolution. In many cases we can document the evolutionary history of certain organisms in great detail, as in the evolution of modern birds with feathers and no teeth from prehistoric toothed featherless reptiles through feathered toothed intermediates. The evolution of the modern horse from remote dog-sized ancestors is represented in another well-documented series of fossils. In sediments or geological layers (**strata**) that are older, the more primitive forms are always found; newer strata yield forms closer to those living to-day. Over 99% of all fossil species of plants and animals are wholly extinct. None of this is to suggest there are not lively controversies concerning the interpretation of the geological record. Many of these come about as a consequence of the inevitable gaps inherent in the nature of the fossilization process (discussed above) and others occur as a consequence of puzzling geological processes that make it unclear which layers are truly the oldest. To an outsider, it also seems at times that evolutionary biologists are simply more contentious than many of their colleagues in other specialties enjoy being.

---

*The British biologist J. Haldane was once asked what the study of biology could tell him about the nature of God. "I'm really not sure," he replied, "except that He must be inordinately fond of beetles." Given that there are estimated to be currently living about 300,000 species of them, it is not a bad inference.

cause of the serendipitous nature of fossil discoveries, important pieces of various evolutionary puzzles are constantly being unearthed and fitted into place. One recent such discovery has been a fossil whale that had well-developed rear legs. It has long been believed that modern whales descended from legged land mammals, but modern whales show in their skeletons only the tiniest traces of hind legs. (The flippers are clearly derived from earlier front legs.) However, until very recently no fossil whales with better developed hind legs had been found. It almost seemed that hind legs had been lost "all at once", something that almost never happens in evolution. (Legs, after all, are difficult things to lose, unnoticed!) But now the gap has been filled with the discovery of a fossil whale skeleton bearing well-developed legs and feet. Since this specimen appears to date from a time about ten million years after whales returned to the water, the persistence of hind legs and feet this late in whale evolution suggest that they continued to have some use, not for locomotion but perhaps in the mating process.[1]

COMPARATIVE ANATOMY

Another very important line of evidence for evolution stems from the study of comparative anatomy. Such studies reveal that structures that outwardly appear very different are often composed from the same basic plan. The arms of humans, the flippers of whales, the forelimbs of horses and cats, the wings of bats all shore a common underlying bone structure, which in turn reflects a common embryological origin of these parts. All these animals are mammals and to that extent related. The similarity of their forelimbs underscores the important evolutionary principle that there is only rarely anything truly "new"[*]. Evolution proceeds by the modification of already existing structures adapted to new uses. This "recycling" of pre-existing structures is unexplainable on the premise that each species was created *de novo*, but is readily explained if evolution of one species into another occurs by the slow accumulation of small inherited changes. Structures that have a common embryological origin but serve different function in the adult are said to be **homologous**. By contrast, there are structures that serve the same function in the adult but are otherwise unrelated. The wings of birds and insects and the eyes of mammals and octopuses, for example, serve the same function but have arisen by the modification of quite different structures during evolution. Such structures are said to be **analogous**.

------

[*]Beware of reading too much into this statement. It refers principally to evolution within related groups such as all mammals or all birds. Obviously there are mechanisms at work that occasionally augment the amount of genetic material available for evolution to work on; otherwise evolution would be largely a down-hill process. It is not possible in this account to discuss in detail the events that can increase the amount of genetic material, but accidental gene or chromosome duplication is an obvious one. When this happens, the duplicated genetic material sometimes becomes free to take on entirely new functions over time. DNA also has an inherent tendency to increase in amount beyond the actual developmental and functional needs of the organism. In humans, for example, almost 90% of the DNA in every cell has no apparent function at all. Hence there is lots of "spare" DNA lying around in the genome, some of which could presumably evolve into functional genetic material.

Comparative anatomy also often reveals the existence of **vestigial structures**, ones that have lingered in rudimentary form within an organism long after the need for them had ceased. Snakes, for example, are often found to have a rudimentary pelvic girdle and even vestigial hind limbs buried within their bodies, although these structures serve no purpose in the snakes. Similarly, the non-functional human appendix survives as rudiment of the caecum found in many mammals as an important digestive organ for leaves and grasses. Again, it is only a knowledge of evolution that makes the existence of such vestigial structures explainable.

EMBRYOLOGY

The study of embryology, besides serving as a guide in interpreting comparative anatomy, sheds light on evolution in other ways. Close study of a developing embryo often reveals in some detail the evolutionary history of the species, a principle sometimes summarized in the phrase "embryology recapitulates phylogeny"—meaning that many embryos sequentially develop and then discard or modify structures that reassemble those that functioned in their remote evolutionary progenitors. The early embryos of birds, reptiles, and mammals all develop a row of vestigial gill slits behind the head, suggesting that all these forms evolved from organisms that once had functional gills; bird embryos develop tooth buds, although the adults are toothless, and human embryos develop for a time tails, although only rarely are babies born with a tail. All these embryological structures did fully develop in the organisms from which the modern ones are descended.

The embryological gill arches mentioned above have an interesting fate as development of a mammalian embryo proceeds. Again showing how evolution remodels pre-existing structures for new uses, the parts of the gill arches that would have provided the bony support for the gills of fishes become instead parts of the inner ear, lower jaw, and air passages in adult humans.

BIOGEOGRAPHY

Biogeography, the study of the distribution of plants and animals, is what first tuned Darwin to the idea of evolution. Why did the Galapagos, a small group of islands off to the west of South America, contain so many different kinds of finches, more than in all of South America? Why were marsupials (pouch-rearing mammals) endemic only to South America and Australia? And why did Australia not have any endemic placental mammals, but had marsupials that in many cases resembled Old World placentals, at least in the ecological niche they occupied and sometimes physically, too? Both Darwin and Wallace, in puzzling over many of these same these facts, came to the conclusion that they were best explained as a consequence of evolution. The many species finches in the Galapagos were descended from a very few that were blown from the mainland of South America and then found their way to neighboring islands. Being isolated in these new niches, they evolved into many distinct species each uniquely adapted to their new habitats, although all are ultimately descended from the very few that first found their errant way from the mainland. This process by which new species develop to occupy different ecological niches is termed **adaptive radiation**. A similar process went on in Australia, where the continent became separated from the rest of the world before the development of placental mammals. The marsupials were free to "radiate" into the ecological niches in Australia that placental mammals in the Old World later occupied there. In some cases the way of life required in the two continents was similar enough, that is to say the evolutionary forces acting on the different organisms were similar enough, that the morphological adaptations in the organisms were also similar. For example, there is a marsupial "wolf", a marsupial "woodchuck", and a marsupial "flying squirrel", as well as many completely unique adaptions, such as the kangaroo.

BIOCHEMISTRY

The evidence from biochemistry is also very impressive. All currently living organisms use DNA as the repository of their genetic instructions and all read it by first transcribing it into RNA and then translating it by means of t-RNA and ribosomes. All use essentially the same genetic code in translation, even though there seems no biochemical reason for assigning any particular codon to a particular amino acid. In all probability, the current genetic code just happened to be the one functioning in the first successful cell, and all descendant cells—that is, all subsequent life—inherited it essentially unchanged. All cells use ATP as an energy transducer...the list of fundamental similarities is almost endless and certainly much greater than the differences among cells, which tend to be merely superficial. This unity of all living things at the biochemical level speaks strongly for their evolutionary relationship and equally strongly against their individual creation.

PLANT AND ANIMAL BREEDING

The final evidence for the plasticity of species comes from the achievements of plant and animal breeders. At the outset, it should be underscored that the efforts of humans to produce different varieties of plants and animals is a conscious, goal-directed process that is quite unlike what happens in nature. For this reason, selective breeding is usually termed **artificial selection**, as opposed to the unconscious, undirected process that occurs in the wild. The different varieties of dogs, cats, cattle, etc. that have been produced from the original domesticated prototypes demonstrate clearly the effects of selection in radically altering over time the appearance of organisms. The results in plant breeding are even more impressive. For example, cabbage, brussels sprouts, kale, and cauliflower are all artificially produced derivatives of the same wild-type plant, *Brassica oleracea*. It could be argued that these plant and animal varieties are nonetheless still the same species and that selection has not really changed their fundamental nature. This is to some extent true, but then conscious selection by humans has likely gone on only very sporadically and for at most a few tens of thousands of years. This compares with the millions of years available to natural selection. It is also the case, as will be discussed presently, that the concept of species, which up to now has not been defined in this account, is not as clear-cut as might be supposed.

REFERENCES

1. Gingerich, P., Smith, B., and Simons, E., 1990. Science 249:154-157.

LECTURE 48

THE ORIGIN OF LIFE

How life originated on earth cannot ever be decided with certainty, and maybe not even with a satisfying degree of probability. This has not prevented the lush growth of a variety of theories about the origin of life, which fossil evidence suggests occurred about 3.5 billion or so years ago. Most scientists think that all living things are descended from a single first-living-thing, which suggests either that life arose only once on this planet or else that only one of the several first-living-things survived to populate the earth. For the sake of economy, this account assumes the former and goes on to suggest how it could have happened.

EARLY EARTH

Earth 3.5 billion years ago was not the accommodating place it has since become. The atmosphere was composed mainly of methane, ammonia, nitrogen, water vapor and carbon monoxide and dioxide. Hydrogen gas may (or may not) have been present in appreciable quantities, and there may also have been hydrogen cyanide. There was no oxygen and no pro-tecting ozone layer. Ultraviolet radiation was intense; so was the sun's heat. Torrential rains and severe electrical storms were commonplace. The earth's surface was rock and water and little else; the developing oceans would contain a variety of inorganic minerals and salts leached from the rocks by water runoff. At first blush, it seems hardly a propitious environment for the origin of life; one also wonders why life would want to live there in the first place.

ABIOTIC SYNTHESIS OF ORGANIC MOLECULES[1]

In 1936 the Russian biochemist A. Oparin proposed that conditions such as these actually were favorable for the origin of life from non-living precursors. He suggested that energy in the form of heat, electricity (from storms) and ultraviolet light could act on the gasses present in the early environment to produce many of the organic molecules essential to life and that in this primordial organic soup that resulted, living, reproducing things evolved. Oparin's ideas were put to a test in a famous experiment by S. Miller and H. Urey in 1953. These investigators combined what they considered the gasses present in the early earth (their guesses were ammonia, methane, water and hydrogen) in a sealed apparatus into which they introduced energy in the form of electrical discharges and heat. The mixture was repeatedly refluxed (i.e., subjected to cycles of evaporation and condensation) and the contents examined after about a week of such treatment. To their surprise, the final mixtures were rich in organic molecules such as acetic, formic, succinic and lactic acids, urea, hydrogen cyanide, and the amino acids glycine, alanine, aspartic and glutamic acid. When they and subsequent investigators repeated this work using other plausible ingredi-ents (e.g., hydrogen cyanide, carbon dioxide, and hydrogen sulphide, which is still found in volcanic outgassing), an even greater variety of molecules accumulated, including the nitrogenous base adenine and some lipids and carbohydrates. Moreover other energy sources such as ultraviolet light would substitute for the electrical energy. Whatever the exact composition of the primitive environment on earth, it is clear that circumstances were very favorable to the abiotic synthesis of many organic molecules that are now fundamental to the metabolism of all living things.*

These molecules could accumulate in the early seas because neither of the two things that currently break down organic molecules existed then: living organisms and oxygen. As the soup grew thicker (one shouldn't take this metaphor too literally), there is an increasing likelihood that the molecules contained therein would interact, either through chance encounters in the sea or, more likely, on the surface of clay and mineral particles. Small drying pools of water containing organic compounds would serve to concentrate them further and increase the likelihood of their interaction. S. Fox has shown, for example, that heating dried amino acids results in the ready formation of polypeptides ("proteinoids", he calls them). These polymers could then wash back into the seas, and the seas in time would become rich in them, too.

Since the fundamental building unit of all current living organisms is the cell, some means had to exist to encourage the aggregation of these disparate macromolecules into small, discrete "packages". As it turns out, there are purely physical inducements for this to happen spontaneously. Fox has shown that the polypeptide-like molecules will aggregate into **microspheres** if they are heated together and slowly cooled. These microspheres exhibit many of the properties associated with cells. They have a double-layered boundary vaguely resembling the cell membrane, but without lipids. They absorb material through this "membrane" and concentrate it internally, shrink or swell according to the tonicity of the surrounding medium, develop an electrical potential across the "membrane", aggregate into groups of microspheres, and even divide into two by a budding process not unlike what happens in yeast cells. They are not cells, and may not even be a forerunner of cells, but do show that many of the properties that are

---

* Some scientists have speculated that amino acids and perhaps other molecules of biological importance could have been seeded on earth by meteorites, particularly the softer type called carbonaceous chondrites. It is certain that meteorites do contain amino acids that are presumably of abiotic origin, or at least not of earthly origin. One meteorite that has been extensively studied is the so-called Murchison meteorite that fell to earth in the late 1960s. It contained at least 74 different amino acids (only about 26 are commonly found in terrestrial life), and these amino acids had a ratio of C^{13}/C^{12} that was much higher than that found in any terrestrial organic compounds. Moreover, the amino acids consisted of an approximately equal mixture of D and L stereoisomers, whereas terrestrial life uses and synthesizes only the L isomer. (D and L refer to the direction in which light is rotated—to the right or left, respectively—when passed through the amino acid solution. Non-biological syntheses usually yield an equal mixture of the two forms.) Hence, contamination from an earthly source seems an unlikely explanation for the presence of amino acids in the Murchison meteorite. However, because of the great heat generated during the track of a meteorite through our atmosphere, it also appears unlikely that amino acids could arrive to earth in quantity by this route.

characteristic of living cells can emerge solely as a consequence of known chemical and physical forces acting on inert matter.

Other investigators have observed similar cell-like spheres called **coacervates** form when molecules from a Urey-Miller experiment are mixed with water under particular conditions of pH, temperature, and ionic concentration. These, too, form membrane-like shells around them containing lipids. They are stabilized by complex hydrophobic and hydrophilic interactions, and demonstrate a non-random arrangement of molecules internally. They have the ability to bring external molecules into the coacervate and concentrate them internally. If they are prepared in such a way that an enzyme is included inside, they carry out a process resembling metabolism. For example, if phosphorylase is included, glucose-1-phosphate will be absorbed from the surrounding medium and converted into starch. If the enzyme amylase had also been included, the starch would be broken down into maltose, which will diffuse back into the medium. These coacervates, then, show one way in which metabolism could originate. At first, chemical energy would be available from the environment in the form of spontaneously generated ATP, but as this supply diminished, some proto-cells might begin to harness the light-induced electrical potential across their membranes to the production of their own ATP, not unlike what certain bacteria do now. These proto-cells would gain an obvious advantage thereby, as would any that developed any chemical process of their own that made them less dependent on the chance environmental provision of nutrients. Natural selection at this point would be promoting a form of abiotic **chemical evolution** among the proto-cells.

It has already been demonstrated that certain catalytic RNA molecules can, in one sense, evolve to take on new catalytic abilities. For example, a ribozyme from *Tetrahymena*, has normally the ability to excise itself from a longer RNA molecule. The wild-type enzyme has no ability to cleave DNA. Investigators then introduced various mutations (base changes) into this ribozyme and tested a large number of such variants for any ability to cut DNA. Eventually, one such variant was discovered, although the cleaving of DNA was very weak and occurred only at high temperature (50°C). Using reverse transcriptase, they then converted this variant RNA into a DNA copy (c-DNA), which they amplified many-fold by the polymerase chain reaction. The polymerase they used in this step, however, was one deliberately chosen because it was slightly inaccurate in copying the template DNA, the result being the generation of a very large number of c-DNA molecules each slightly different from each other. When these were c-DNA molecules were copied back into RNA by RNA polymerase, the new generation of ribozymes would also be slightly different from each other and from the one used in the previous generation to produce the c-DNA. This second generation of ribozymes was then tested for their ability to cleave DNA, and those that did so most efficiently and at the lowest temperature were selected as the progenitors of the next round of amplification, again with an inaccurate polymerase...and so on. The end result was that after ten such round of amplification and selection, the ability of the ribozyme to cleave DNA had increased 100 times and the temperature at which it performed was reduced to 37°C, i.e., physiological temperature.[2]

True evolution, however, is not possible until some enterprising proto-cell devises a means to reproduce with reasonable accuracy. To-day all living cells use DNA as their genetic storage material, but because recent studies have shown that RNA can both store genetic information (as in RNA viruses) and act as a enzyme to chemically alter its own sequence, current speculations focus on RNA as the first genetic material of cells. Some even suggest that the earliest cells contained only RNA catalysts and that protein enzymes evolved later as a more efficient means of catalyzing metabolic reactions. It has proved successful in the laboratory to manufacture an artificial RNA molecule that has limited ability to catalyze its own replication.[3]

There is little to be gained by further speculating here on how coacervate droplets, proteinoid microspheres or whatever may have diversified their metabolic repertoire and acquired the ability to reproduce accurately, but it is clear that once this giant step had been taken, it is a mere hop, skip and a jump to bacteria-like cells, amoeba-like eukaryotes, then multicellular forms, and soon shopping malls and supermarkets.[4*]

Who knows whether life really did come about in this way? The point of these speculations is not so much to advocate one scheme or another, but to point out that the origin of life on earth is not so utterly implausible an event that it cannot be seriously entertained and studied scientifically. The laws of chemistry and physics acting mindlessly by themselves can conjure up some pretty remarkable (and remarkably pretty) things, given sufficient time and maybe a modicum of luck. The "luck" is mainly cosmological—the fact that the planet earth is favorably situated where it is relative to its star, the sun. Given that, it's just a matter of patience.

Darwin, incidentally, avoided all speculations about the ultimate origin of life in *The Origin of Species*, noting that such musings required extravagant extrapolations from what was knowable in his time and were not, therefore, worth the effort. In his private correspondence he was not so reticent, and his speculations are not too unlike those we could subscribe to to-day.

> It has often been said that all the conditions for the first production of a living organism are now present which could ever have been present. But if (and oh! what a big if!) we could conceive in some warm little pond, with all sorts of ammonia and phosphoric salts, light, heat, electricity, etc., present, that a protein compound was chemically formed ready to develop still more complex changes, at the present day such matter would be instantly devoured or absorbed, which would not have been the case before living creatures were formed.[5]

REFERENCES

1. Dickerson, R., 1978. Chemical evolution and the origin of life. Sci. Amer. (Sept.).

2. Beaudry, A. and G. Joyce, 1992. Science 257: 635-641.

3. Doudna, J., et al., 1991. Science 251:1605-1608.

4. Schopf, J., 1978. The evolution of the earliest cells. Sci. Amer. (Sept.).

5. Darwin, F. *The Life and Letters of Charles Darwin, Including an Autobiographical Chapter.* Murray, London (1887).

---

*As the Nobel laureate, G. Wald, has put it, "Given so much time, the 'impossible' becomes possible, the possible probable, and the probable virtually certain."

INDEX

This index generally indicates only the first or the defining use of the term.

301

NOTES

NOTES

NOTES

NOTES

NOTES

NOTES

NOTES

NOTES

NOTES

NOTES

$$\frac{c^2}{c^2} + \frac{s^2}{c^2} = \frac{1}{c^2}$$

NOTES

$$1 + t^2 = \sec^2$$

$$\sec^2 - t^2 = 1$$

$$1 + \tan^2\theta = \sec^2\theta$$